Process Metallurgy 12

EXPLORATORY ANALYSIS OF METALLURGICAL PROCESS DATA WITH NEURAL NETWORKS AND RELATED METHODS

Process Metallurgy
Advisory Editors: A.W. Ashbrook and G.M. Ritcey

1
G.M. RITCEY and A.W. ASHBROOK
Solvent Extraction: Principles and Applications to Process Metallurgy,
Part I and Part II

2
P.A. WRIGHT
Extractive Metallurgy of Tin (Second, completely revised edition)

3
I.H. WARREN (Editor)
Application of Polarization Measurements in the Control of Metal Deposition

4
R.W. LAWRENCE, R.M.R. BRANION and H.G. EBNER (Editors)
Fundamental and Applied Biohydrometallurgy

5
A.E. TORMA and I.H. GUNDILER (Editors)
Precious and Rare Metal Technologies

6
G.M. RITCEY
Tailings Management

7
T. SEKINE
Solvent Extraction 1990

8
C.K. GUPTA and N. KRISHNAMURTHY
Extractive Metallurgy of Vanadium

9
R. AMILS and A. BALLESTER (Editors)
Biohydrometallurgy and the Environment Toward the Mining of the 21st Century
Part A: Bioleaching, Microbiology
Part B: Molecular Biology, Biosorption, Bioremedation

10
P. BALÁŽ
Extractive Metallurgy of Activated Minerals

11
V.S.T. CIMINELLI and O. GARCIA Jr. (Editors)
Biohydometallurgy: Fundamentals, Technology and Sustainable Development
Part A: Bioleaching, Microbiology and Molecular Biology
Part B: Biosorption and Bioremediation

Process Metallurgy 12

EXPLORATORY ANALYSIS OF METALLURGICAL PROCESS DATA WITH NEURAL NETWORKS AND RELATED METHODS

C. Aldrich
University of Stellenbosch, South Africa

2002

ELSEVIER
Amsterdam • Boston • London • New York • Oxford • Paris
San Diego • San Francisco • Singapore • Sydney • Tokyo

ELSEVIER SCIENCE B.V.
Sara Burgerhartstraat 25
P.O. Box 211, 1000 AE Amsterdam, The Netherlands

First edition 2002

Library of Congress Cataloging in Publication Data

Exploratory analysis of metallurgical process data with neural networks and related methods / edited by C. Aldrich.-- 1st ed.
p. cm. -- (Process metallurgy ; 12)
ISBN 0-444-50312-9
1. Metallurgy. 2. Metallurgical research. I. Aldrich, C. II. Series.

TN673 .E86 2002
669'.07'2--dc21

2002016356

British Library Cataloguing in Publication Data

Exploratory analysis of metallurgical process data with
neural networks and related methods. - (Process metallurgy
; 12)
1.Metallurgy - Data processing 2.Neural networks (Computer
science)
I .Aldrich, C.
669'. 0285

ISBN: 0 444 50312 9

∞The paper used in this publication meets the requirements of ANSI/NISO Z39.48-1992 (Permanence of Paper).
Printed in The Netherlands.

Preface

This book is concerned with the analysis and interpretation of multivariate measurements commonly found in the mineral and metallurgical industries, with the emphasis on the use of neural networks. Methods of multivariate analysis deal with reasonably large numbers of measurements (i.e. variables) made on each entity in one or more samples *simultaneously* (Dillon and Goldstein, 1984). In this respect, multivariate techniques differ from univariate or bivariate techniques, in that they focus on the covariances or correlations of three or more variables, instead of the means and variances of single variables, or the pairwise relationship between two variables.

Neural networks can be seen as a legitimate part of statistics that fits snugly in the niche between parametric and non-parametric methods. They are non-parametric, since they generally do not require the specification of explicit process models, but are not quite as unstructured as some statistical methods in that they adhere to a general class of models. In this context, neural networks have been used to extend, rather than replace, regression models, principal component analysis (Kramer, 1991, 1992), principal curves (Dong and McAvoy, 1996), partial least squares methods (Qin and McAvoy, 1992), as well as the visualization of process data in several major ways, to name but a few. In addition, the argument that neural networks are really highly parallelized neurocomputers or hardware devices and should therefore be distinguished from statistical or other patter recognition algorithms is not entirely convincing. In the vast majority of cases neural networks are simulated on single processor machines. There is no reason why other methods cannot also be simulated or executed in a similar way (and are indeed).

Since the book is aimed at the practicing metallurgist or a process engineer in the first place, a considerable part of it is of necessity devoted to the basic theory, which is introduced as briefly as possible within the large scope of the field. Also, although the book focuses on neural networks, they cannot be divorced from their statistical framework and the author has gone to considerable lengths to discuss this. For example, at least a basic understanding of fundamental statistical modelling (linear models) is necessary to appreciate the issues involved in the development of nonlinear models. The book is therefore a blend of the basic theory and some of the most recent advances in the practical application of neural networks.

Naturally, this preface would not be complete without my expressing my gratitude to the many people who have been involved in the writing of this book in one way or another, my graduate students who have performed many of the experiments described in this book, Juliana Steyl, who has helped in the final stages of the preparation of the book and last but not least, Annemarie and Melissa, who have had to bear with me, despite the wholly underestimated time and effort required to finish this book.

Chris Aldrich

Stellenbosch

November 2001

Table of Contents

Chapter 1
Introduction to Neural Networks

1.1. BACKGROUND

The technological progress of humanity throughout the ages can be summarized as a perpetual cycle of observing nature, interpreting these observations until the system or phenomenon being observed is understood sufficiently well to modify or redesign the system. Clearly man has made spectacular progress in all four areas. Our understanding of nature is reaching new depths at an ever-increasing pace, while we only need to look around us to appreciate the role of engineering and technology in every day life. The growth of each stage in the cycle depends on the previous stages; for example, heavier-than-air flight only became possible when the laws of physics governing flight were understood sufficiently well. The same applies to many of the recent advances in biotechnology, which are contingent upon detailed knowledge of the human genome, etc.

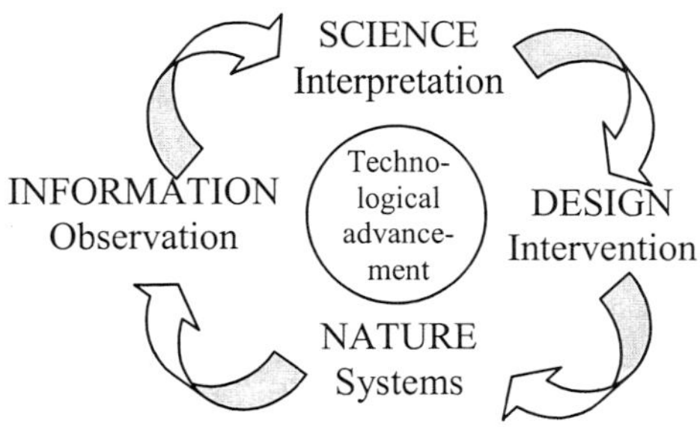

Figure 1.1. The cycle of technological progress.

However, in recent years, the advent of the computer has upset the balance between the elements of the cycle of technological progress portrayed in Figure 1.1. For example, although measurements of process variables on metallurgical plants have been logged for decades, it is only relatively recently, with the large-scale availability of inexpensive computing facilities, that the large historic databases of plant behaviour have become established. These databases can contain tens of thousands of variables and hundreds of thousands or millions of observations, constituting a rich repository of data, detailing the historic behaviour of a plant.

For example, in the production of ferrochrome in a submerged arc furnace, the specific energy consumption, metal production, percentage of valuable metal lost to the slag, etc., may depend on hundreds of other process variables, such as the composition and particulate state of the feed, as well as the electrical configuration of the furnace. These variables are likely to interact in a complex way to influence the quality of the product and the cost of production. Derivation of an empirical model of such a system is unlikely to be successful if at least a sizeable subset of the explanatory variables is not considered simultaneously.

Unfortunately, this proliferation of plant data does not always lead to a concomitant increase in knowledge or insight into process dynamics or plant operation. In fact, on many metallurgical plants personnel have probably experienced a net loss in understanding of the complexities of the behaviour of the plant, owing to increased turnover, rationalisation, etc. This has resulted in younger, less experienced plant operators sometimes having to cope with the unpredictable dynamics of nonlinear process systems. To aggravate the situation, a steady increase in the demand for high quality products at a lower cost, owing to global competition, as well as environmental and legislative constraints, require substantial improvements in process control.

In addition, automated capture of data has not only led to large data sets, but also data sets that can contain many more variables than observations. One such example pertains to spectroscopic data, where observations comprise a function, rather than a few discrete values. The data are obtained by exposing a chemical sample to an energy source, and recording the resulting absorbance as a continuous trace over a range of wavelengths. Such a trace is consequently digitised at appropriate intervals (wavelengths) with the digitised values forming a set of variables. Pyrolysis mass spectroscopy yields, near infrared spectroscopy and infrared spectroscopy yield approximately 200, 700 and 1700 such variables for each chemical sample (Krzanowski and Marriot, 1994). In these cases the number of variables usually exceed the number of samples by far. Similar problems are encountered with the measurement of acoustic signals, such as may be the case in on-line monitoring of process equipment (Zeng and Forssberg, 1992), potentiometric measurements to monitor corrosion, or image analysis, where each pixel in the image would represent a variable. In the latter case, high-resolution two-dimensional images can yield in excess of a million variables.

It is therefore not surprising that exploratory data analysis, multivariate analysis or data mining, is seen as a key enabling technology, and the topic of this book. Many of the techniques for the efficient exploration of data have been around for decades. However, it is only now with the growing availability of processing power that these techniques have become sophisticated instruments in the hands of metallurgical engineers and analysts.

Artificial neural networks represent a class of tools that can facilitate the exploration of large systems in ways not previously possible. These methods have seen explosive growth in the last decade and are still being developed at a breath-taking pace. In many ways neural networks can be viewed as nonlinear approaches to multivariate statistical methods, not bound by assumptions of normality or linearity. Although neural networks have originated outside the field of statistics, and have even been seen as an alternative to statistical methods in some circles, there are signs that this viewpoint is making way for an appreciation of the way in which neural networks complement classical statistics.

1.2. ARTIFICIAL NEURAL NETWORKS FROM AN ENGINEERING PERSPECTIVE

By the end of World War II several groups of scientists of the United States and England were working on what is now known as a computer. Although Alan Turing (1912-1954), the principal British scientist at the time, suggested the use of logical operators (such as OR, AND, NOT, etc.) as a basis for fundamental instructions to these machines, the majority of investigators favoured the use of numeric operators (+, -, <, etc.). It was only with the shifting emphasis on methods to allow computers to behave more like humans, that the approach advocated by Turing had begun to attract new interest. This entire research effort and its commercial repercussions are known as *artificial intelligence* (AI), and comprise many aspirations, ranging from the design of machines to do various things considered to be intelligent, to machines which could provide insight into the mental faculties of man.

Although different workers in the field have different goals, all seek to design machines that can solve problems. In order to achieve this goal, two basic strategies can be pursued. The first strategy or top-down approach has been developed productively for several decades and entails the reduction of large complex systems to small manipulable units. These techniques encompass heuristic programming, goal-based reasoning, parsing and causal analysis and are efficient systematic search procedures, capable of the manipulation and rearrangement of elements of complex systems or the supervision or management of the interaction between subsystems interacting in intricate ways. The disadvantages of symbolic logic systems such as these are their inflexibility and restricted operation, which limits them to very narrow domains of knowledge.

Bottom-up strategies (i.e. connectionist procedures) endeavour to build systems with as little architecture as possible. These systems start off with simple elements (such as simplified models, small computer programs, elementary principles, etc.) and move towards more complex systems by connecting these units to produce large-scale phenomena. As a consequence, these systems are very versatile and capable of the representation of uncertain approximate relations between elements or the solution of problems involving large numbers of weak interactions (such as found in pattern recognition and knowledge retrieval problems). On the other hand, connectionist systems cannot reason well and are not capable of symbolic manipulation and logic analyses.

With the exception of basic arithmetic, it is quite obvious that the human brain is superior to a digital computer at many tasks. Consider for example the processing of visual information. A one-year old baby is much better and faster at recognizing objects and faces than the most advanced supercomputer.

Historically the cost of computing has been directly related to the energy consumed in the process, and not the computation itself. Computational systems are limited by the system overheads required to supply the energy and to get rid of the heat, i.e. the boxes, the heaters, the fans, the connectors, the circuit boards, and the other superstructure that is required to make the system work. Technology development has therefore always been in the direction of the lowest energy consumed per unit computation. For example, an ordinary wristwatch today does far more computation than the ENIAC[1] did in the 1940s.

From this perspective it is interesting to consider the capability of biological systems in computation. Contrary to the myth that nervous systems are slow an inefficient, we can still not match the capabilities of the simplest insects, let alone handling tasks routinely performed by humans, despite a 10 000 000 fold increase in computational capability over the last few decades. With silicon technology we can envision today will dissipate in the order of 10^{-9} J of energy per operation on the chip level, and will consume approximately 100 to 1000 times more energy on a box level.

---

[1]Electronic Numerical Integrator And Computer, the first general purpose electronic computer built in the Moore School of Electrical Engineering of the University of Pennsylvania during 1943-1946.

Compared to the energy requirements of the human brain, with approximately 10^{15} synapses, each of which receives a nerve pulse roughly 10 times per second, the brain accomplishes approximately 10^{16} complex operations per second. All this is accomplished while dissipating only a few watts (i.e. an energy dissipation of about 10^{-16} J per operation), making the brain roughly 10 000 000 times more efficient than the best supercomputer today. From a different perspective, it is estimated that state-of-the-art digital technology would require about 10 MW of power to process information at the rate at which the human brain is capable. It is therefore clear that much can be gained from emulation of biological computational systems in the quest for more advanced computing systems.

Apart from its computational efficiency, the human brain also excels in many other respects. It differs significantly from current digital hardware with regard to its bottom-level elementary functions, representation of information, as well as top-level organizing principles. More specifically,

- It does not have to be programmed, but is flexible and can easily adjust to its environment by learning.
- It is robust and fault tolerant. An estimated 10 000 nerve cells in the brain die daily without perceptibly affecting its performance.
- It can deal with information of various kinds, be it fuzzy, probabilistic, noisy or inconsistent.
- It is highly parallel, small and compact.

Neural computation can therefore be seen as an alternative to the usual one based on a programmed instruction sequence introduced by Von Neuman. Neural networks are useful mathematical techniques inspired by the study of the human brain. Although the brain[2] is a very complex organ that is still largely an enigma, despite considerable advances in the neurosciences, it is clear that it operates in a massively parallel mode. Artificial systems inspired by the basic architecture of the brain emerged under various names, such as connectionist systems artificial neural networks, parallel distributed processing and Boltzmann machines. These systems differ in many subtle ways, but share the same general principles. Unlike traditional expert systems, where knowledge is stored explicitly in a database or as a set of rules or heuristics, neural networks generate their own implicit rules by learning from examples. Items of knowledge are furthermore distributed across the network and reasonable responses are obtained when the network is presented with incomplete, noisy or previously unseen inputs. From the perspective of cognitive modelling of process systems know-how, these pattern recognition and generalization capabilities of neural networks are much more attractive than the symbol manipulation methodology of expert systems, especially as far as complex, ill-defined systems are concerned.

Many parallels can be drawn between the development of knowledge-based systems and that of neural networks. Both had suffered from an overzealous approach in the early stages of their development. In the mid-1980s for example, a common perception had temporarily made its way into the process engineering community that knowledge-based systems had

[2]The typical human brain contains between 10^{10} and 10^{11} neurons each of which can be connected to as many as 10 000 other neurons. The neuron, which is the basic processing unit in the brain, is very slow, with a switch time in the order of milliseconds and it has to operate in parallel to achieve the performance observed.

failed to live up to expectations. Like their rule-based counterparts, neural networks are also sometimes seen as 'solutions looking for problems'. Although the application of neural networks in the process engineering industry has not matured yet, there is every reason to believe that like other computational methods it will also find a solid niche in this field. A closer look at the historic development of neural networks will underpin the analogous paths of these two branches of artificial intelligence.

1.3. BRIEF HISTORY OF NEURAL NETWORKS

The modern era of neural networks had its inception in the 1940s, when the paper of McCulloch (a psychiatrist and neuroanatomist) and Pitts (a mathematician) on the modelling of neurons appeared (McCulloch and Pitts, 1943). The McCulloch-Pitts model contained all the necessary elements to perform logic operations, but implementation was not feasible with the bulky vacuum tubes prevalent at the time. Although this model never became technically significant, it laid the foundation for future developments.

Donald Hebb's book *The Organization of Behaviour* first appeared in the late 1940s (Hebb, 1949), as well as his proposed learning scheme for updating a neuron's connections, presently referred to as the Hebbian learning rule. During the 1950s the first neurocomputers that could adapt their connections automatically were built and tested (Minsky, 1954). Minsky and Edmond constructed an analog synthetic brain at Harvard in 1951, to test Hebb's learning theory. Referred to as the *Snark*, the device consisted of 300 vacuum tubes and 40 variable resistors, which represented the weights of the network. The *Snark* could be trained to run a maze.

The interest sparked by these ideas was further buoyed when Frank Rosenblatt invented his Mark I Perceptron in 1958. The perceptron was the world's first practical neurocomputer, used for the recognition of characters mounted on an illuminated board. It was built by Rosenblatt, Wightman and Martin in 1957 at the Cornell Aeronautics Laboratory and was sponsored by the US Office of Naval Research. A 20 x 20 array of cadmium sulphide photo-sensors provided the input to the neural network. An 8 x 8 array of servomotor driven poten-tiometers constituted the adjustable weights of the neural network.

This was followed by Widrow's ADALINE (ADAptive LINEar combiner) in 1960, as well as the introduction of a powerful new learning rule called the Widrow-Hoff learning rule developed by Bernard Widrow and Marcian Hoff (Widrow and Hoff, 1960). The rule minimized the summed square error during training associated with the classification of patterns. The ADALINE network and its MADALINE (Multiple ADALINEs) extension were applied to weather forecasting, adaptive controls and pattern recognition.

Although some success was achieved in this early period, the machine learning theorems were too limited at the time to support application to more complicated problems. This, as well as the lack of adequate computational facilities resulted in stagnation of the research in the neural network field or cybernetics, as it was known at the time. The early development of neural networks came to a dramatic end when Minsky and Papert (1969) showed that the capabilities of the linear networks studied at the time were severely limited. These revelations caused a virtually total cessation in the availability of research funding and many talented researchers left the field permanently.

This growth has in part been fomented by improvements in very large scale integration (VLSI) technology, as well as the efforts of a small number of investigators who had continued to work during the 1970s and early 1980s, despite a lack of funds and public interest. For example, in Japan, Sun-Ichi Amari (1972, 1977) pursued the investigation of neural networks with threshold elements and the mathematical theory of neural networks. His compatriot, Kunihiko Fukushima developed a class of neural networks known as neocognitrons (Fukushima, 1980). The neocognitrons were biologically inspired models for visual pattern recognition, that emulated retinal images, and processed them by use of two-dimensional layers of neurons. In Finland, Teuvo Kohonen (1977, 1982, 1984, 1988, 1990, 1995) developed unsupervised neural networks capable of feature mapping into regular arrays of neurons, while James A. Anderson conducted research into associative memories (Anderson et al., 1977). Stephen Grossberg and Gail Carpenter introduced a number of neural architectures and theories, while developing the theory of adaptive resonance theory (ART) neural networks (Grossberg, 1976, 1982; Carpenter and Grossberg, 1990).

However, the initial interest in neural networks was only revived again in the early 1980s, and since then the field of neural networks has seen phenomenal growth, passing from a research curiosity to commercial fruition. Several seminal publications saw the light from 1982 to 1986. John J. Hopfield initiated the new renaissance of neural networks through the introduction of a recurrent neural network for associative memories (Hopfield, 1982, 1984). Further revitalisation of the field occurred with the publication of James McClelland and David Rumelhart's (Rumelhart and McClelland, 1986) two volumes on distributed parallel processing. In this publication, the earlier barriers that had led to the slump in the mainstream neural network development in the 1960s were circumvented. Much of this work concerning the training of neural networks with multiple layers had actually been discovered and rediscovered earlier by Dreyfus (1962), Bryson (Bryson and Ho, 1969), Kelley (1969), as well as Paul Werbos (Werbos, 1974), but it went largely unnoticed at the time. However, by the mid-1980s, neural network business has soared from an approximately $7 million industry in 1987, to an estimated $120 million industry in 1990.

1.4. STRUCTURES OF NEURAL NETWORKS

Although much of the development of neural networks has been inspired by biological neural mechanisms, the link between artificial neural networks and their biological neural systems is rather tenuous. Biological organisms are simply not understood sufficiently to allow any meaningful emulation. As a result, artificial neural networks are better interpreted as a class of mathematical algorithms (since a network can essentially be regarded as a graphical notation for a large class of algorithms), as opposed to synthetic networks capable of competing with their biological equivalents. In general terms neural networks are therefore simply computers or computational structures consisting of large numbers of primitive process units connected on a massively parallel scale. These units, nodes or artificial neurons are relatively simple devices by themselves, and it is only through the collective behaviour of these nodes that neural networks can realize their powerful ability to form generalized representations of complex relationships and data structures. A basic understanding of the structure and functioning of a typical neural network node is therefore necessary for a better understanding of the capabilities and limitations of neural networks. The basic model of an artificial neuron that will be used throughout this text is consequently presented in more detail below.

1.4.1. Models of single neurons

Each node model consists of a processing element with a set of input connections, as well as a single output connection, as illustrated in Figure 1.1. Each of these connections is characterized by a numerical value or weight, which is an indication of the strength of the connection. The flow of information through the node is unidirectional, as indicated by the arrows in this figure.

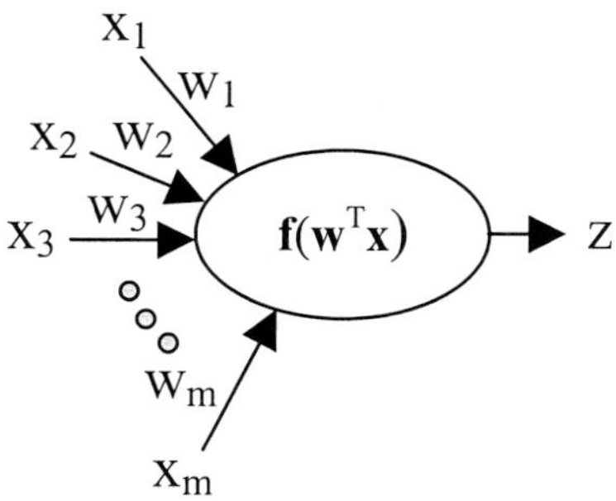

Figure 1.1. Model of a single neuron.

The output of the neuron can be expressed as follows.

$$z = f(\Sigma_{i=1}^{m} w_i x_i), \text{ or}$$

$$z = f(\mathbf{w}^T\mathbf{x}) \qquad (1.1)$$

where **w** is the weight vector of the neural node, defined as

$$\mathbf{w} = [w_1, w_2, w_3, \ldots w_m]^T$$

and **x** is the input vector, defined as

$$\mathbf{x} = [x_1, x_2, x_3, \ldots x_m]^T$$

Like all other vectors in the text, these vectors are column vectors, of which the superscript T denotes the transposition.

The function $f(\mathbf{w}^T\mathbf{x})$ is referred to as the activation function of the node, defined on the set of activation values, which are the scalar product of the weight and input vectors of the node. The argument of the activation function is sometimes referred to as the potential of the node, in analogy to the membrane potentials of biological neurons.

An additional input can be defined for some neurons, i.e. x_0, with associated weight w_0. This input is referred to as a bias and has a fixed value of -1. Like the other weights $w_1, w_2, w_3, \ldots w_m$, the bias weight is also adaptable. The use of a bias input value is sometimes necessary to enable neural networks to form accurate representations of process trends, by offsetting the output of the neural network. Although the above model is used commonly in the application of neural networks, some classes of neural networks can have different definitions of potential

($\neq \mathbf{w}^T\mathbf{x}$). Also, in some neural networks, nodes can perform different functions during different stages of the application of the network.

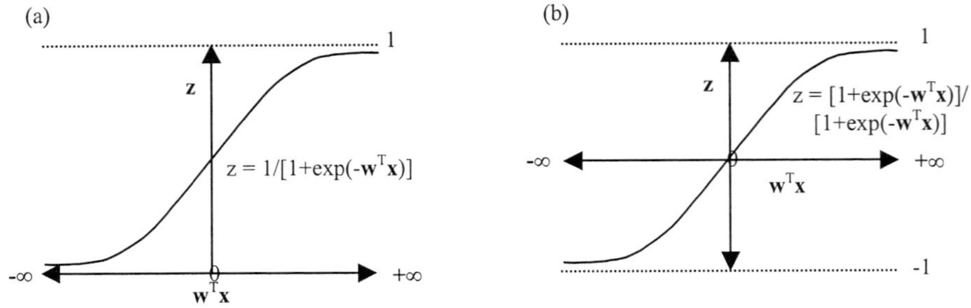

Figure 1.2. Sigmoidal activation functions, (a) unipolar, $z = 1/[1+\exp(-\mathbf{w}^T\mathbf{x})]$ and (b) bipolar, that is $z = [1-\exp(-\mathbf{w}^T\mathbf{x})]/[1+\exp(-\mathbf{w}^T\mathbf{x})]$.

Sigmoidal activation functions are used widely in neural network applications, that is bipolar sigmoidal activations functions (with $\phi = \mathbf{w}^T\mathbf{x}$ and $\lambda > 0$)

$$f(\phi) = 2/(1 + e^{-\lambda\phi}) - 1 \tag{1.2}$$

and their hard-limiting equivalent (bipolar sign functions, or bipolar binary functions)

$$f(\phi) = \text{sgn}(\phi) = +1, \text{ if } \phi > 0 \tag{1.3}$$

$$f(\phi) = \text{sgn}(\phi) = -1, \text{ if } \phi < 0 \tag{1.4}$$

and unipolar sigmoidal activation functions

$$f(\phi) = 1/(1 + e^{-\lambda\phi}) \tag{1.5}$$

with their hard-limiting version (unipolar sign functions, or unipolar binary functions) the same as for the bipolar activations functions (equations 1.3-1.4). The parameter λ is proportional to the gain of the neuron, and determines the steepness of the continuous activation function. These functions are depicted graphically in Figure 1.12. For obvious reasons, the sign function is also called a bipolar binary function.

1.4.2. Models of neural network structures

Neural networks consist of interconnections of nodes, such as the ones described above. These processing nodes are usually divided into disjoint subsets or layers, in which all the nodes have similar computational characteristics. A distinction is made between input, hidden and output layers, depending on their relation to the information environment of the neural network. The nodes in a particular layer are linked to other nodes in successive layers by means of the weighted connections discussed above. An elementary feedforward neural network can therefore be considered as a structure with n neurons or nodes receiving inputs (x), and m nodes producing outputs (z), that is

$$\mathbf{x} = [x_1, x_2, x_3, \ldots x_n]^T$$

$$\mathbf{z} = [z_1, z_2, z_3, \ldots z_m]^T$$

The potential for a particular node in this single layer feedforward neural network is similar to that for the neuron model, except that a double index i,j notation is used to describe the destination (first subscript) and source (second subscript) nodes of the weights. The activation value or argument of the i'th neuron in the network is therefore

$$\phi_i = \Sigma_{j=1}^{n} \mathbf{w}_{ij}\mathbf{x}_i, \text{ for } i = 1, 2, \ldots m \tag{1.6}$$

This value is subsequently transformed by the activation function of the i'th output node of the network

$$z_i = f(\mathbf{w}_i^T\mathbf{x}), \text{ for } i = 1, 2, 3, \ldots m \tag{1.7}$$

Neural networks with multiple layers are simply formed by cascading the single-layer networks represented by equations 1.6-1.7. Once the structure of the network (number of layers, number of nodes per layer, types of nodes, etc.) is fixed, the parameters (weights) of the network have to be determined. This is done by *training* (optimization) of the weight matrix of the neural network. Feedforward neural networks, like the one discussed above, learn by repeatedly attempting to match sets of input data to corresponding sets of output data or target values (a process called supervised learning). The optimized weights constitute a distributed internal representation of the relationship(s) between the inputs and the outputs of the neural network.

Learning typically occurs by means of algorithms designed to minimize the mean square error between the desired and the actual output of the network through incremental modification of the weight matrix of the network. In feedforward neural networks, information is propagated back through the network during the learning process, in order to update the weights. As a result, these neural networks are also known as back propagation neural networks. Training of the neural network is terminated when the network has learnt to generalize the underlying trends or relationships exemplified by the data. Generalization implies that the neural network can interpolate sensibly at points not contained in its training set, as indicated in Figure 1.3. The ability of the neural network to do so is typically assessed by means of cross-validation, where the performance of the network is evaluated against a novel set of test data, not used during training. Modes of training other than the basic approach mentioned above are possible as well, depending on the function and structure of the neural network. These algorithms are considered in more detail below.

1.5. TRAINING RULES

As was mentioned above, neural networks are essentially data driven devices that can form internal distributed representations of complex relationships. These relationships are typically specified implicitly by means of examples, i.e. input-output data. Without loss of generality, a modelling problem (the behaviour of a process, plant unit operation, etc.) can be treated as follows, if the behaviour of the process is characterized by data of the following form

$$\mathbf{Y} = \begin{bmatrix} y_{1,1} & y_{1,2} & .. & y_{1,p} \\ y_{2,1} & y_{2,2} & .. & y_{2,p} \\ .. & .. & .. & .. \\ .. & .. & .. & .. \\ y_{n,1} & y_{n,2} & .. & y_{n,p} \end{bmatrix} \in \Re^{nxp} \qquad (1.8)$$

$$\mathbf{X} = \begin{bmatrix} x_{1,1} & x_{1,2} & .. & x_{1,m} \\ x_{2,1} & x_{2,2} & .. & x_{2,m} \\ .. & .. & .. & .. \\ .. & .. & .. & .. \\ x_{n,1} & x_{n,2} & .. & x_{n,m} \end{bmatrix} \in \Re^{nxm} \qquad (1.9)$$

where $y_{i,k}$ ($i = 1, 2, \ldots p$) represent p variables dependent on m causal or independent variables $x_{j,k}$ ($j = 1, 2, \ldots m$), based on n observations ($k = 1, 2, \ldots n$). The variables $y_{i,k}$ are usually parameters which provide a measure of the performance of the plant, while the $x_{j,k}$ variables are the plant parameters on which these performance variables are known to depend.

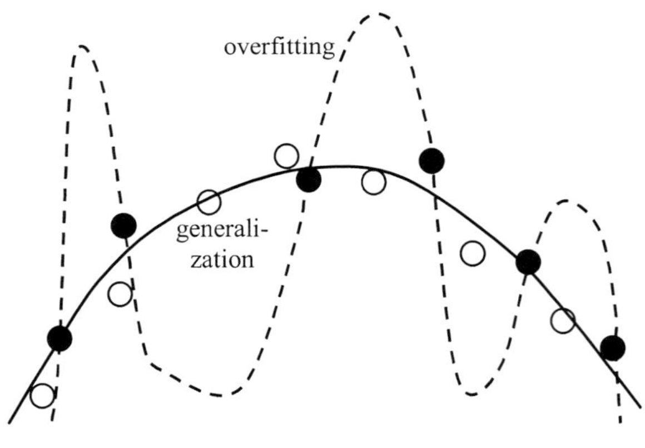

Figure 1.3. Overfitting of data (broken line), compared with generalization (solid line) by a neural network. The solid and empty circles indicate training and test data respectively.

The problem is then to relate the matrix **Y** to some set of functions **Y** = **f(X)** of matrix **X**, in order to predict **Y** from **X**. The main advantage of modelling techniques based on the use of neural networks, is that a priori assumptions with regard to the functional relationship between **X** and **Y** are not required. The network learns this relationship instead, on the basis of examples of related **x-y** vector pairs or exemplars and forms an internal distributed implicit model representation of the process.

In supervised training the weights of neural network nodes can be modified, based on the inputs received by the node, its response, as well as the response of a supervisor, or target value. In unsupervised learning, the response of a target value to guide learning is not available. This can be expressed by a generalized learning rule (Amari, 1990) where the weight vector of a node increases proportional to the product of the input vector **x**, and a learning signal r.

The learning signal r is generally a function of the weight vector, $\mathbf{w}_i \in \Re^m$, the input $\mathbf{x} \in \Re^m$ and a target signal, $d_i \in \Re$, where applicable, that is

$$r = r(\mathbf{w}_i, \mathbf{x}, d_i) \quad (1.12)$$

The weight vector at time t is incremented by

$$\Delta\mathbf{w}_i(t) = \beta r[\mathbf{w}_i(t), \mathbf{x}(t), d_i(t)]\mathbf{x}(t) \quad (1.13)$$

The parameter β determines the learning rate, so that the weight vector is updated at discrete time steps as follows

$$\mathbf{w}_i(t+1) = \mathbf{w}_i(t) + \beta r[\mathbf{w}_i(t), \mathbf{x}(t), d_i(t)]\mathbf{x}(t) \quad (1.14)$$

Different learning rules can be distinguished on the basis of their different learning signals, as considered in more detail below.

1.5.1. Supervised training

a) *Perceptron learning rule*

The perceptron learning rule is characterized by a learning signal that is the difference between the desired or target response of the neuron, and the neuron's actual response.

$$r = d_i - z_i \quad (1.15)$$

The adjustment of the weights in this supervisory procedure takes place as follows

$$\Delta\mathbf{w}_i = \beta[d_i - \text{sgn}(\mathbf{w}_i^T\mathbf{x})]\mathbf{x} = \Delta\mathbf{w}_i = \beta[d_i - \text{sgn}(\mathbf{w}_i^T\mathbf{x})]\mathbf{x} \quad (1.16)$$

The perceptron rule pertains to binary node outputs only, and weights are only adjusted if there is a difference between the actual and target response of the neural network. The weights of the network can assume any initial values. The method is elucidated by the example below.

Assume the set of training vectors to be $\mathbf{x}_1 = [1, 2, 3 \mid 1]$, $\mathbf{x}_2 = [-1, -2, -1 \mid -1]$ and $\mathbf{x}_3 = [-3, -1, 0 \mid -1]$. The learning rate is assumed to be β = 0.1, and the initial weight vector is arbitrarily assumed to be $\mathbf{w}(0) = [1, 0, -1]$. As before, the input or training vectors are presented to the network sequentially.

Step 1: For the first input x_1, with desired output d_1, the activation of the node is

$$\mathbf{w}(0)^T\mathbf{x}_1 = [1, 0, -1][1, 2, 3]^T = -2$$

and the output of the node is $z_1 = \text{sgn}(-2) = -1$.

Since z_1 is not equal to the target $d_1 = 1$, the weights of the network have to be adjusted.

$\mathbf{w}(1) = \mathbf{w}(0) + \beta[d_i - \mathrm{sgn}(\mathbf{w}(0)^T\mathbf{x}_1]\mathbf{x}_1$

$\mathbf{w}(1) = [1, 0, -1]^T + 0.1[1-(-1)][1, 2, 3]^T = [1, 0, -1]^T + [0.2, 0.4, 0.6]^T$

$\mathbf{w}(1) = [1.2, 0.4, -0.4]^T$

Step 2: For the second input $\mathbf{x}_2$, with desired output d_2, the activation of the node is

$\mathbf{w}(1)^T\mathbf{x}_1 = [1.2, 0.4, -0.4][-1, -2, -1]^T = -1.6$

and the output of the node is $z_2 = \mathrm{sgn}(-1.6) = -1$.

Since z_2 is equal to the target $d_2 = -1$, adjustment of the weights of the network is not required, so that

$\mathbf{w}(2) = [1.2, 0.4, -0.4]^T$

Step 3: For the second input x_3, with desired output d_3, the activation of the node is

$\mathbf{w}(2)^T\mathbf{x}_3 = [1.2, 0.4, -0.4][-3, -1, 0]^T = -4$

and the output of the node is $z_1 = \mathrm{sgn}(-4) = -1$.

Since z_3 is equal to the target $d_3 = -1$, adjustment of the weights of the network is again not required, so that

$\mathbf{w}(3) = [1.2, 0.4, -0.4]^T$

Training of the network can therefore be terminated, since the network (node) has learnt to reproduce all the targeted outputs correctly.

b) Delta and generalized delta rules

The delta learning rule pertains to nodes with continuous activation functions only, and the learning signal of the rule is defined as

$$r = [d_i - f(\mathbf{w}_i^T\mathbf{x})]f'(\mathbf{w}_i^T\mathbf{x}) \quad (1.17)$$

The term $f'(\mathbf{w}_i^T\mathbf{x})$ is the derivative of the activation function $f(\mathbf{w}_i^T\mathbf{x})$. The rule can be derived from the least squared error of the difference between the actual output of the network and the desired output, in the form

$$E = ½(d_i - z_i)^2 = ½[d_i - f(\mathbf{w}_i^T\mathbf{x})]^2 \quad (1.18)$$

Calculation of the gradient vector of the squared error in equation, with regard to w_i, gives

$$\nabla E = -[d_i - f(\mathbf{w}_i^T\mathbf{x})]f'(\mathbf{w}_i^T\mathbf{x})\mathbf{x} \quad (1.19)$$

The adjustment of the weights in this supervisory procedure takes place as follows

$$\Delta\mathbf{w}_i = -\beta\nabla E \quad (1.20)$$

or

$$\Delta \mathbf{w}_i = \beta[d_i - f(\mathbf{w}_i^T\mathbf{x})]f'(\mathbf{w}_i^T\mathbf{x})\mathbf{x} \tag{1.21}$$

for a single weight, this is equivalent to

$$\Delta \mathbf{w}_{ij} = \beta[d_i - f(\mathbf{w}_i^T\mathbf{x})]f'(\mathbf{w}_i^T\mathbf{x})\mathbf{x}_{j}, \text{ for } j = 0, 1, 2, \ldots N \tag{1.22}$$

As an example, assume the set of training vectors to be $\mathbf{x}_1 = [1, 2, 3 \mid 1]$, $\mathbf{x}_2 = [-1, -2, -1 \mid 0]$ and $\mathbf{x}_3 = [-3, -1, 0 \mid 0]$. The learning rate is assumed to be $\beta = 0.1$, and the initial weight vector is arbitrarily assumed to be $\mathbf{w}(0) = [1, 0, -1]$. The conditions are the same as for the perceptron rule considered above, except that the target vectors are different. As before, the input or training vectors are presented to the network sequentially. The node is assumed to have a continuous unipolar activation function of the form $f(x) = 1/[1 + e^{-x}]$. In this case both the value of the activation of the neuron, as well as the derivative of the activation have to be computed. One of the reasons for using sigmoidal activation functions is that their derivatives can be calculated easily. For a continuous unipolar sigmoidal activation function, the derivative is

As an example, assume the set of training vectors to be $\mathbf{x}_1 = [1, 2, 3 \mid 1]$, $\mathbf{x}_2 = [-1, -2, -1 \mid 0]$ and $\mathbf{x}_3 = [-3, -1, 0 \mid 0]$. The learning rate is assumed to be $\beta = 0.1$, and the initial weight vector is arbitrarily assumed to be $w(0) = [1, 0, -1]$. The conditions are the same as for the perceptron rule considered above, except that the target vectors are different. As before, the input or training vectors are presented to the network sequentially. The node is assumed to have a continuous unipolar activation function of the form $f(x) = 1/[1 + e^{-x}]$. In this case both the value of the activation of the neuron, as well as the derivative of the activation have to be computed. One of the reasons for using sigmoidal activation functions is that their derivatives can be calculated easily. For a continuous unipolar sigmoidal activation function, the derivative is

$$\begin{aligned}
d/dx[f(x)] &= d/dx[1 + \exp(-x)]^{-1} \\
&= (-1)[1 + \exp(-x)]^{-2}\exp(-x)(-1) \\
&= [1 + \exp(-x)]^{-2}\exp(-x) \\
&= [f(x)]^2[1 + \exp(-x) - 1] \\
&= [f(x)]^2[1/f(x) - 1] \\
&= [f(x)]^2\{[1-f(x)]/f(x)\} \\
&= f(x)[1-f(x)]
\end{aligned}$$

The initial activation of the node is $\mathbf{w}(0)^T\mathbf{x}_1 = -2$. The output of the node is $z(0) = 1/[1 + \exp(-\mathbf{w}(0)^T\mathbf{x}_1)] = 1/[1 + e^{-(-2)}] = 0.119$ (which differs from the desirable output $d_1 = 1$). The derivative of the output node is $z'(0) = 0.119(1-0.119) = 0.1048$, and

$$\mathbf{w}(1) = \mathbf{w}(0) + \beta[d_i - f(\mathbf{w}_i^T\mathbf{x})]f'(\mathbf{w}_i^T\mathbf{x})\mathbf{x}$$

$$\mathbf{w}(1) = [1, 0, -1]^T + 0.1[1 - 0.119]0.1048[1, 2, 3]^T$$

$$\mathbf{w}(1) = [1, 0, -1]^T + 0.0092[1, 2, 3]^T = [1.0092, 0.0185, -0.9723]^T$$

This adjustment in the weight vector of the neuron results in a smaller error on the first exemplar, reducing it from 0.879 to approximately 0.867.

For step 2:

The activation of the node is $\mathbf{w}(1)^T\mathbf{x}_2 = [1.0092, 0.0185, -0.9723][-1, -2, -1]^T = -0.074$

The output of the node is $z(1) = 1/[1 + \exp(-\mathbf{w}(1)^T\mathbf{x}_2)] = 1/[1 + e^{-(-0.074)}] = 0.4815$ (which differs from the desirable output $d_1 = 0$).

The derivative of the output node is $z'(1) = 0.4815(1\text{-}0.4815) = 0.2497$, and

$$\mathbf{w}(1) = \mathbf{w}(0) + \beta[d_i - f(\mathbf{w}_i{}^T\mathbf{x})]f'(\mathbf{w}_i{}^T\mathbf{x})\mathbf{x}$$

$$\mathbf{w}(1) = [1, 0, -1]^T + 0.1[1 - 0.4815]0.2497[1, 2, 3]^T$$

$$\mathbf{w}(1) = [1, 0, -1]^T + 0.0129[1, 2, 3]^T = [1.0129, 0.0258, -0.9613]^T$$

This adjustment in the weight vector of the node results in a smaller error on the second exemplar, reducing it from 0.482 to approximately 0.474.

The delta rule requires a small learning rate (approximately $0.001 \leq \beta \leq 0.1$), since the weight vector is moved in the negative error gradient direction in the weight space. Since the error is only reduced with small increments at a time, the set of training data has to be presented to the neural network repeatedly, in order to reduce the output errors of the network satisfactorily.

The best value for the learning rate τ depends on the error surface, i.e. a plot of E versus $\mathbf{w}_{ji}$ (which is rarely known beforehand). If the surface is relatively smooth, a larger learning rate will speed convergence. On the other hand, if the error surface changes relatively rapidly, a smaller learning rate would clearly be desirable. As a general rule of thumb, the largest learning rate not causing oscillation should be used.

Note that with this training scheme, if the desired output of the j'th unit is less than the actual output of the neural network, the weight $\mathbf{w}_{kj}$ connecting input unit k with output unit j is increased. This does not take into account the response of the network to other training patterns. Moreover, zero inputs do not result in any adjustment, not even for non-zero unit errors.

A simple method of increasing the learning rate without risking instability is to modify the delta rule through inclusion of a momentum term ($\alpha > 0$), that is

$$\Delta\mathbf{w}_{ji}(t) = \alpha\Delta\mathbf{w}_{ji}(t-1) + \tau\delta_j(t)y_i(t) \tag{1.23}$$

Equation (1.23) is known as the *generalized delta rule*, since it includes the delta rule ($\alpha = 0$).

The inclusion of a momentum term has the following benefits.

- When the partial derivative $\partial E(k)/\partial \mathbf{w}_{ji}(k)$ has the same algebraic sign in consecutive in iterations, the weighted sum $\Delta \mathbf{w}_{ji}(t)$ grows, resulting in large adjustments to the weight $\mathbf{w}_{ji}(k)$. This tends to result in *accelerated descent* in steady downhill directions.

- When the partial derivative $\partial E(k)/\partial \mathbf{w}_{ji}(k)$ has alternating algebraic signs in consecutive iterations, the weighted sum $\Delta \mathbf{w}_{ji}(t)$ is reduced, resulting in small adjustments to the weight $\mathbf{w}_{ji}(k)$. The inclusion of the momentum term therefore has a stabilizing effect in directions that tend to produce oscillation.

- In addition, the momentum term can have the advantage of preventing the learning process of getting trapped in shallow local minima on the error surface.

When all the network weights are adjusted for the k'th exemplar (i.e. for all i and j) as indicated above, it is referred to as per sample training or pattern training. An alternative is to train per epoch, by accumulating weight changes prior to adjustment, i.e. $\Delta^k \mathbf{w}'_{ji} = \Sigma_n \Delta^k \mathbf{w}_{ji}$. The weights in the network are thus only adjusted after each presentation of all the exemplars in the training base, or a subset (epoch) of these exemplars.

The supervised training of back propagation neural networks may be viewed as a global identification problem, which requires the minimization of a cost function. The cost function (E) can be defined in terms of the discrepancies between the outputs of the neural network and the desired or target output values. More specifically, the cost function can be expressed in terms of the weight matrix of the network (which has an otherwise fixed configuration). The purpose of training is to adjust these free parameters (weights) to enable the outputs of the neural network to match the target outputs more closely.

The standard back propagation algorithm modifies a particular parameter based on an instantaneous estimate of the gradient ($\partial E/\partial \mathbf{w}_i$) of the cost function with respect to the parameter (weight). This is an efficient method of training, although it uses a minimum amount of information in the process. As a consequence, the use of the algorithm becomes unpractical with large networks (which require excessively long training times). The problem can be alleviated by making better use of available information during training, e.g. by incorporating training heuristics into the algorithm.

A wide variety of approaches to the optimization of the weight matrices of neural networks have been documented to date. In practice, gradient descent methods, such as the generalized delta rule have proved to be very popular, but other methods are also being used to compensate for the disadvantages of these methods (chiefly their susceptibility towards entrapment in local minima). These methods include 2nd order gradient descent methods, such as conjugate gradients, Newton and Levenberg-Marquardt methods (Reklaitis et al., 1983), as well as genetic algorithms, among other, as discussed in Chapter 2.

c) Widrow-Hoff learning rule

Like the perceptron and delta rules, the Widrow-Hoff rule applies to the supervised training of neural networks. The Widrow-Hoff rule does not depend on the activation function of the node, since it minimizes the squared error between the target value and the activation of the node, that is

$$r = d_i - \mathbf{w}_i^T\mathbf{x} \tag{1.24}$$

and

$$\Delta\mathbf{w}_i = \beta[d_i - \mathbf{w}_i^T\mathbf{x}]\mathbf{x} \tag{1.25}$$

which his equivalent to

$$\Delta\mathbf{w}_{ij} = \beta[\, d_i - \mathbf{w}_i^T\mathbf{x}]\mathbf{x}_j, \text{ for } j = 0, 1, 2, \ldots n \tag{1.26}$$

for the adjustment of a single weight. This rule is clearly a special case of the delta rule, where the activation function is the identity function, $f(\mathbf{w}_i^T\mathbf{x}) = \mathbf{w}_i^T\mathbf{x}$, and $f'(\mathbf{w}_i^T\mathbf{x}) = 1$. As is the case with the delta function, the weights of the neural network are also initialized to arbitrary (small) values.

d) *Correlation learning rule*

The correlation rule is obtained by substituting $r = d_i$ in the general learning rule (equation 1.12), so that

$$\Delta\mathbf{w}_i = \beta d_i\mathbf{x}, \text{ or} \tag{1.27}$$

$$\Delta\mathbf{w}_{ij} = \beta d_i\mathbf{x}_j, \text{ for } j = 0, 1, 2, \ldots n \tag{1.28}$$

The increment in the weight vector is directly proportional to the product of the target value of a particular exemplar, and the exemplar (input) itself. The rule is similar to the Hebbian rule discussed below (with a binary activation function and $z_i = d_i$), except that it is a supervised learning rule. Similar to Hebbian learning, the initial weight vector should also be zero.

1.5.2. Unsupervised training

a) *Hebbian and anti-Hebbian learning rule*

For the Hebbian learning rule, the weight vector is initialized to small random values. Subsequent change in the weight vector is related to the product of the input and outputs of the neural network. The learning signal is simply the output of the network, and therefore dependent on the activation of the nodes, as well as the activation functions of the nodes, only. Since the Hebbian learning rule is not dependent on a target value, it is an unsupervised learning rule, i.e.

$$r = f(\mathbf{w}_i^T\mathbf{x}) \tag{1.29}$$

$$\Delta\mathbf{w}_i = \beta f(\mathbf{w}_i^T\mathbf{x})\mathbf{x}, \text{ or} \tag{1.30}$$

$$\mathbf{w}_i(t+1) = \mathbf{w}_i(t) + \beta f[\mathbf{w}_i(t)^T\mathbf{x}(t)]\mathbf{x}(t) \tag{1.31}$$

As can be seen from equation 1.31, the weights of a node are increased when the correlation between the input and the output of the node is positive, and decreased when this correlation is negative. Also, the output is progressively strengthened for each presentation of the input.

Frequent exemplars will therefore tend to have a larger influence on the node's weight vector and will eventually produce the largest output.

A simple example demonstrates the dynamics of the Hebbian learning rule for a single node, with a bipolar binary activation function. Since there is only one weight vector, the subscript of the weight vector has been dropped.

Let $\mathbf{w}(0) = [0.1,-0.1,0]^T$, and the two input vectors available for training be $\mathbf{x}_1 = [1,-1,1]^T$ and $\mathbf{x}_2 = [1,-0.5,-2]^T$. A unity value for the learning parameter β is assumed, while the input vectors are presented sequentially to the neural network.

Step 1: Calculation of the activation and output of the neural network, based on input of first exemplar.

$$\mathbf{w}(0)^T\mathbf{x}_1 = [0.1, -0.1, 0][1, -1, 1]^T = 0.2$$

$$z(0) = \text{sgn}[\mathbf{w}(0)^T\mathbf{x}_1] = +1$$

Step 2: Modification of weights based on input of first exemplar.

$$\mathbf{w}(1) = \mathbf{w}(0) + \text{sgn}[\mathbf{w}(0)^T\mathbf{x}_1]\mathbf{x}_1 = [0.1, -0.1, 0]^T + (+1)[1, -1, 1]^T = [1.1, -1.1, 1]^T$$

Step 3: Calculation of the activation and output of the neural network, based on input of first exemplar.

$$\mathbf{w}(1)^T\mathbf{x}_2 = [1.1, -1.1, 1][1, -0.5, -2]^T = -0.35$$

$$z(1) = \text{sgn}[\mathbf{w}(1)^T\mathbf{x}_2] = -1$$

Step 4: Modification of weights based on input of second exemplar.

$$\mathbf{w}(2) = \mathbf{w}(1) + \text{sgn}[\mathbf{w}(1)^T\mathbf{x}_2]\mathbf{x}_2 = [1.1, -1.1, 1]^T + (-1)[1, -0.5, -2]^T = [0.1, -0.6, 3]^T$$

The procedure can be repeated with additional presentation of the weight vectors, until the weights of the neural network had stabilized, or until some other termination criterion had been satisfied. From the above example it can be seen that learning with a discrete activation function $f(\mathbf{w}^T\mathbf{x}) = \pm 1$, and learning rate $\beta = 1$, results in addition or subtraction of the entire input vector to and from the weight vector respectively. Note that with continuous activation functions, such as unipolar or bipolar sigmoids, the change in the weight vector is some fraction of the input vector instead.

b) Winner-takes-all rule

Like the outstar rule, the winner-takes-all rule can be explained best by considering an ensemble of neurons, arranged in a layer or some regular geometry. No output data are required to train the neural network (i.e. unsupervised training). Each node in the ensemble measures the (Euclidean) distance of its weights to the input values (exemplars) presented to the nodes.

For example, if the input data consist of m-dimensional vectors of the form $\mathbf{x} = [x_1, x_2, \ldots x_m]^T$, then each node will have M weight values, which can be denoted by $\mathbf{w}_i = [w_{i1}, w_{i2} \ldots w_{im}]$. The Euclidean distance $D_i = ||\mathbf{x} - \mathbf{w}_i||$ between the input vectors and the weight vectors of the neural network is then computed for each of the nodes and the winner is determined by the minimum Euclidean distance. This is equivalent to

$$\mathbf{w}_p{}^T\mathbf{x} = \max_{i = 0, 1, 2, \ldots m}(\mathbf{w}_p{}^T\mathbf{x}) \qquad (1.32)$$

The weights of the winning node (assumed to be the p'th node), as well as its neighbouring nodes (constituting the adaptation zone associated with the winning node) are subsequently adjusted in order to move the weights closer to the input vector, as follows

$$\Delta\mathbf{w}_p = \alpha(\mathbf{x} - \mathbf{w}_p), \text{ or}$$

$$\Delta\mathbf{w}_{ip} = \alpha(x_j\text{-}w_{ip,old}) \qquad (1.33)$$

where α is an appropriate learning coefficient which decreases with time (typically starting at 0.4 and decreasing to 0.1 or lower).

The adjustment of the weights of the nodes in the immediate vicinity of the winning node is instrumental in the preservation of the order of the input space and amounts to an order preserving projection of the input space onto the ensemble of nodes (typically a one- or two-dimensional layer). As a result similar inputs are mapped to similar regions in the output space, while dissimilar inputs are mapped to different regions in the output space. The winner-takes-all rule is especially important in the multivariate analysis of data, and will be considered in more detail at a later stage.

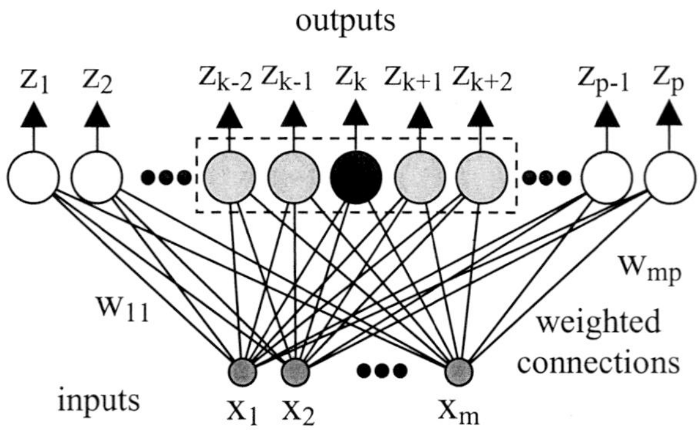

Figure 1.4. One-dimensional array of p competitive nodes ($z_1, z_2, \ldots z_p$), each receiving m inputs ($x_1, x_2, \ldots x_m$), showing the winning node (z_k), surrounded by neighbouring nodes (z_{k-2}, z_{k-1}, z_{k+1} and z_{k+2}) in its neighbourhood (broken lines).

c) Outstar learning rule

The outstar rule is best explained in the context of an ensemble of nodes, arranged in a layer. The rule applies to supervised learning, but is also designed to allow the network to extract statistical features from the inputs and outputs of the neural network. Weight adjustments are computed as follows

$$\Delta \mathbf{w}_j = \beta(d - \mathbf{w}_j) \tag{1.34}$$

or in terms of the adjustments of the individual weights of the nodes, for i = 1, 2, … m

$$\Delta \mathbf{w}_{ij} = \beta(d_i - \mathbf{w}_{ij}) \tag{1.34}$$

In contrast with the previous rules, the weight vector that is updated fans out of the j'th node. The learning rate β is typically a small positive parameter that decreases as learning progresses.

Although the above rules are not the only ones used to train neural networks, they are widely used, although their application does not guarantee convergence of the neural network being trained. More detailed discussions are available in other sources, such as Zurada (1992), Bishop (1995) and Haykin (1999).

1.6. NEURAL NETWORK MODELS

1.6.1. Multilayer Perceptrons

Since multilayer perceptron neural networks are by far the most popular (and simple) neural networks in use in the process engineering industries, they are considered first.

a) Basic structure

As mentioned previously, in multilayer perceptron neural networks the processing nodes are usually divided into disjoint subsets or layers, in which all the nodes have similar computational characteristics.

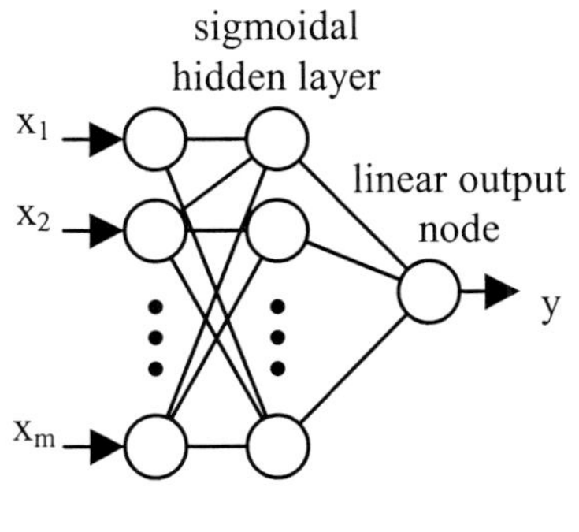

Figure 1.5. Structure of a typical multilayer perceptron neural network.

A distinction is made between input, hidden and output layers depending on their relation to the information environment of the network. The nodes in a particular layer are linked to other nodes in successive layers by means of artificial synapses or weighted connections (adjustable numeric values), as shown in Figure 1.5. These weights form the crux of the model, in that they define a distributed internal relationship between the input and output activations of the neural network.

The development of neural network models thus consists of first determining the overall structure of the neural network (number of layers, number of nodes per layer, types of nodes, etc.). Once the structure of the network is fixed, the parameters (weights) of the network have to be determined. Unlike the case with a single node, a network of nodes requires that the output error of the network be apportioned to each node in the network.

b) Back propagation algorithm

The back propagation algorithm can be summarized as follows, for a network with a single hidden layer with q nodes and an output layer with p nodes, without loss in generalization. Before training, the learning rate coefficient (η) and the maximum error (E_{max}) are specified by the user.

i) Initialize the weight matrices of the hidden layer (**V**) and the output layers (**W**) to small random values.

ii) Compute the response of the hidden layer (**y**) and the output layer (**z**) of the neural network, when presented with an exemplar of the form [**x**|**t**], where $\mathbf{x} \in \Re^m$, and $\mathbf{t} \in \Re^p$, i.e.

$$\mathbf{y} = \phi(\mathbf{Vx})$$

$$\mathbf{z} = \phi(\mathbf{Wy})$$

iii) Compute the error, $E_{new} = E_{old} + \frac{1}{2}||\mathbf{z} - \mathbf{t}||^2$, for k = 1, 2, … p.

iv) Calculate the error terms δ_z and δ_y, with $\delta_z \in \Re^p$ and $\delta_y \in \Re^q$, i.e.

$$\delta_z = \tfrac{1}{2}[(t_k - z_k)(1 - z_k^2)], \text{ for } k = 1, 2, \dots p$$

$$\delta_y = \mathbf{w}_j^T \delta_z f'y = \tfrac{1}{2}(1 - y_j^2)\Sigma_{k=1}\delta_{zk}{}^p w_{kj})], \text{ for } j = 1, 2, \dots q$$

v) Adjust the weights of the output layer

$$\mathbf{w}_{new} = \mathbf{w}_{old} + \eta\delta_z\mathbf{y}^T \text{ or}$$

$$\mathbf{w}_{kj,new} = \mathbf{w}_{kj,old} + \eta\delta_{zk}\mathbf{y}_j, \text{ for } k = 1, 2, \dots p \text{ and } j = 1, 2, \dots q)$$

vi) Adjust the weights of the hidden layer

$$\mathbf{v}_{new} = \mathbf{v}_{old} + \eta\delta_y\mathbf{z}^T \text{ or}$$

$$\mathbf{v}_{ji,new} = \mathbf{w}_{ji,old} + \eta\delta_{yj}\mathbf{z}_i, \text{ for } j = 1, 2, \dots q \text{ and } i = 1, 2, \dots m)$$

vii) If there are more exemplars in the training set, go to step ii)

viii) If $E < E_{max}$, stop, otherwise go to step ii)

The back propagation algorithm highly efficient, as the number of operations required to evaluate the derivatives of the error function scales with $O(N_W)$, for a sufficiently large number of weights N_W. Since the number of weights are usually much larger than the number of nodes in the neural network, most of the computational effort in the forward cycle during training is devoted to the evaluation of the weighted sums (one multiplication and one addition per connection) in order to determine the activations of the nodes, while the evaluation of the activation functions can be seen as a small computational overhead (Bishop, 1995). Normally, numerical evaluation of the derivatives would each be $O(N_W)$, so that the evaluation of the derivatives would scale to $O(N_W^2)$. However, with the back propagation algorithm, evaluation for each exemplar scales to $O(N_W)$ only. This is a crucial gain in efficiency, since training can involve a considerable computational effort.

1.6.2. Kohonen self-organizing mapping (SOM) neural networks

Self-organizing neural networks or self-organizing (feature) maps (SOM) are systems that typically create two- or three-dimensional feature maps of input data in such a way that order is preserved. This characteristic makes them useful for cluster analysis and the visualization of topologies and hierarchical structures of higher-dimensional input spaces.

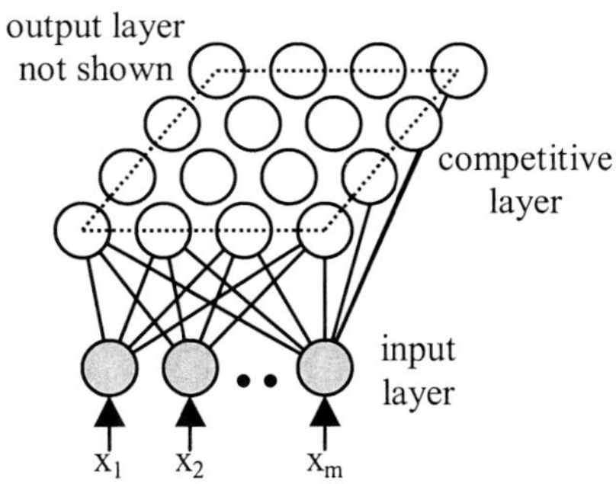

Figure 1.6. A self-organizing mapping neural network (Kohonen).

Self-organizing systems are based on competitive learning, that is the outputs of the network compete among themselves to be activated or fired, so that only one node can win at any given time. These nodes are known as winner-takes-all nodes.

Self-organizing or Kohonen networks do not require output data to train the network (i.e. training is unsupervised). Such a network typically consists of an input layer, which is fully connected to one or more two-dimensional Kohonen layers, as shown in Figure 1.6. Each node in the Kohonen layer measures the (Euclidean) distance of its weights to the input values (exemplars) fed to the layer. For example, if the input data consist of m-dimensional vectors of the form $\mathbf{x} = \{x_1, x_2, \dots x_m\}$, then each Kohonen node will have m weight values, which can be denoted by $\mathbf{w}_i = \{w_{i1}, w_{i2}, \dots w_{im}\}$. The Euclidean distances $D_i = \| \mathbf{x} - \mathbf{w}_i \|$ between the input vectors and the weights of the network are then computed for each of the Kohonen nodes and the winner is determined by the minimum Euclidean distance.

The weights of the winning node, as well as its neighbouring nodes, which constitute the adaptation zone associated with the winning node are subsequently adjusted in order to move the weights closer to the input vector.

The adjustment of the weights of the nodes in the immediate vicinity of the winning node is instrumental in the preservation of the order of the input space and amounts to an order preserving projection of the input space onto the (typically) two-dimensional Kohonen layer. As a result similar inputs are mapped to similar regions in the output space, while dissimilar inputs are mapped to different regions in the output space.

One of the problems that have to be dealt with as far as the training of self-organizing neural networks is concerned, is the non-participation of neurons in the training process. This problem can be alleviated by modulation of the selection of (a) winning nodes or (b) learning rates through frequency sensitivity.

Frequency sensitivity entails a history-sensitive threshold in which the level of activation of the node is proportional to the amount by which the activation exceeds the threshold. This threshold is constantly adjusted, so that the thresholds of losing neurons are decreased, and those of winning neurons are increased. In this way output nodes which do not win sufficiently frequently, become increasingly sensitive. Conversely, if nodes win too often, they become increasingly insensitive. Eventually this enables all neurons to be involved in the learning process. Training of Kohonen self-organised neural networks can be summarized as follows.

a) *Summary of the SOM algorithm (Kohonen)*

Initialization: Select small random values for the initial weight vector $\mathbf{w}_j(0)$, so that the $\mathbf{w}_j(0)$ are different or $j = 1, 2, \ldots p$, where p is the number of neurons in the lattice.

i. *Sampling*: Draw a sample $\mathbf{x}$ from the input distribution with a certain probability.

ii. *Similarity matching*: Find the winning neuron $I(\mathbf{x})$ at time t, using the minimum distance Euclidean or other criterion:

iii. $I(\mathbf{x}) = \arg_j\min\|\mathbf{x}(t) - \mathbf{w}_j\|$, for $j = 1, 2, \ldots p$ (1.35)

iv. *Updating*: Modify the synaptic weights of the neurons in the lattice as follows:

$$\mathbf{w}_j(t+1) = \mathbf{w}_j(t) + \eta(t)[\mathbf{x}(t) - \mathbf{w}_j(t)],\ j \in \Lambda_{I(x)}(t) \quad (1.36)$$

$$\mathbf{w}_j(t+1) = \mathbf{w}_j(t), \text{ otherwise.}$$

where $\eta(t)$ is a time-variant learning rate parameter, $\Lambda_{I(\mathbf{x})}(t)$ is the neighbourhood function centred around the winning neuron $I(\mathbf{x})$, all of which are varied dynamically. These parameters are often allowed to decay exponentially, for example $\eta(t) = \eta_0 e^{-t/\tau}$.

For example, for Gaussian-type neighbourhood functions the modification of the synaptic weight vector $\mathbf{w}_j$ of the j'th neuron at a lateral distance d_{ji} from the winning neuron $I(\mathbf{x})$ is

$$\mathbf{w}_j(t+1) = \mathbf{w}_j(t) + \eta(t)\pi_{j,i(x)}(t)[\mathbf{x}(t) - \mathbf{w}_j(t)] \quad (1.37)$$

v. *Continuation*: Continue with step 2 until the feature map has stabilized.

b) Properties of the SOM algorithm

The feature map displays important statistical properties of the input.

- *Approximation of the input space*: The self-organised feature map Φ presented by the set of synaptic weight vectors $\{\mathbf{w}_j \mid j = 1, 2, \ldots p\}$ in the output space Θ, provides a good approximation of the input space **X**.
- *Topological ordering*: The feature map Φ is topologically ordered in that the spatial location of a neuron in the lattice corresponds to particular domain of input patterns.
- *Density matching*: Variations in the distribution of the input are reflected in the feature map, in that regions in the input space **X** from which samples are drawn with a higher probability than other regions, are mapped with better resolution and onto larger domains of the output space Θ than samples drawn with a lower probability.

If f(**x**) denotes the multidimensional probability density function of the input vector x, then the probability density function integrated over the entire input space should equal unity, that is

$$\int_{-\infty}^{\infty} f(\mathbf{x})d\mathbf{x} = 1$$

Let m(x) denote the magnification factor of the SOFM, defined as the number of neurons associated with a small volume d**x** of the input space **X**. Then the magnification factor, integrated over the entire input space should equal the total number of neurons in the network lattice, that is

$$\int_{-\infty}^{\infty} M(\mathbf{x})d\mathbf{x} = p \qquad (1.38)$$

In order to match the input density exactly in the feature map

$$M(\mathbf{x}) \propto f(\mathbf{x})$$

In other words, if a particular region in the input space occurs more often, that region is mapped onto a larger region of neurons in the network lattice. For 2D- and higher-D maps the magnification factor M(**x**) is generally not expressible as a simple function of the probability density function f(**x**) of the input vector **x**. In fact, such a relationship can only be derived for 1D-maps, and even then the magnification factor is usually not proportional to the probability density function.

Generally speaking, the map tends to underrepresent regions of high input density and conversely, to overrepresent regions of low input density. It is for this reason that heuristics (such as the conscience mechanism) are sometimes included in the SOM algorithm, i.e. to force a more exact density matching between the magnification factor of the map and the probability density function of the input. Alternatively, an information-theoretic approach can be used to construct the feature map.

1.6.3. Generative topographic maps

Generative topographic maps (Bishop et al., 1997) are density models of data based on the use of a constrained mixture of Gaussians in the data space in which the model parameters (**W** and β) are determined by maximum likelihood using the expectation-maximization algorithm.

Generative topographic maps are defined by specifying a set of points $\{x_i\}$ in a latent space, together with a set of basis functions $\{\phi_j(\mathbf{x})\}$. A constrained mixture of Gaussians is defined by adaptive parameters $\mathbf{W}$ and β, with centres $\mathbf{W}\phi(x_i)$ and common covariance $\beta^{-1}\mathbf{I}$.

As a latent variable model, a generative topographic map represents a distribution $p(\mathbf{t})$ of data in an m-dimensional space, $\mathbf{t} = (t_1, t_2, \ldots t_m)$ in terms of a set of p latent variables $\mathbf{x} = (x_1, x_2, \ldots x_p)$. The mapping between points in the p-dimensional latent space and the m-dimensional data space is represented by a function $\mathbf{y}(\mathbf{x},\mathbf{W})$, as indicated in Figure 1.7, for $p = 2$ and $m = 3$. The matrix of parameters determines the mapping (which represents the weights and biases in the case of a neural network model).

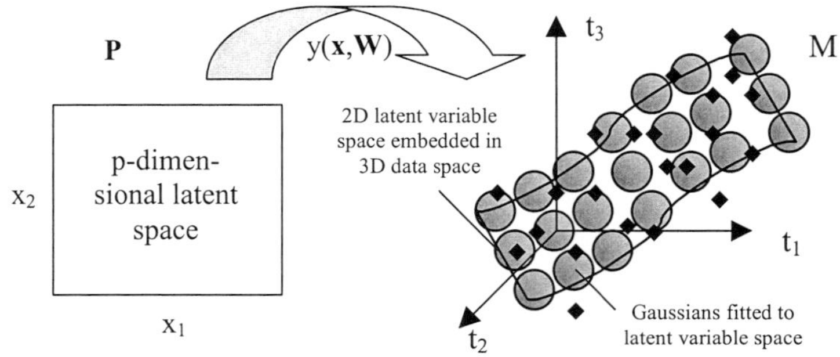

Figure 1.7. A manifold M embedded in data space is defined by the function $\mathbf{y}(\mathbf{x},\mathbf{W})$ given by the image of the latent variable space under the mapping $\mathbf{x}\rightarrow\mathbf{y}$.

If a probability distribution $p(\mathbf{x})$ is defined on the latent variable space P, this will induce a corresponding distribution $p(\mathbf{y}|\mathbf{W})$ in the data space M. If $p < m$, then the distribution in the data space will be confined to a P-dimensional manifold and would therefore be singular. However, since this manifold will only approximate the actual distribution of the data in the data space, it is appropriate to include a noise model with the $\mathbf{t}$ vector. The distribution of $\mathbf{t}$ is can be represented by spherical Gaussians[3] centred on $\mathbf{y}(\mathbf{x},\mathbf{W})$, with a common inverse variance β^{-1}, i.e.

$$p(\mathbf{t} \mid \mathbf{x}, \mathbf{W}, \beta) = (\beta/2\pi)^{m/2}\exp\{-\beta/2 \, \| \mathbf{y}(\mathbf{x},\mathbf{W}) - \mathbf{t} \|^2\} \quad (1.39)$$

The distribution in the $\mathbf{t}$ space (M), for a given matrix of parameters $\mathbf{W}$, is obtained by integration over the $\mathbf{x}$ space (P), that is

$$p(\mathbf{t} \mid \mathbf{W}, \beta) = \int p(\mathbf{t} \mid \mathbf{x}, \mathbf{W}, \beta)\, p(\mathbf{x})\, d\mathbf{x} \quad (1.40)$$

For a given data set of size n, the parameter matrix $\mathbf{W}$ and the inverse variance β^{-1} can be determined by use of maximum likelihood. In practice, the log likelihood is maximized, i.e.

$$L(\mathbf{W},\beta) = \ln \prod_{k=1}^{N} p(\mathbf{t}_k|\mathbf{W},\beta) \quad (1.41)$$

[3] Of course, other models for $p(\mathbf{t}|\mathbf{x})$ may also be appropriate.

Unlike the SOM algorithm, the GTM algorithm defines an explicit probability density given by the mixture distribution (Bishop et al., 1998). Consequently, there is a well-defined objective function given by the log likelihood and convergence to a local maximum of the objective function is guaranteed by use of the expectation maximization (EM) algorithm (Dempster et al., 1977). In contrast, the SOM algorithm does not have an explicit cost function. Moreover, conditions under which self-organisation occurs in SOM neural networks are not quantified and in practice it is necessary to validate the spatial ordering of trained SOM models.

1.6.4. Learning vector quantization neural networks

Vector quantization for data compression is an important application of competitive learning, and is used for both the storage and transmission of speech and image data.

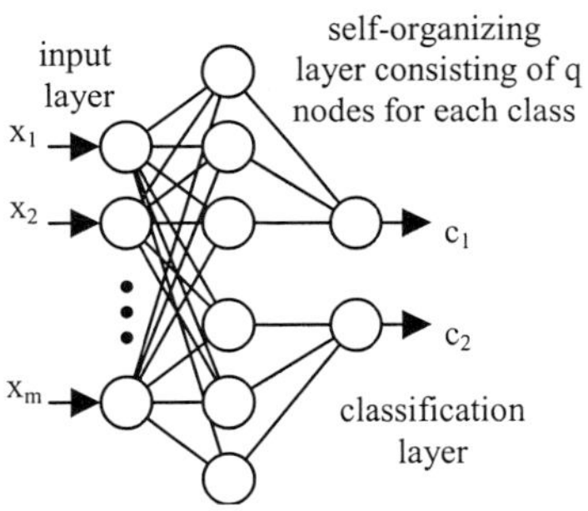

Figure 1.8. A learning vector quantization neural network with two classes.

The essential concept on which learning vector quantization networks (see Figure 1.8) is based, is that a set of vectors can be distributed across a space in such a way that their spatial distribution corresponds with the probability distribution of a set of training data. The idea is to categorise the set of input vectors into c classes $C = c_1, c_2, \ldots c_C$, each class of which is characterized by a class or prototype vector. Each of the original input vectors is subsequently represented by the class of which it is a member, which allows for high ratios of data compression. The components of the vectors usually have continuous values, and instead of storing or transmitting the prototype vectors, only their indices need to be handled, once a set of prototype vectors or a codebook has been defined.

The class of a particular input can be found by locating its nearest prototype vector, using an ordinary (Euclidean) metric, i.e. $| w_i^* - c_i | \leq | w_i - c_i |$. This divides the vector space into a so-called Voronoi (or Dirichlet) tesselation, as indicated in Figure 1.9. Learning vector quantization networks differ from supervised neural networks, in that they construct their own representations of categories among input data.

A learning vector quantization network contains an input layer, a Kohonen layer which performs the classification based on the previously learned features of the various classes, and an output layer, as shown in Figure 1.8. The input layer is comprised of one node for each feature or input parameter of the various classes, while the output layer contains one node for each class. Although the number of classes is predefined, the categories (q) assigned to these classes are not. During training the Euclidean distance (d_i) between a training exemplar ($\mathbf{x}$) and the weight vector ($\mathbf{w}_i$) of each node is computed. That is

$$d_i = \text{norm}(\mathbf{w}_i-\mathbf{x}) = [\Sigma_j,n(w_{ij}-x_j)^2\}]^{1/2} \qquad (1.55)$$

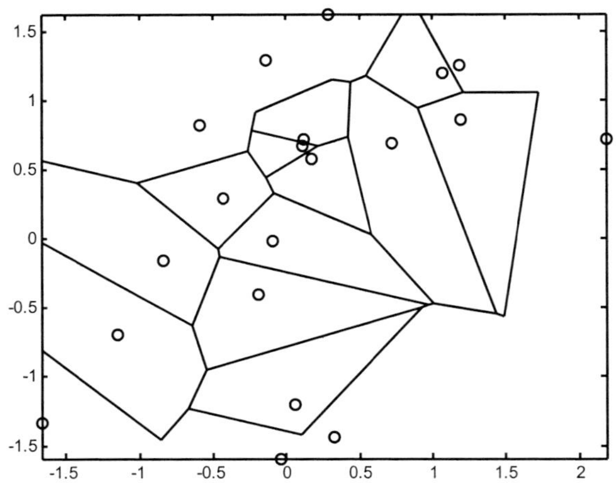

Figure 1.9. An example of Voronoi tessellation of data in two dimensions.

If this winning node and the training vector share the same class, the winning node is moved towards the training vector, otherwise it is moved away, or repulsed, i.e.

◊ $w_p = w_p + \beta(x_p-w_p)$, if winning node is in the correct class (1.56)

◊ $w_p = w_p - \beta(x_p-w_p)$, if winning node is not in the correct class (1.57)

As a consequence, the nodes assigned to a class migrate to a region associated with their class. In the classification mode, the distance of the input vector to each node is determined and the vector is assigned to the class of the winning node.

1.6.5. Probabilistic neural networks

Many methods of pattern classification and feature evaluation presuppose complete knowledge of class conditional probability density functions $p(\mathbf{x}|c_j)$, with $j = 1, 2, \ldots m$. In practice the actual probability structure of the classes are usually unknown and the only information available to the analyst is a set of exemplars with known class memberships.

It is therefore necessary to infer the unknown probability density functions from the data. One way to accomplish this is to make use of kernels. For example, given a particular exemplar $\mathbf{x}_i$ for a class c_j, we can assert that $p(\mathbf{x}|c_j)$ assumes a non-zero value at the point $\mathbf{x}_i$. Moreover, assuming $p(\mathbf{x}|c_j)$ to be continuous, it can be inferred that $p(\mathbf{x}|c_j)$ will assume non-zero values in the immediate vicinity of observation $\mathbf{x}_i$. The information about $p(\mathbf{x}|c_j)$ gained by observing $\mathbf{x}_i \in c_j$ can be repesented by a function $K(\mathbf{x},\mathbf{x}_i)$ centred at $\mathbf{x}_i$. This function (known as a kernel function) attains a maximum value at $\mathbf{x}_i$ and decreases monotonically as the distance from $\mathbf{x}_i$ increases, as indicated in Figure 1.10(a).

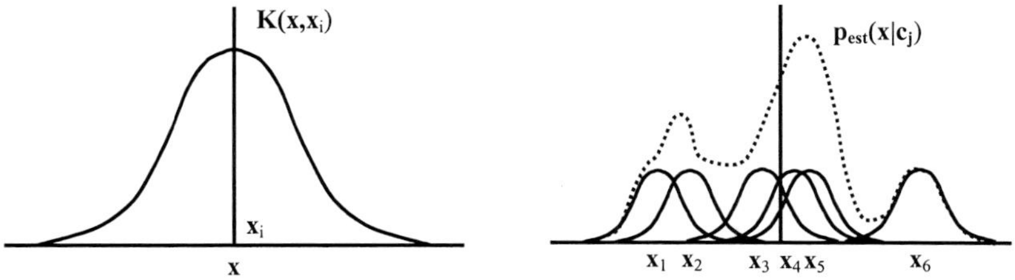

Figure 1.10(a) A one-dimensional Gaussian kernel function, and (b) estimation of the probability density function $p_{est}(\mathbf{x}|c_j)$ of class c_j by summing the contributions of the kernels centred on the exemplars for the class.

With a set of n exemplars for a given class c_j, $p(\mathbf{x}|c_j)$ can be estimated by calculating the average of all the contributions of all the exemplars, i.e.

$$p_{est}(\mathbf{x}|c_j) = \Sigma_{i=1}^{n} K(\mathbf{x},\mathbf{x}_i)/n \tag{1.58}$$

As indicated in Figure 1.10(b), exemplars close to each other give rise to larger values of $p_{est}(\mathbf{x}|c_j)$ than exemplars situated further apart.

Clearly, the contributions of the kernels to $p_{est}(\mathbf{x}|c_j)$ also depend on their range of influence. If this is very small, the estimate of the probability density function will be spiky, while too large a range of influence will miss local variations in $p(\mathbf{x}|c_j)$. Intuition dictates that as the number of exemplars increase, the influence of $K(\mathbf{x},\mathbf{x}_i)$ should decrease progressively, and conversely, if few exemplars are available, the influence of $K(\mathbf{x},\mathbf{x}_i)$ should be large to smooth out sampling effects. The kernel function reflecting this arrangement should have the form

$$K(\mathbf{x},\mathbf{x}_i) = \rho^{-\lambda} h[d(\mathbf{x},\mathbf{x}_i)/\rho] \tag{1.59}$$

Where ρ is a parameter of the estimator that depends on the sample size and satisfies

$$\lim_{n\to\infty} \rho^{\lambda}(n) = 0 \tag{1.60}$$

$d(\mathbf{x},\mathbf{x}_i)$ is a suitable metric and $h(\bullet)$ is a function attaining a maximum at $d(\mathbf{x},\mathbf{x}_i) = 0$ and decreasing monotonically as $d(\mathbf{x},\mathbf{x}_i)$ increases. Provided that $h(\bullet)$ is a non-negative function, the only constraint which $h(\bullet)$ has to satisfy is

$$\int K(\mathbf{x},\mathbf{x}_i) d\mathbf{x} = 1 \qquad (1.61))$$

Conditions (1.60) and (1.61) guarantee that $p_{est}(\mathbf{x}|c_j)$ is a density function providing an unbiased and consistent estimate of $p(\mathbf{x}|c_j)$. The most important kernels are hyperspheric kernels (Figure 1.11(a) and equation 1.62), hypercubic kernels (Figure 1.11(b) and equation 1.63) and Gaussian kernels (Figure 1.11(c) and equation 1.64).

$$K(\mathbf{x},\mathbf{x}_i) = 1/v, \text{ if } \{\mathbf{x}|d_E(\mathbf{x},\mathbf{x}_i) \le \rho\}$$

$$= 0, \text{ if } \{\mathbf{x}|d_E(\mathbf{x},\mathbf{x}_i) > \rho\} \tag{1.62}$$

where $d_E(\mathbf{x},\mathbf{x}_i) = [(\mathbf{x} - \mathbf{x}_i)^T(\mathbf{x} - \mathbf{x}_i)]^*$ is the Euclidean distance metric and v is the volume of hypersphere with radius ρ.

$$K(\mathbf{x},\mathbf{x}_i) = (2\rho)^{-\lambda}, \text{ if } \{\mathbf{x}|d_T(\mathbf{x},\mathbf{x}_i) \leq \rho\}$$

$$= 0, \text{ if } \{\mathbf{x}|d_T(\mathbf{x},\mathbf{x}_i) > \rho\} \tag{1.63}$$

where $d_T(\mathbf{x},\mathbf{x}_i) = \max_j|(\mathbf{x} - \mathbf{x}_i)|$ is the Chebyshev distance metric. Unlike the Gaussian estimator, hypercubic kernels are easy to calculate.

$$K(\mathbf{x},\mathbf{x}_i) = [\rho^2 2\pi)^{\lambda}|\mathbf{Q}|]^{-½}\exp[-½d_Q(\mathbf{x},\mathbf{x}_i)/\rho^2] \tag{1.64}$$

where $d_Q(\mathbf{x},\mathbf{x}_i) = (\mathbf{x} - \mathbf{x}_i)^T Q(\mathbf{x} - \mathbf{x}_i)$ is a quadratic distance and $\mathbf{Q}$ is a positive definite scaling matrix, typically the sampling covariance matrix $\mathbf{S}_j$ of class c_j.

Probabilistic neural networks (see Figure 1.11) are based on the use of Bayesian classification methods and as such provide a powerful general framework for pattern classification problems. These networks use exemplars to develop distribution functions as outlined above, which in turn are used to estimate the likelihood of a feature vector belonging to a particular category.

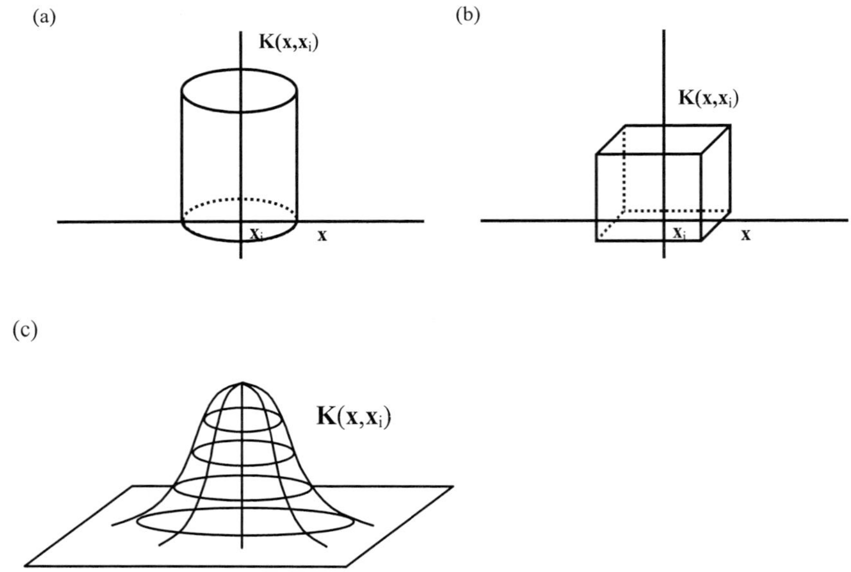

Figure 1.11(a). A hyperspheric, (b) cubic and c) Gaussian kernel.

These estimates can be modified by the incorporation of *a priori* information. Suppose a classification problem consists of p different classes c_1, c_2, ... c_p, and that the data on which the classification process are based can be represented by a feature vector with m dimensions

$$\mathbf{x} = [x_1, x_2, \ldots x_m]^T$$

If $F(\mathbf{x}) = [F_1(\mathbf{x}), F_2(\mathbf{x}), \ldots F_p(\mathbf{x})]^T$ is the set of probability density functions of the class populations and $\mathbf{A} = [a_1, a_2, \ldots a_p]^T$ is the set of *a priori* probabilities that a feature vector belongs to a particular class, then the Bayes classifier compares the p values $a_1.F_1(\mathbf{x}), a_2.F_2(\mathbf{x}), \ldots a_p.F_p(\mathbf{x})$ and determines the class with the highest value.

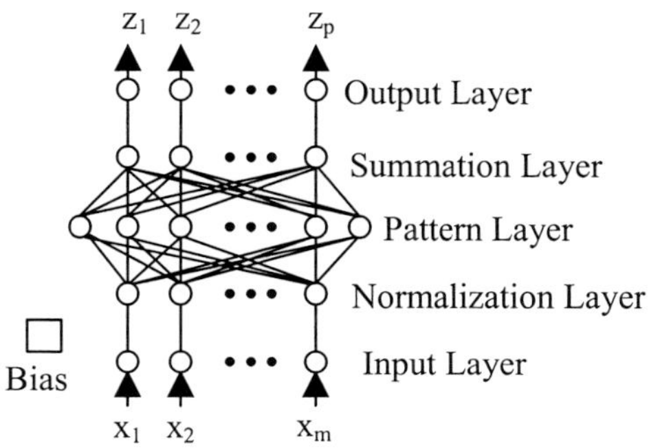

Figure 1.12. Structure of a probabilistic neural network.

Before this decision rule (in which the multivariate class of probability density functions are evaluated, weighted and compared) can be implemented, the probability density functions have to be constructed. Parzen estimation is a non-parametric method of doing so, in which no assumption is made with regard to the nature of the distributions of these functions, that is

$$F_k(\mathbf{x}) = [B/m_k]\Sigma_j \exp[-(\mathbf{x}-\mathbf{x}_{kj})^T(\mathbf{x}-\mathbf{x}_{kj})/4\sigma^2] \quad (1.65)$$

where $B = 1/(2\pi^{p/2}\sigma^p)$.

The Parzen estimator is constructed from n training data points available. As explained above, the exponential terms or Parzen kernels are small multivariate Gaussian curves that are added together and smoothed (B-term). As shown in Figure 1.12, the neural network version of this Bayesian classifier consists of an input layer, a normalizing layer (which normalizes the feature vector $\mathbf{x}$, so that $\mathbf{x}^T\mathbf{x} = 1$), a pattern or exemplar layer, which represents the Parzen kernels, a summation layer in which the kernels are summed, and a competitive output layer. The weights associated with the nodes in the output (class) layer constitute the *a priori* probabilities a_k ($k = 1, 2, \ldots p$) of the occurrence of the classes, and usually assume equal values unless specified otherwise.

Probabilistic neural networks are also useful for pattern recognition and classification problems, especially where the probabilities of some events or classes are known in advance, since these can be incorporated directly into the network.

1.6.6. Radial basis function neural networks

In can be shown that in solving problems concerning nonlinearly separable patterns, there is practical benefit to be gained in mapping the input space into a new space of sufficiently high dimension. This nonlinear mapping in effect turns a nonlinearly separable problem into a linearly separable one. The idea is illustrated in Figure 1.13, where two interlocked two-dimensional patterns are easily separated by mapping them to three dimensions where they can be separated by a flat plane. In the same way it is possible to turn a difficult nonlinear approximation problem into an easier linear approximation problem.

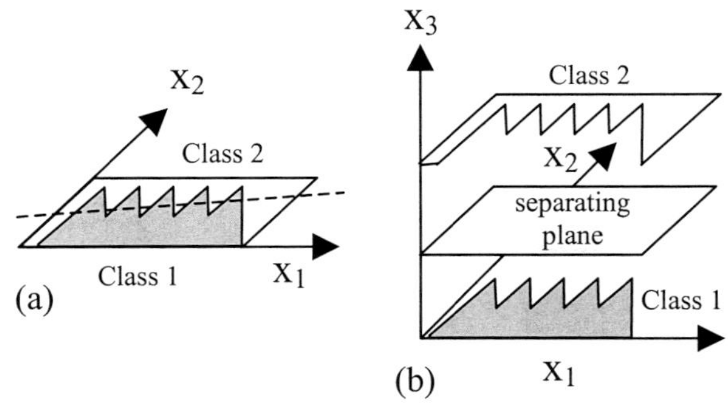

Figure 1.13. Linear separation of two nonlinearly separable classes, after mapping to a higher dimension.

Consider therefore without loss of generality a feedforward neural network with an input layer with p input nodes, a single hidden layer and an output layer with one node. This network is designed to perform a nonlinear mapping from the input space to the hidden space, and a linear mapping from the hidden space to the output space.

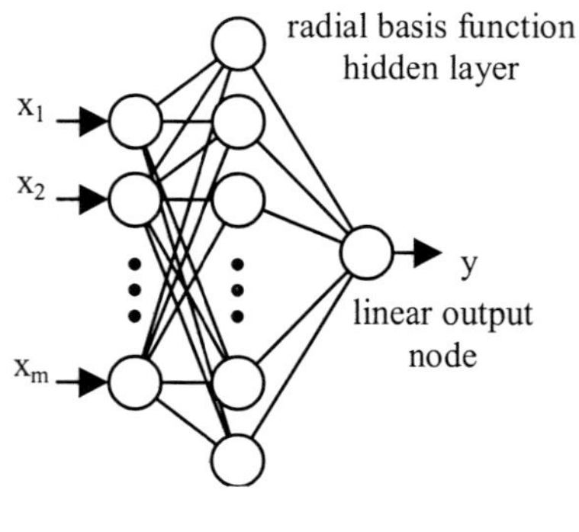

Figure 1.14. Structure of a radial basis function neural network.

Overall the network represents a mapping from the m-dimensional input space to the one-dimensional output space, as follows, $s:\Re^m \rightarrow \Re^1$ and the map s can be thought of as a hyper-surface $\Gamma \subset \Re^{m+1}$ in the same way as we think of the elementary map $s: \Re^1 \rightarrow \Re^1$, where $s(x) = x^2$, as a parabola drawn in $\Re^2$-space. The curve Γ is a multidimensional plot of the output as a function of the input. In practice the surface Γ is unknown, but exemplified by a set of training data (input-output pairs).

As a consequence, training constitutes a fitting procedure for the hypersurface Γ, based on the input-output examples presented to the neural network. This is followed by a generalization phase, which is equivalent to multivariable interpolation between the data points, with interpolation performed along the estimated constrained hypersurface (Powell, 1985).

In a strict sense the interpolation problem can be formulated as follows. Given a set of n different observations on m variables $\{\mathbf{x}_i \in \Re^m \mid i = 1, 2, \ldots n\}$ and a corresponding set of n real numbers $\{z_i \in \Re^1 \mid i = 1, 2, \ldots n\}$, find a function F: $\Re^m \rightarrow \Re^1$ that complies with the interpolation condition $F(\mathbf{x}_i) = z_i$, for i = 1, 2, ... n. Note that in the strict sense specified, the interpolation surface is forced to pass through all the training data points.

Techniques based on radial basis functions are based on the selection of a function F of the following form

$$F(\mathbf{x}) = \Sigma_1^n w_i \varphi (||\mathbf{x}-\mathbf{x}_i||) \tag{1.66}$$

where $\{\varphi(||\mathbf{x}-\mathbf{x}_i||) \mid i = 1, 2, \ldots n\}$is a set of n arbitrary functions, known as radial basis functions. $||\cdot||$ denotes a norm that is usually Euclidean. The known data points x_i typically form the centres of the radial basis functions. Examples of such functions are multiquadrics, $\varphi(r) = (r^2 + c^2)^{1/2}$, inverse multiquadrics, $\varphi(r) = (r^2 + c^2)^{-1/2}$, Gaussian functions $\varphi(r) = \exp\{-r^2/(2\sigma^2)\}$ and thin-plate splines $\varphi(r) = (r/\sigma)\log(r/\sigma)$, where c and σ are postive constants and $r \in \Re$.

By use of the interpolation condition $F(\mathbf{x}_i) = z_i$ and equation (1.66), a set of simultaneous linear equations for the unknown coefficients or weights (w_i) of the expansion can be obtained.

$$\begin{bmatrix} \varphi_{11} & \varphi_{12} & \cdots & \varphi_{1n} \\ \varphi_{21} & \varphi_{22} & \cdots & \varphi_{2n} \\ \cdots & \cdots & \cdots & \cdots \\ \varphi_{n1} & \varphi_{n2} & \cdots & \varphi_{nn} \end{bmatrix} \begin{bmatrix} w_1 \\ w_2 \\ \cdots \\ w_n \end{bmatrix} = \begin{bmatrix} z_1 \\ z_2 \\ \cdots \\ z_n \end{bmatrix}$$

where $\varphi_{ij} = \varphi(||x_i - x_j||)$, for i, j = 1, 2, ... n. Moreover, the n x 1 vectors $\mathbf{w} = [w_1, w_2, \ldots w_n]^T$ and $\mathbf{z} = [z_1, z_2, \ldots z_n]^T$ represent the linear weight vector and target or desired response vector respectively. With $\Phi = \{\varphi_{ij} \mid i, j = 1,2, \ldots n\}$ the n x n interpolation matrix, $\mathbf{\Phi w} = \mathbf{z}$ represents a more compact form of the set of simultaneous linear equations.

For a certain class of radial basis functions, such as inverse multiquadrics (equation 1.67) and Gaussian functions (equation 1.68) the n x n matrix Φ is positive definite.

$$\varphi(r) = (r^2 + c^2)^{-1/2} \tag{1.67}$$

for $c > 0$ and $r \geq 0$, and

$$\varphi(r) = \exp\{- r^2/(2\sigma^2)\} \tag{1.68}$$

for $\sigma > 0$ and $r \geq 0$

If all the data points are distinct, and the matrix Φ is positive definite, then the weight vector can be obtained from $\mathbf{w} = \Phi^{-1}\mathbf{z}$. If the matrix is arbitrarily close to singular, perturbation of the matrix can help to solve for $\mathbf{w}$. These radial basis functions are used for interpolation, where the number of basis functions is equal to the number of data points.

Although the theory of radial basis function neural networks is intimately linked with that of radial basis functions themselves (a main field of study in numerical analysis), there are some differences. For example, with radial basis function neural networks, the number of basis functions need not be equal to the number of data points and is typically much less. Moreover, the centres of the radial basis functions need not coincide with the data themselves and the widths of the basis functions also do not need to be the same. The determination of suitable centres and widths for the basis functions is usually part of the training process of the network. Finally, bias values are typically included in the linear sum associated with the output layer to compensate for the difference between the average value of the targets and the average value of the basis functions over the data set (Bishop, 1995).

In its most basic form, the construction of a radial basis function neural network involves three different types of layers. These networks typically consist of input layers, hidden (pattern) layers, as well as output layers, as shown in Figure 1.14. The input nodes (one for each input variable) merely distribute the input values to the hidden nodes (one for each exemplar in the training set) and are not weighted. In the case of multivariate Gaussian functions[4], the hidden node activation functions can be described by

$$z_{i,j}(\mathbf{x}_j,\alpha_i,\beta_i) = \exp(-||\alpha_i - \mathbf{x}_j||^2/\beta_i^2) \tag{1.69}$$

where $\mathbf{x}_j = \{x_1, x_2, \ldots x_m\}_j$ is the j'th input vector of dimension m presented to the network, $z_{i,j}(x_j,\alpha_i,\beta_i)$ is the activation of the i'th node in the hidden layer in response to the j'th input vector $\mathbf{x}_j$. m+1 parameters are associated with each node, viz. $\alpha_i = \{\alpha_1, \alpha_2, \ldots \alpha_m\}_i$, as well as β_i, a distance scaling parameter which determines the distance in the input space over which the node will have a significant influence.

The parameters α_i and β_i function in much the same way as the mean and standard deviation in a normal distribution. The closer the input vector to the pattern of a hidden unit (i.e. the smaller the distance between these vectors, the stronger the activity of the unit. The hidden layer can thus be considered to be a density function for the input space and can be used to derive a measure of the probability that a new input vector is part of the same distribution as the training vectors. Note that the training of the hidden units is unsupervised, i.e. the pattern layer representation is constructed solely by self-organisation.

Whereas the α_i vectors are typically found by vector quantization, the β_i parameters are usually determined in an ad hoc manner, such as the mean distance to the first k nearest α_i centres. Once the self-organizing phase of training is complete, the output layer can be trained using standard least mean square error techniques.

Each hidden unit of a radial basis function network can be seen as having its own receptive field, which is used to cover the input space. The output weights leading from the hidden units

---

[4] The use of Gaussian radial basis functions is particularly attractive in neural networks, since these are the only functions that are factorizable, and can thus be constructed from 1- and 2-dimensional radial basis functions.

to the output nodes subsequently allow a smooth fit to the desired function. Radial basis function neural networks can be used for classification, pattern recognition and process modelling and can model local data more accurately than multilayer perceptron neural networks. They perform less well as far as representation of the global properties of the data are concerned.

The classical approach to training of radial basis function neural networks consists of unsupervised training of the hidden layer, followed by supervised training of the output layer, which can be summarized as follows.

i) *Estimation of clusters centres in hidden layer*

- Start with a random set of cluster centres $c = \{c_1, c_2, \ldots c_k\}$.
- Read r'th input vector $\mathbf{x}_r$.
- Modify closest cluster centre (the learning coefficient $\eta$is usually reduced with time:

$$\mathbf{c}_k^{new} = \mathbf{c}_k^{old} + \eta(\mathbf{x}_r - \mathbf{c}_k^{old}) \quad (1.70)$$

- Terminate after a fixed number of iterations, or when $\eta = 0$.

ii) *Estimation of width of activation functions*

- The width of the transfer functions of each of the Gaussian kernels or receptive fields is based on a P nearest neighbour heuristic.

$$\sigma_k = \{1/P\Sigma_p||\mathbf{c}_k - \mathbf{c}_{kp}||^2\}^{1/2} \quad (1.71)$$

where $\mathbf{c}_{kp}$ represent the p'th nearest neighbour of the k'th cluster c_k.

iii) *Training of the output layer*

The output layer is trained by minimization of a least squares criterion and is equivalent of parameter estimation in linear regression, i.e. it does not involve a lengthy process, since there is only one linear (output) layer.

In summary, when compared with multilayer perceptrons

- Radial basis function neural networks have single hidden layers, whereas multilayer perceptrons can have more than one hidden layer. It can be shown that radial basis function neural networks require only one hidden layer to fit an arbitrary function (as opposed to the maximum of two required by multilayer perceptrons). This means that training is considerably faster in radial basis function networks.

- In contrast with radial basis function neural networks, a common neuron model can be used for all the nodes in an multilayer perceptron. In radial basis function networks the hidden layer neurons differ markedly from those in multilayer perceptrons.

- In radial basis function networks the hidden nodes are nonlinear and the output nodes linear, while in multilayer perceptrons the hidden and output nodes can be linear or nonlinear.

One of the drawbacks of kernel-based approximations (such as radial basis function neural networks) is that they suffer from the so-called curse of dimensionality, which is associated with the exponential increase in the required number of hidden nodes with an increase in the dimensionality of the input space. These problems are particularly acute in large-scale problems, such as those concerned with image analysis and speech recognition.

Other approaches to the training of radial basis function neural networks include one-phase learning, where only the output weights are adjusted through some kind of supervised optimization. The centres are subsampled from the input data and the widths of the Gaussians are all equal and predefined. Support vector learning (see 1.5.8) is a special example of one-phase training. Variants of two-phase learning differ mainly in the way that the radial basis functions are determined. For example, Kubat (1998) has proposed a method to transform the disjoint hyperectangular regions represented by the leaves of a decision tree into a set of centres and scaling parameters in order to initialize radial basis function neural networks. Similarly, Kohonen's learning vector quantization can also be used to determine prototypes for basis functions. Finally, three-phase learning (Schwenker et al., 2001) entails separate training of the hidden and output layers of the neural network, followed by a third phase of optimization of the entire network architecture.

1.6.7. Adaptive resonance theory neural networks

Adaptive resonance theory neural networks (Carpenter and Grossberg, 1987; 1995) are designed to discover clusters in an unsupervised mode, similar to the Kohonen neural network. Unlike the way in which other neural networks are used, clusters are discovered in a controlled way and new clusters can be accommodated without disruption of previously established cluster patterns.

In essence, the first exemplar defines the first cluster, after which the network generates new clusters if the distances of subsequent exemplars from established cluster centres exceed a certain threshold. The network has a two-layered architecture, as shown in Figure 1.15.

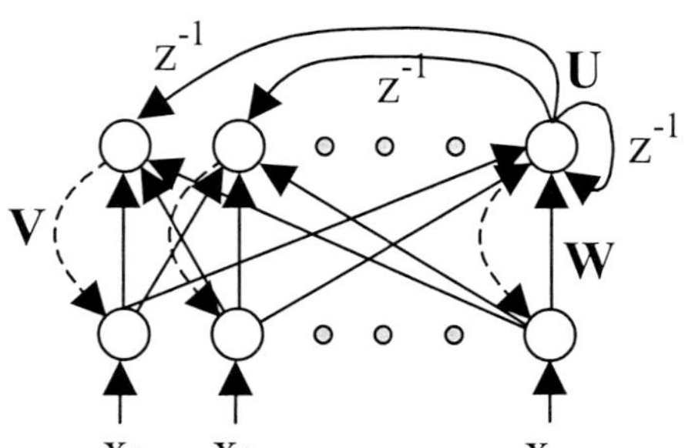

Figure 1.15. Architecture of an adaptive resonance theory neural network. z^{-1} indicates unit delays.

The network is defined by three weight matrices, **U**, **V** and **W**, of which only **V** and **W** are trained, since **U**, the matrix of recurrent connections in the top layer, is fixed. The top layer computes the matching score, reflecting the degree of similarity of a given exemplar to previously encoded clusters. This similarity of the input with the best matching cluster (winning node in then top layer) is then checked against a predefined threshold value or vigilance parameter and if sufficiently similar, the best matching cluster is updated. If not sufficiently similar, the search for a new cluster continues and if none of the existing cluster match the input sufficiently closely, a new cluster is created. The architecture and algorithm are described in more detail below for a relatively simple ART-1 system, which deals with binary inputs only. However, extension of the same idea to continuous-valued inputs is straight-forward.

a) *Network architecture*

Consider an ART neural network such as the one shown in Figure 1.15, with m nodes in the bottom layer and q nodes in the top layer. The weight matrix **W** represents the bottom-up connections from the bottom layer to the top layer, while the matrix **V** represents the top-down connections from the top layer to the bottom layer. As mentioned previously, the matrix **U** represents lateral feedback connections from the top layer and can actually be omitted in simplified versions of ART neural networks, since these connections affect neither the identification of the best matching nodes in the top layer, nor the assessment of the similarity of the input exemplars to the existing set of clusters.

Inputs are passed through the bottom layer to the top layer, so that the initializing input for the k'th node of the top layer is the scalar product similarity measure between the exemplar x and the weight vector of the k'th node (with weights $\mathbf{w}_k = [w_{1k}, w_{2k} \ldots w_{mk}]$ and $k = 1, 2, \ldots q$).

$$d^0_k = \mathbf{w}_k^T\mathbf{x}, \text{ for } k = 1, 2, \ldots q \qquad (1.72)$$

The first index in the double subscript denotes the input node in the bottom layer of the network, while the second index denotes the node in the top layer of the network. For bipolar binary exemplars, this amounts to the number of positions of agreement minus the number of positions of disagreement (matching or non-matching zeros and ones).

With the input to the k'th node (with weight vector w_k) in the top layer denoted by y, the activations of the nodes in the top layer are computed as follows.

$$\phi = \tfrac{1}{2}\mathbf{y}^T\mathbf{w}_k + m/2 \qquad (1.73)$$

The value m/2 is a fixed bias added to the input of each node and the factor ½ is convenient for scaling. The activation functions of the nodes in the top layer are

$$f(\phi) = 0, \text{ if } \phi < 0$$
$$f(\phi) = \phi, \text{ if } \phi \geq 0 \qquad (1.74)$$

The nodes in the top layer have both excitatory and inhibitory connections. The excitatory connections are implemented in the form of single positive feedback loops with unitary weight coefficients for each node. All the other (lateral) connections in this fully coupled layer are inhibitory, each with a weight coefficient of $-\varepsilon$, where $0 < \varepsilon < 1/q$. The weight matrix of these feedback connections is therefore

$$\mathbf{U} = \begin{bmatrix} 1 & -\varepsilon & -\varepsilon & \dots & -\varepsilon \\ -\varepsilon & 1 & \dots & \dots & -\varepsilon \\ -\varepsilon & -\varepsilon & \dots & \dots & \dots \\ \dots & \dots & \dots & \dots & -\varepsilon \\ -\varepsilon & -\varepsilon & \dots & -\varepsilon & 1 \end{bmatrix} \qquad (1.75)$$

b) Training of the network

When initialized with the input vector $\mathbf{y}(0)$, the network starts to add positive self-feedback and negative cross-feedback, so that finally the only node not suppressed (i.e. with non-zero output) will be the winning node in the first place.

The vigilance test on the input vector (equation 1.78) assesses the ratio of the scalar product of the best matching cluster weights $\mathbf{v}_j$ with the input vector to the number of unitary bits in the input vector. If the ratio exceeds the vigilance threshold, the input is considered similar to the cluster prototype and the input is then assigned to the cluster represented by the weight vector $\mathbf{w}_j$. If the ratio is less than the vigilance threshold, the input is considered different from the previously stored cluster prototypes and a new cluster is then added to the existing set of clusters. The system can resonate in step iv) when the input exemplar can resonate with one of the exemplars of class j, reinforcing the storage of the class.

The training algorithm can be summarized as follows (Figure 1.16)

i) The vigilance threshold ρ is selected ($0 < \rho < 1$) and for the m-dimensional input exemplars, $\mathbf{x} \in \Re^m$ the weight matrices $\mathbf{V} = [1] \in \Re^{qxm}$ and $\mathbf{W} = [1/(1+m)] \in \Re^{qxm}$ are initialized with identical entries.

ii) A binary unipolar exemplar $\mathbf{x}$ is presented at the input layer of the network.

iii) Matching scores are computed from equations

$$y_k(0) = \Sigma_{i=1}^{m} w_{ik} x_i, \text{ for } k = 1, 2, \dots q \qquad (1.76)$$

Selection of the best matching existing cluster (k) is determined by

$$y_j(0) = \max\{y_k(0)\}, \text{ for } k = 1, 2, \dots q \qquad (1.77)$$

iv) The similarity of the winning node j is tested, with the norm defined as $||\mathbf{x}|| = \Sigma_{i=1}^{m}|x_i|$

$$||\mathbf{x}||^{-1}\Sigma_{i=1}^{m} v_{ij} x^i > \rho \qquad (1.78)$$

If the test is passed, the algorithm continues from step v). If the test is failed, the algorithm goes to step vi), only if the top layer has more than one single active node left. Otherwise, the algorithm goes to step v)

v) The weight matrices are updated only for index j passing the test in step iv), which updates the weights of the j'th cluster. The algorithm then returns to step ii)

$$w_{ij}(t+1) = v_{ij}(t)x_i/\{½ + \sum_{i=1} v_{ij}(t)x_i\} \quad (1.79)$$

$$v_{ij}(t+1) = x_i v_{ij}(t) \quad (1.80)$$

vi) The j'th node is deactivated by setting $y_j = 0$, so that its participation in the current cluster search is terminated. The algorithm returns to step iii), where it will attempt to establish a new cluster different from the j'th pattern being tested.

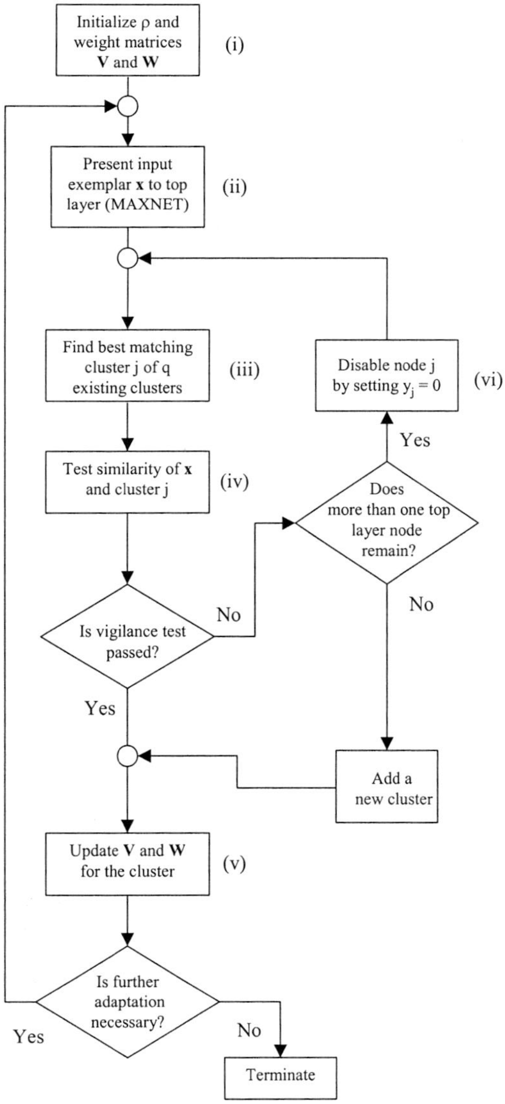

Figure 1.16. Flowchart of the ART training algorithm for unipolar binary exemplars.

1.6.8. Support vector machines

Support vector machines are linear machines derived from the principle of structural risk minimization embedded in the Vapnik-Chervonenkis (VC) dimension theory (Vapnik, 1992). The design of these machines is based on the extraction of a subset of the training data serving as *support vectors*. The support vector learning algorithm operates in batch mode only.

The basic concept can be explained by considering a pattern recognition problem, where two linearly separable patterns have to be recognised. The idea behind support vector machines is to construct an optimal hyperplane separating the two classes. The optimal hyperplane is defined as the plane or decision surface that maximizes the margin of separation between the two classes.

a) Linearly separable patterns

Consider a set of training exemplars of the form $\{\mathbf{x}_j, t_j\}$, for j = 1, 2, 3, … n), where $\mathbf{x}_j$ denotes the j'th input vector and t_j ($= \pm 1$) the target or class membership associated with the j'th input vector. Assuming that the two classes are linearly separable, the hyperplane or decision surface separating the two classes, can be defined as

$$\mathbf{w}^T\mathbf{x} + w_0 = 0 \tag{1.81}$$

where $\mathbf{w}$ is some weight vector and w_0 is a bias. That is

$$\mathbf{w}^T\mathbf{x}_j + w_0 \geq 0, \text{ for } t_j = +1 \tag{1.82}$$

$$\mathbf{w}^T\mathbf{x}_j + w_0 < 0, \text{ for } t_j = -1$$

For a given weight vector $\mathbf{w}$ and bias w_0, the margin of separation (ρ) is defined as the distance between the separating hyperplane and the closest data point. The hyperplane is considered to be *optimal* when this distance ρ is maximized. If the optimal hyperplane is defined by

$$\mathbf{w}^{*T}\mathbf{x} + w_0^* = 0 \tag{1.83}$$

then the discriminant function $g(\mathbf{x}) = \mathbf{w}^{*T}\mathbf{x} + w_0^*$ gives an algebraic measure of the distance between the optimal hyperplane and an observation $\mathbf{x}$ (Duda and Hart, 1973). If equation (1.73) holds, $\mathbf{w}^*$ and w_0^* can be rescaled so that equation (1.75) also holds, i.e. the parameters $\mathbf{w}^*$ and w_0^* must satisfy the constraints

$$\mathbf{w}^{*T}\mathbf{x}_j + w_0^* \geq 1, \text{ for } t_j = +1 \tag{1.84}$$

$$\mathbf{w}^{*T}\mathbf{x}_j + w_0^* < 1, \text{ for } t_j = -1$$

Note that this scaling does not affect equation (1.74). The data points ($\mathbf{x}_j$, t_j) satisfying equation (1.75) with equality signs are called *support vectors* ($\mathbf{x}^{(s)}$). By definition

$$g(\mathbf{x}^{(s)}) = \mathbf{w}^{*T}\mathbf{x}^{(s)} \pm w_0^* = \pm 1, \text{ for } t^{(s)} = \pm 1 \tag{1.85}$$

Conceptually, the support vectors are those observations lying closest to the decision surface and which are consequently the most difficult to classify, as indicated in Figure 1.17, which shows the optimal separation plane for two classes depending on two features x_1 and x_2. As such they have a direct influence on the location of the optimal decision surface.

It can also be shown (Haykin, 1999) that the algebraic distances of the support vectors to the optimal hyperplane are

$$1/||\mathbf{w}^*||, \text{ if } t^{(s)} = +1, \text{ and} \tag{1.86}$$

$$-1/||\mathbf{w}^*||, \text{ if } t^{(s)} = -1$$

The '+' sign indicates that $\mathbf{x}^{(s)}$ lies on the positive side of the optimal hyperplane, while the '-' sign indicates that $\mathbf{x}^{(s)}$ lies on the negative side. The margin of separation between the two classes constituting the training data set is $2/||\mathbf{w}^*||$. This means that maximizing the margin of separation between the different exemplars is equivalent to minimizing the Euclidean norm of the weight vector **w**.

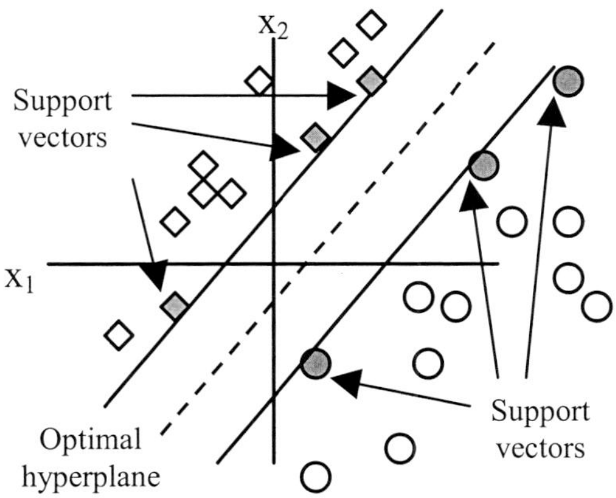

Figure 1.17. Optimal hyperplane for linearly separable patterns. Shaded markers indicate the elements of the support vectors.

b) Nonlinearly separable patterns

With nonlinearly separable patterns, it is not possible to construct a hyperplane without incurring classification errors. In this case, the hyperplane is constructed so as to minimize the probability of classification errors, averaged over the training data set. The margin of separation is referred to as *soft*, if the following condition is violated

$$t_j(\mathbf{w}^T\mathbf{x}_j + w_0) \geq +1, \text{ for } j = 1, 2, 3, \ldots n \tag{1.87}$$

As indicated in Figure 1.18, the constraint can be violated, if

- An observation $(\mathbf{x}_j, t_j)$ is located inside the region of separation, but on the right side of the decision surface. This still leads to correct classification.

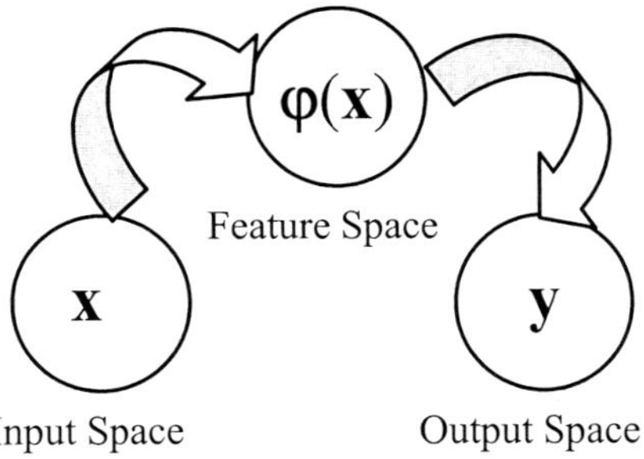

Figure 1.19. Mapping of the observations in the input space to a higher-dimensional feature space, followed by mapping onto the target vectors in the output space.

d) Location of the optimal hyperplane

The constrained optimization problem that has to be solved can be summarized as follows. Given a training sample $(\mathbf{x}_j,t_j)$, for j = 1, 2, 3, ... n, find the optimal values of the weight vector **w**, so that the cost function

$$\omega(\mathbf{w}) = \tfrac{1}{2}\mathbf{w}^T\mathbf{w} \qquad (1.93)$$

is minimized, subject to

$$t_j(\mathbf{w}^T\mathbf{x}_j + w_0) \geq 1, \text{ for } j = 1, 2, 3, \ldots n \qquad (1.94)$$

This constrained optimization problem is referred to as the *primal* problem. The constraints are linear in **w** and the cost function $\omega(\mathbf{w})$ is a convex function of **w**. The problem can be solved by the method of Lagrangian multipliers, that is by constructing a Lagrangian function

$$J(\mathbf{w},w_0,\alpha) = \tfrac{1}{2}\mathbf{w}^T\mathbf{w} - \Sigma_{j=1}^{n}\alpha_j t_j[(\mathbf{w}^T\mathbf{x}_j + w_0)-1] \qquad (1.95)$$

with (nonnegative) Lagrange multipliers α_j. The Lagrangian function has to be minimized with respect to w and w_0, i.e.

$$\partial J(\mathbf{w},w_0,\alpha)/\partial\mathbf{w} = \mathbf{0} \qquad (1.96)$$

and

$$\partial J(\mathbf{w},w_0,\alpha)/\partial w_0 = 0 \qquad (1.97)$$

This gives (first condition)

$$\mathbf{w} = \Sigma_{j=1}^{n}\alpha_j t_j\mathbf{x}_j \qquad (1.98)$$

and (second condition)

$$\Sigma_{j=1}^{n}\alpha_j t_j = 0 \qquad (1.99)$$

The solution vector **w** is thus defined in terms of an expansion that involves all the training exemplars and is unique, since the Lagrangian is unique. The α_j-values (Lagrangian coefficients) on the other hand, are not uniquely defined.

From the optimal Lagrangian multipliers (α_j*), the optimal weight vector **w*** can be computed, i.e.

$$\mathbf{w}^* = \Sigma_{j=1}^{n}\alpha_j^* t_j \mathbf{x}_j \tag{1.100}$$

while w_0 can be calculated as

$$w_0 = 1 - \mathbf{w}^{*T}\mathbf{x}^{(s)}, \text{ for } t^{(s)} = 1 \tag{1.101}$$

e) Inner-product kernels

If **x** represents a vector drawn from an input space of dimension m and $\varphi_j(\mathbf{x})$ (j = 1, 2, ... p) denote a set of a priori defined nonlinear transformations from the input space to the feature space of dimension p, then a decision surface can be defined as

$$\Sigma_{j=1}^{p} w_j\varphi_j(\mathbf{x}) = 0 \text{ or } \mathbf{w}^T\varphi(\mathbf{x}) = \mathbf{0} \tag{1.102}$$

In this case $\varphi_0(\mathbf{x}) = 1$ for all **x** (i.e. w_0 is a bias). The vector $\varphi(\mathbf{x})$ can be seen as the image produced by the input vector **x** in the feature space. This means that equation (1.89) can be adapted, so that

$$\mathbf{w} = \Sigma_{j=1}^{n}\alpha_j t_j \varphi(\mathbf{x}_j) \tag{1.103}$$

In other words, the feature pattern is used, instead of the corresponding input vector **x**. By substituting equation (1.94) in equation (1.93), the decision surface can be defined in the feature space as

$$\Sigma_{j=1}^{n}\alpha_j t_j \varphi(\mathbf{x}_j)^T\varphi(\mathbf{x}) = 0 \tag{1.104}$$

The term $\varphi(\mathbf{x}_j)^T\varphi(\mathbf{x})$ represents the inner product of the two vectors in the feature space corresponding to the input vector **x** and the input pattern $\mathbf{x}_i$. The model can also be expressed as

$$\Sigma_{j=1}^{n}\alpha_j t_j K(\mathbf{x},\mathbf{x}_j) = 0 \tag{1.105}$$

where $K(\mathbf{x},\mathbf{x}_j)$ represents the inner product kernel $K(\mathbf{x},\mathbf{x}_j) = \varphi(\mathbf{x}_j)^T\varphi(\mathbf{x}) = \Sigma_{i=0}^{m}\varphi_i(\mathbf{x})\varphi_i(\mathbf{x}_j)$[6], for j = 1, 2, ,3, ... n. This inner product kernel can be used to construct the optimal hyperplane, *without having to consider the feature space explicitly.*

f) Building support vector machines for nonlinear regression problems

In the application of support vector machines to nonlinear regression problems, the following model can be considered

$$t = f(\mathbf{x}) + e \tag{1.106}$$

---

[6] Known as Mercer's condition.

where the scalar-valued nonlinear function f($\mathbf{x}$) is defined by the conditional expectation E{T|$\mathbf{x}$}, where T is a random variable with a realization denoted by t and e is an unknown additive noise term, independent of the input vector $\mathbf{x}$. The problem is to estimate the dependence of t on $\mathbf{x}$, given a set of training data $\{\mathbf{x}_j|t_j\}$, for j = 1, 2, 3, ... n.

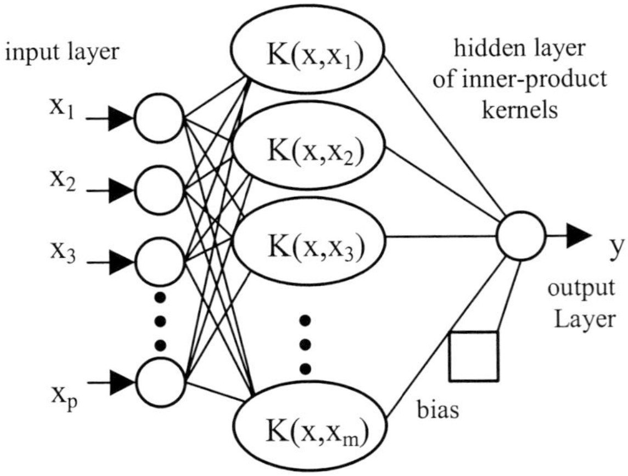

Figure 1.20. Generic structure of a support vector machine.

This is accomplished by postulating an estimate of t, denoted by y, which is subsequently expanded in terms of a set of nonlinear basis functions $\varphi_i(\mathbf{x})$, i = 1, 2, 3, ... p, as follows

$$y = \Sigma_{i=0}^{p}\varphi_i(\mathbf{x}) \tag{1.107}$$

$$y = \mathbf{w}^T\varphi(\mathbf{x}) \tag{1.108}$$

where

$$\varphi(\mathbf{x}) = [\varphi_0(\mathbf{x}), \varphi_1(\mathbf{x}), \varphi_2(\mathbf{x}), \ldots \varphi_p(\mathbf{x})]^T \tag{1.109}$$

and

$$\mathbf{w} = [w_0, w_1, w_2, \ldots w_p]^T \tag{1.110}$$

In this case it is assumed that $\varphi_0(\mathbf{x}) = 1$, so that w_0 is a bias term. The model can be optimized by minimization of a least squares criterion

$$R = 1/n\Sigma_{j=1}^{n}(y_j-t_j)^2 \tag{1.111}$$

subject to the inequality

$$||\mathbf{w}||^2 \leq c_0 \tag{1.112}$$

where c_0 is a constant. If more robust estimation is required, e.g. where the data contain influential outliers or when the underlying distribution has log-tailed additive noise, other criteria are more suitable. This includes the use of an absolute error criterion, $R = 1/n\Sigma_{j=1}^{n}|y_j-t_j|$ or the use of an ε-insensitive loss function, $R_\varepsilon(t,y) = |t - y| - \varepsilon$, for $|t - y| \geq \varepsilon$ and 0 otherwise, as originally proposed by Vapnik (1995, 1998). The absolute error criterion is simply a special case of the ε-insensitive loss function, with $\varepsilon = 0$.

As shown by Haykin (1999), optimization of the model can be formulated as a constrained (dual) optimization problem. Given the training exemplar $[\mathbf{x}_j,t_j]$, for $j = 1, 2, 3, \ldots n$, find the Lagrangian multipliers α_j and α_j' that maximize the objective function

$$Q(\alpha_j,\alpha_j') = \Sigma_{j=1}^{n}t_j(\alpha_j- \alpha_j') - \varepsilon\Sigma_{j=1}^{n}t_j(\alpha_j+ \alpha_j')- \tfrac{1}{2}\Sigma_{j=1}^{n}\Sigma_{i=1}^{n}(\alpha_j- \alpha_j')(\alpha_i- \alpha_i')K(x_j,x_i) \qquad (1.113)$$

subject to

$$\Sigma_{j=1}^{n}(\alpha_j- \alpha_j') = 0 \qquad (1.114)$$

$$0 \leq \alpha_j \leq C, \text{ for } j = 1, 2, 3, \ldots n$$

$$0 \leq \alpha_j' \leq C, \text{ for } j = 1, 2, 3, \ldots n$$

The parameters ε and C are free parameters (to be selected by the user) that control the VC dimension of the approximating function

$$F(\mathbf{x},\mathbf{w}) = \mathbf{w}^T\mathbf{x} = \Sigma_{j=1}^{n}(\alpha_j- \alpha_j')K(\mathbf{x}_j,\mathbf{x}_i) \qquad (1.115)$$

This makes the regression problem more difficult, since the parameters ε and C have to be tuned simultaneously. The generic structure of a support vector machine is shown in Figure 1.20.

1.7. NEURAL NETWORKS AND STATISTICAL MODELS

From the preceding discussion, it should be clear that despite their different origins, many neural networks are closely related to, or equivalent to statistical models, especially popular techniques such as generalized linear models, regression methods, discriminant analysis, principal component analysis and cluster analysis. A few examples will illustrate these similarities.

For example, consider a simple multilayer perceptron with no hidden layer and an output layer with linear activation functions, as shown in Figure 1.21(a), (for m input variables and a single output y). This network is equivalent to a multiple linear regression model, while the almost identical network shown in Figure 1.21(b), with a sigmoidal output node, is equivalent to a logistic regression model. The activation functions in multilayer perceptrons are analogous to the inverse of the link functions used in generalized linear models (McCullagh and Nelder, 1989). Activation functions are usually bounded, while inverse link functions, such as identity, reciprocal or exponential functions tend not to be.

If the activation function of the network is changed to a threshold function, $f(x) = 0$, if $x < 0$, or $f(x) = 1$ otherwise, then the network is equivalent to a linear discriminant model, suitable for the classification of data into two classes, as indicated in Figure 1.21(c). With only one output, these networks are also referred to as ADALINE networks, while MADALINE networks can accommodate more than one output. Note that instead of threshold functions, logistic functions can also be used to estimate the conditional probabilities of each class.

Figure 1.21(d) shows a functional link neural network, with a hidden layer that expands its single output into polynomials of orders 0 to p (the bias node is not shown). This neural network is clearly equivalent to a polynomial regression model. In general, functional links can be any transformation not requiring additional parameters, i.e. only one layer of weights need to be determined. Finally, in Figure 1.21(e), a multilayer perceptron with a sigmoidal hidden layer is shown.

This network is equivalent to multiple nonlinear regression. If the input nodes are also connected direct tot the output nodes (sometimes referred to as a fully connected neural network), these connections can be seen as equivalent to what are referred to as main effects in the statistical literature. Since multilayer perceptrons with nonlinear activation functions are nonlinear models in the true sense of the word (i.e. nonlinear in their parameters), they tend to take more computer time to fit than polynomials or splines. On the other hand, they may also be more stable numerically, than high-order polynomials.

Moreover, multilayer perceptrons do not require the specification of knot positions and also have different extrapolation properties than their statistical counterparts. Polynomials extrapolate to infinity, whereas multilayer perceptrons tend to flatten out. Nonetheless, this does not somehow make them more suitable for extrapolation than polynomial models.

Also important, is the fact that nonlinear statistical models are supported by extensive statistical theory, allowing the estimation of various diagnostics, confidence intervals, prediction intervals, etc. These options are generally not readily available when multilayer perceptron models are developed.

The same parallels can be drawn with principal component analysis. The extraction of features from data by use of neural networks is typically accomplished via unsupervised learning. Unsupervised learning entails the formulation of a suitable criterion, which has to be optimized during the training of the neural network. For example, the extraction of features from the data should be such that the original observations can be reconstructed from the features.

In the simple neural network structure shown in Figure 1.21(f), a set of variables x_1, x_2, ... x_m is presented to a multilayer perceptron with linear activation nodes. The multilayer perceptron has a hidden layer with a single node, and the target variables are identical to the input variables. The neural network is therefore forced to reconstruct the original variables from a single feature. As will be shown in more detail later on, this feature is equivalent to the first principal component of the data. Although this is not a particularly efficient way of extracting principal components, the addition of hidden layers before and after the hidden layer shown in Figure 1.21(g), allows the extraction of nonlinear principal components from the data.

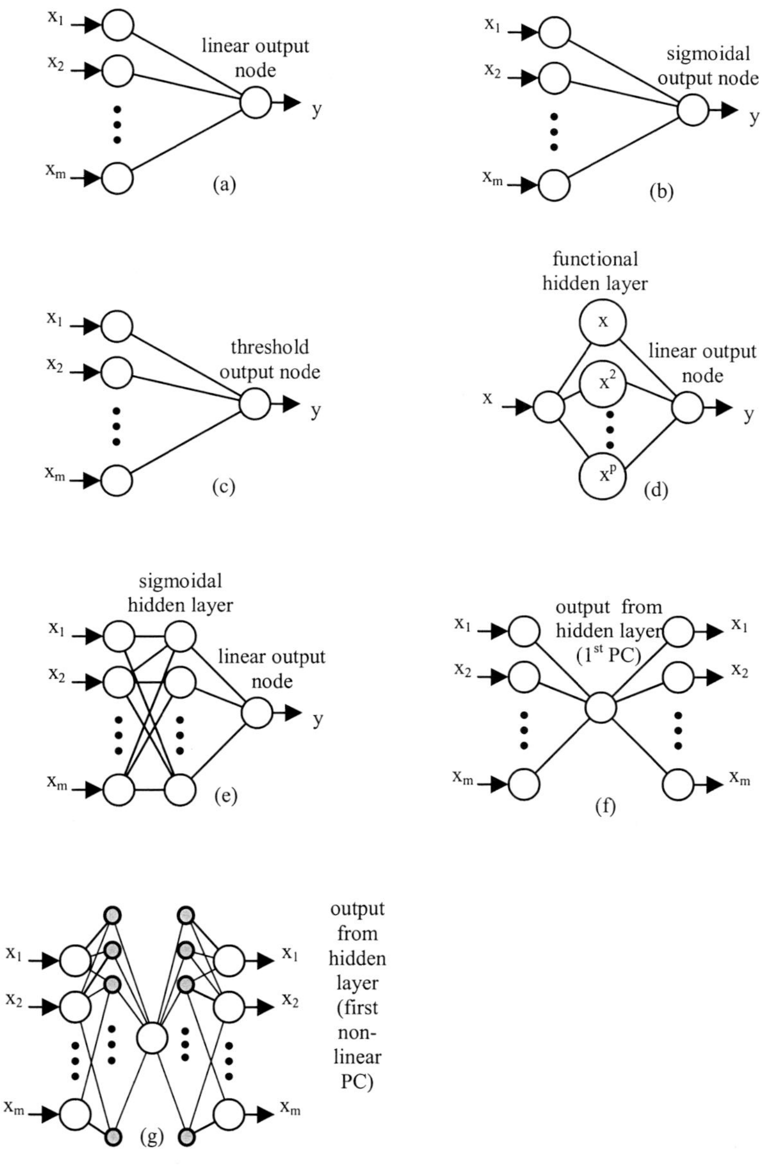

Figure 1.21. Regression and classification with neural networks, i.e. the neural network equivalent of a) linear regression, b) logistic regression, c) linear discriminant analysis, d) polynomial regression, e) nonlinear (nonparametric) regression, f) principal component analysis and g) nonlinear principal component analysis.

These examples can be extended to cluster analysis, where competitive neural networks, such the self-organizing map of Kohonen (1995) and adaptive resonance theory neural networks can be shown to be closely related to k-means cluster analysis and the leader algorithm, respectively. The same goes for kernel-based methods, such as embodied in the probabilistic neural network and the general regression neural network (Specht, 1991), that are direct derivatives of their statistical counterparts.

1.8. APPLICATIONS IN THE PROCESS INDUSTRIES

Since the late 1980s, researchers at the Siemens Corporate Research and Development Division (ZFE) in Munich have worked on advanced distributed processing systems, together with outside partners which included the Bavarian Research Centre for Knowledge-based Systems and several European universities. As a result of these research efforts, a number of mills have been equipped with neurocontrol systems. One of the contracts (worth approximately DM100 million in 1996 concerned the process automation of a rolling mill for EKO Stahl of Eisenhüttenstadt in eastern Germany. Other applications by this group is also reported, viz. neurocontrol systems in strip processing lines, electric arc furnaces and aluminium cold strip tandem mills, while the Siemens group also has research contracts with a number of other European mill operators, such as Thyssen and Krupp-Hoesch Stahl in Germany, Rautaruukki in Finland, VOEST-Stahl in Austria and SSAB Oxelösund in Sweden.

The early 1990s saw the implementation of a hybrid neuro-fuzzy control system from Pavilion Technologies in Eastman Kodak's refinery in Texas, among other. In the pyrometallurgical industries, commercial applications included hybrid control systems sold by Neural Applications Corporation, consisting of neural networks as well as expert systems, used in arc furnaces. These systems were used to optimize the positions of the electrodes of the arc furnaces used for the smelting of scrap metal in steel plants, and were reported to have saved an estimated $US 2 000 000 annually on the operating costs of each furnace.

Kawasaki Steel Corporation in Japan has applied neural networks for control of the burden distribution at their Mizushima No 3 blast furnace. The system consists of three neural networks operating in series. The first neural network characterizes the gas distribution in the furnace, and its outputs are fed to a second neural network, which computes appropriate changes in the charging patterns of the furnace. The third neural network predicts the effect of charge patterns on gas flow in the furnace, and serves as a guide to the implementation of control decisions.

One of the early applications of neural networks in the nuclear industry was related to the control of a nuclear fusion reactor at AEA Technology's Culham Laboratory in Oxfordshire in the United Kingdom. The optimal conditions for fusion in the Compass tokamak reactor occur where the turbulence in the plasma is minimal and cannot be calculated sufficiently fast by conventional computers, which can take hours or even days to compute the setup of the magnetic fields needed to produce suitable plasma shapes in the reaction chamber. The problem is solved by making use of a neural network that can do the necessary calculations in approximately ten microseconds (significantly faster than the fluctuations in the plasma that typically last for a few hundred milliseconds). The Compass network obtains data from 16 magnetic field sensors inside the chamber and has four output nodes linked to the magnetic controls of the system. An added advantage is the flexibility of the network, which can be retrained (by sets of approximately 2000 exemplars at a time) when the implementation of different control strategies are warranted. In contrast, conventional controllers can only cope with narrow ranges of process conditions.

In 1994, Star Enterprise installed NeuCOP (Neural Control and Optimization Package) jointly developed by NeuralWare and Texaco to control its multi-product atmospheric distillation tower. The multivariable, adaptive, closed-loop controller runs on an IBM RS-6000, while integrating Gensym's G2 to provide sensor validation, alarm filtering, access to plant-wide cost information and an interactive graphical user interface. The system reportedly reduced product variability and increases controller up-time. NeuralWare and Gensym Corporation's G2/NeuCOP system has also been tested by Texaco in the USA, at one of its debutaniser units at the Puget Sound refinery, as well as other Texaco sites. NeuralWare has since been acquired by Aspen.

CEMQUEST (Cement quality estimation) is a Gaussian radial basis function (localized receptive field) neural network developed and commercialised by Schlumberger Cambridge Research for use in the oil industry. It is currently used by Schlumberger Dowell to predict the quality of cement being used in oil wells. CEMQUEST can predict the quality of the cement linings (composition, particle size distributions, and thickening time for certain cement slurry formations) from the Fourier Transform infrared (FTIR) spectra of the dry cement powders. The original training data of the system consisted of FTIR spectra of more than 160 different cements sourced from all over the world. These spectra were digitised, resulting in more than 2000 points per spectrum, which were subsequently compressed to 35 inputs by means of principal component analysis. Development between 1991 and 1993 took approximately 12 man-years, and the system is reported to save clients an estimated $US 3 000 000 to $US 5 000 000 annually, through detection of potentially major operating failures.

In South Africa, neural network systems are currently used in the petrochemical industries for data validation. Similar systems are being undergoing plant trials in other sectors of the mineral processing industries in South Africa. On a number of flotation plants, self-organizing neural networks are used to visualize plant conditions, and serve as a decision support system for plant operators. Prototypes of these systems are still being evaluated and they cannot be considered fully commercialised yet.

Despite these successes, the on-line implementation of neural networks in advanced control systems is not widespread in the mineral and metallurgical process industries. This can be attributed to the high cost associated with the maintenance of the plant models, which have to cope with large disturbances over short time intervals.

Chapter 2
Training of Neural Networks

2.1. GRADIENT DESCENT METHODS

The essence of the back propagation algorithm is the credit assignment problem, i.e. the evaluation of the contribution of each particular weight to the output error. This is readily accomplished for neural networks with objective functions containing continuously differentiable functions of the network weights.[7] Although the development of the back propagation algorithm has contributed significantly to the renewed interest in neural networks in the last few decades, it suffers from various drawbacks in practice.

For example, gradient descent methods, such as those implemented with back propagation are prone to get stuck in local minima, although this problem is ameliorated by various factors. For example, the initialization of the weights of the neural network strongly affects its training. These and other stochastic elements in the algorithm, such as stochastic data, the way in which the training error is calculated, etc., make it more likely to find a good solution than would have been possible with a purely deterministic approach. Nonetheless, the use of such gradient information is of critical importance in algorithms that are sufficiently fast to handle large-scale applications. In recent years, various modifications have been made to gradient descent methods in order to enhance their efficiency.

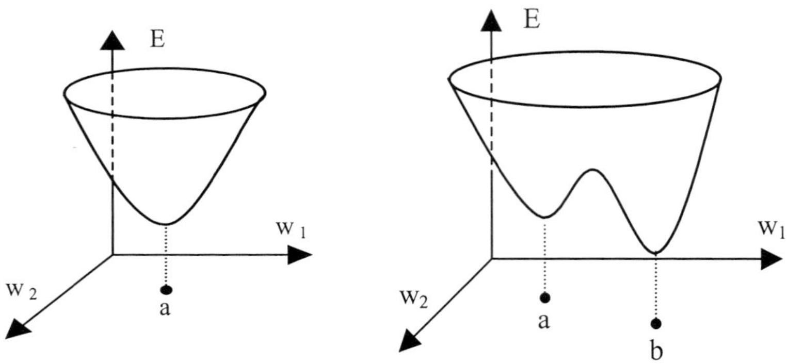

Figure 2.1. Examples of error surfaces E(**w**) located above a two-dimensional weight space (w_1,w_2).

Figure 2.1 shows examples of error surfaces positioned above a weight space (w_1, w_2), with points a and b representing the minima of these surfaces, which can be described by error functions E(w_1,w_2) or more generally the error surfaces E(**w**) that are positioned above weight spaces (**w**). The problem is therefore to find the weight vectors **w*** that minimize these error functions. For simple networks with linear activation functions, a single layer of weights and a sum-of-squares error, the error surfaces will have a general multidimensional quadratic form. In this case the minimum (or degenerate minima) can be found by solution of a set of coupled linear equations. However, in general (for neural networks with more than one layer,

[7] For example, it is not possible for perceptrons with hard-limiting (discrete-valued) activation functions.

nonlinear activation functions, etc.) the error surfaces will be highly nonlinear and may contain many minima. In this case, closed form solutions of the minima are not always possible and algorithms involving a search through weight space need to be considered. These algorithms typically consist of a series of steps j of the form

$$\mathbf{w}_{(j+1)} = \mathbf{w}_{(j)} + \Delta\mathbf{w}_{(j)} \tag{2.1}$$

Algorithms differ in the way in which the weight increments are chosen. For some algorithms, such as the quasi-Newton and conjugate gradient algorithms, the weight increment is guaranteed not to increase the error function, while in others this is not the case.

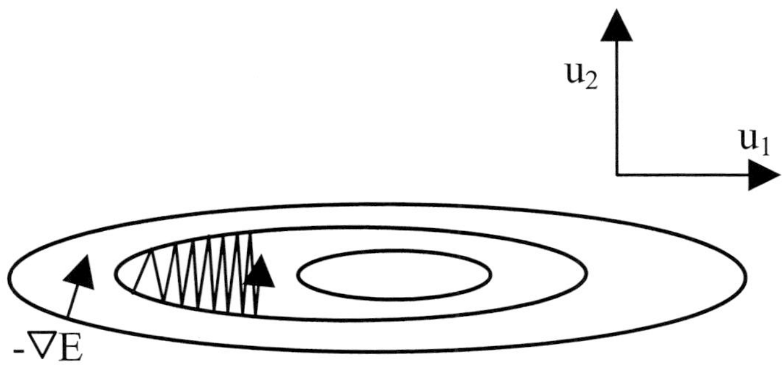

Figure 2.2. Fixed step gradient descent for an error function with different curvatures along different directions. The ellipses represent constant error values, while the vectors u_1 and u_2 depict the eigen-vectors of the Hessian matrix ($\mathbf{H} = \partial^2 E/\partial w_i \partial w_j$). Successive steps oscillate across the valley, with slow convergence, since the negative gradient does not point to the minimum of the surface.

In the steepest gradient descent methods discussed previously, movement in the weight space is in the direction of the largest rate of decrease of the error, i.e.

$$\Delta\mathbf{w}_{(j)} = -\eta\nabla\mathbf{E}(\mathbf{w}_{(j)}) \tag{2.2}$$

η denotes the learning rate and provided that it is sufficiently small, the value of the error can be expected to decrease at each successive step. However, this approach, even with the addition of momentum, is not particularly efficient and numerous attempts have been made in recent years to improve this approach by means of ad hoc modifications. A review of the extensive literature on this topic is beyond the scope of this book and instead a few robust and theoretically well-founded algorithms will be considered briefly below.

The algorithms considered below all involve a sequence of movements in the weight space. First, the direction of the move has to be determined, and second, the distance to be moved along this direction. With simple gradient descent the direction is given by the negative gradient of the error function and the distance moved in this direction is arbitrarily determined by the learning rate. Line search methods improve on this approach, by moving along the direction of the negative gradient to the point where the error function is minimized, that is

$$\mathbf{w}_{(j+1)} = \mathbf{w}_{(j)} + \rho\mathbf{d}_{(j)} \quad (2.3)$$

where $\mathbf{d}_{(j)}$ is the direction of search through the weight space at the j'th iteration and the parameter ρ is selected to minimize

$$E(\rho) = E(\mathbf{w}_{(j)} + \rho\mathbf{d}_{(j)}) \quad (2.4)$$

Moreover, these algorithms are based on quadratic models, since these models are the simplest nonlinear models to minimize (linear functions do not contain interior optima) and any technique has to work satisfactorily on quadratics if it hopes to succeed with more general functions. Just as important, near the optimum, all nonlinear functions can be approximated by a quadratic (in the Taylor expansion, the linear part must vanish). Therefore, the behaviour of the algorithm on the quadratic will also be an indication of how the algorithm will converge for general functions.

2.2. CONJUGATE GRADIENTS

The use of successive gradient vectors generally does not represent the best choice of search direction. At the minimum of the search line

$$\partial E(\rho)/\partial\rho = \partial/\partial\rho\{E(\mathbf{w}_{(j)} + \rho\mathbf{d}_{(j)})\} = 0 \quad (2.5)$$

so that

$$\mathbf{g}_{(j+1)}{}^{T}\mathbf{d}_{(j)} = 0 \quad (2.6)$$

where $g \equiv \nabla E$. This means that the gradient at the new minimum is orthogonal to the previous search direction. This can lead to problems if the search directions are always chosen to coincide with the negative gradients of the error function and minimization may proceed very slowly.

This problem can be circumvented by choosing the successive search directions so that the component of the gradient parallel to the previous search direction, which has been made zero, is unaltered. For example, if line minimization has already taken place along the direction $\mathbf{d}_{(j)}$, starting from the point $\mathbf{w}_{(j)}$ to give the new point $\mathbf{w}_{(j+1)}$, then at $\mathbf{w}_{(j+1)}$

$$\mathbf{g}[\mathbf{w}_{(j+1)}]^{T}\mathbf{d}_{(j)} = 0 \quad (2.7)$$

The new search direction $\mathbf{d}_{(j+1)}$ is then chosen, so that the component of the gradient parallel to the previous search direction remains zero to the lowest order, i.e.

$$\mathbf{g}[\mathbf{w}_{(j+1)} + \rho\mathbf{d}_{(j+1)})]^{T}\mathbf{d}_{(j)} = 0 \quad (2.8)$$

This can be expanded to

$$\mathbf{d}_{(j+1)}{}^{T}\mathbf{H}\mathbf{d}_{(j)} = 0 \quad (2.9)$$

where **H** is the Hessian matrix evaluated at the point $\mathbf{w}_{(j+1)}$. Search directions that satisfy the above relationship are known as non-interfering or conjugate. In fact, it can be shown that a sequence of search directions $\mathbf{d}_{(j)}$ can be constructed, so that each direction is conjugate to all the previous directions, up to the dimensionality of the search space.

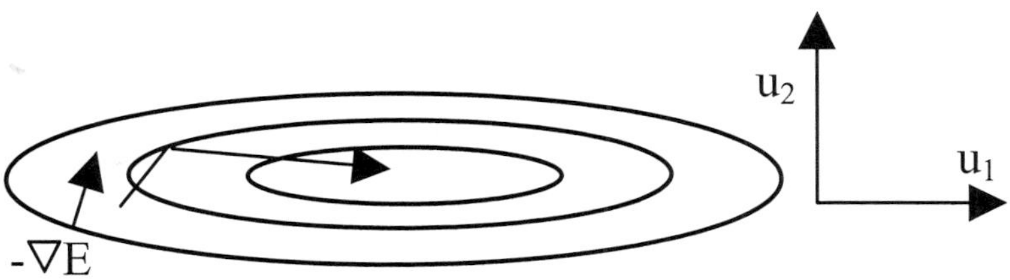

Figure 2.3. Minimization of a two-dimensional quadratic error function by use of a conjugate gradient algorithm, which locates the minimum after two steps.

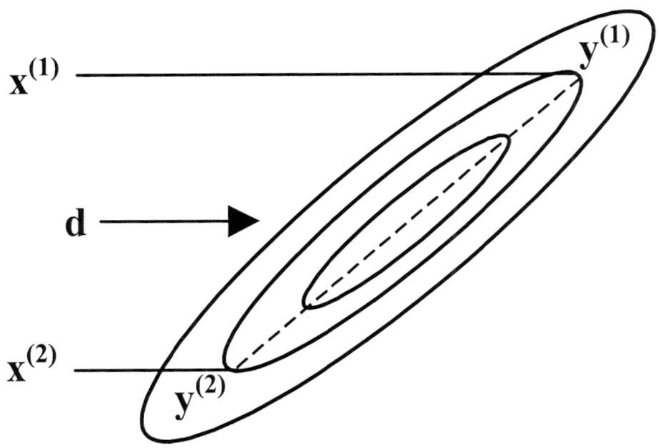

Figure 2.4. Conjugacy in two dimensions.

In general, if $q(\mathbf{x}) = \mathbf{a} + \mathbf{b}^T\mathbf{x} + \frac{1}{2}\mathbf{x}^T\mathbf{C}\mathbf{x}$, and $Q(\mathbf{x}) = \mathbf{x}^T\mathbf{C}\mathbf{x}$, $\mathbf{x}$ is transformed to $\mathbf{x} = \mathbf{T}\mathbf{z}$, the quadratic part of the function is diagonalized, that is $Q(\mathbf{x}) = \mathbf{z}^T\mathbf{T}^T\mathbf{C}\mathbf{T}\mathbf{z} = \mathbf{z}^T\mathbf{D}\mathbf{z}$,

A simple example illustrates the basic idea. Consider the quadratic function

$$f(\mathbf{x}) = 4x_1^2 + 3x_2^2 - 4x_1x_2 + x_1$$

with transformation $\mathbf{x}_1 = \mathbf{z}_1 + \frac{1}{2}\mathbf{z}_2$ and $\mathbf{x}_2 = \mathbf{z}_2$ or $\mathbf{x} = \mathbf{T}\mathbf{z} = [(1,0)^T, (\frac{1}{2},1)^T]\mathbf{z}$ giving the transformed[8] quadratic

---

[8] This transformation is not unique in that other transformations will also diagonalize the quadratic function.

$f(\mathbf{z}) = 4z_1^2 + 2x_2^2 + z_1 + ½z_2$

Starting the search for the minimum of $f(\mathbf{x})$ with $\mathbf{x}^{(0)} = [0,0]^T$, gives $\mathbf{t}_1 = [1,0]^T$ and $\mathbf{t}_2 = [½,1]^T$. Searching along $\mathbf{t}_1$ first, gives $\mathbf{x}^{(1)} = \mathbf{x}^{(0)} + \lambda\mathbf{t}_1$, yielding $\lambda = 1/8$ and $\mathbf{x}^{(1)} = [-1/8,0]^T$. With the second search along $\mathbf{t}_2$ from $\mathbf{x}^{(1)}$, again $\lambda = -1/8$ and $\mathbf{x}^{(2)} = [-3/16,-1/8]^T$, which is the solution found by two successive single variable searches in the $\mathbf{t}_1$ and $\mathbf{t}_2$ directions.

2.3. NEWTON'S METHOD AND QUASI-NEWTON METHOD

Unlike the conjugate gradient approach, where second-order information about the error surface is used *implicitly*, as represented by the Hessian matrix, the Newton method makes explicit use of this information. Since the gradient at any point is given by

$$\mathbf{g} = \nabla E = \mathbf{H}(\mathbf{w}\text{-}\mathbf{w}^*) \quad (2.10)$$

the weight vector corresponding to the minimum of the error function satisfies

$$\mathbf{w}^* = \mathbf{w} - \mathbf{H}^{-1}\mathbf{g} \quad (2.11)$$

The vector $-\mathbf{H}^{-1}\mathbf{g}$ is known as the *Newton step* or *Newton direction* and forms the basis of the Newton and quasi-Newton approaches. At any $\mathbf{w}$, the Newton direction for a quadratic error surface points directly at the minimum of the error function. This can be illustrated by way of a simple example. Let

$$f(\mathbf{x}) = 4x_1^2 + 2x_1x_2 + 3x_2^2$$

$$\nabla f(\mathbf{x}) = [8x_1^2 + 2x_2, 6x_2 + 2x_1]^T$$

$$\nabla^2 f(\mathbf{x}) = \mathbf{H} = [(8, 2)^T, (2, 6)^T]$$

$$\mathbf{H}^{-1} = [(0.1364, -0.0455)^T, (-0.0455, 0.1818)^T]$$

starting with $\mathbf{x}^{(0)} = [10,10]^T$

$$\mathbf{x}^{(1)} = [10,10]^T - [(0.1364, -0.0455)^T, (-0.0455, 0.1818)^T]*[100,80]^T$$

$$\mathbf{x}^{(1)} = [0,0]^T$$

which is the exact solution to the problem.

The Newton method will therefore minimize a quadratic function from any starting point in exactly one step. In general, the Newton step would have to be applied iteratively, requiring the Hessian to be evaluated at every new search point. This is computationally demanding, requiring $O(nm^2)$ steps, where n is the number of patterns in the training data set and m the number of weights in the neural network. Inversion of the Hessian requires $O(m^3)$ steps, which adds to the computational burden. Moreover, the Newton step may result in a move towards a maximum or a saddle point, rather than a minimum, if the Hessian is not positive definite. Lastly, the step size predicted by the Newton step might be sufficiently large to invalidate the quadratic approximation, which may lead to an unstable algorithm.

Despite these shortcomings, the full Newton rule can be modified to make its application practical. For example, the Hessian matrix can be changed to ensure that it is positive definite, by adding the unit matrix scaled by a constant factor λ, that is

$$\mathbf{H}' = \mathbf{H} + \lambda\mathbf{I} \tag{2.12}$$

The corresponding step direction thus becomes a compromise between the Newton direction (favoured by small values of λ) and the negative gradient direction (favoured by large values of λ), i.e. $-(\mathbf{H}+\lambda\mathbf{I})^{-1}\mathbf{g} \approx -\mathbf{g}/\lambda$. Likewise, the computational burden associated with the computation and neglecting the off-diagonal elements of the Hessian matrix can ease inversion of the Hessian matrix[9]. As a consequence, computation of the inverse of the Hessian becomes trivial and the Newton update equations decouple into separate equations for each weight of the network.

In quasi-Newton approaches, the Hessian and its inverse are not calculated directly. Instead, an approximation to the inverse is built up over several steps. This is done by generating a sequence of matrices $\mathbf{G}_{(j)}$ that are increasingly accurate approximations of the inverse Hessian[10]. This is accomplished by using information on the first derivatives of the error function. The quasi-Newton condition can be expressed as

$$\mathbf{w}_{(j+1)} - \mathbf{w}_{(j)} = -\mathbf{H}^{-1}(\mathbf{g}_{(j+1)} - \mathbf{g}_{(j)}) \tag{2.13}$$

The approximation of the inverse Hessian (**G**) is constructed so that the above condition is satisfied

$$\mathbf{G}_{(j+1)} = \mathbf{G}_{(j)} + \mathbf{p}\mathbf{p}^T/\mathbf{p}^T\mathbf{v} - (\mathbf{G}_{(j)}\mathbf{v})\mathbf{v}^T\mathbf{G}_{(j)}/\mathbf{v}^T\mathbf{G}_{(j)}\mathbf{v} + (\mathbf{v}^T\mathbf{G}_{(j)}\mathbf{v})\mathbf{u}\mathbf{u}^T \tag{2.14}$$

where

$$\mathbf{p} = \mathbf{w}_{(j+1)} - \mathbf{w}_{(j)} \tag{2.15}$$

$$\mathbf{v} = \mathbf{g}_{(j+1)} - \mathbf{g}_{(j)} \tag{2.16}$$

$$\mathbf{u} = \mathbf{p}/(\mathbf{p}^T\mathbf{v}) - \mathbf{G}_{(j)}\mathbf{v}/(\mathbf{v}^T\mathbf{G}_{(j)}\mathbf{v}) \tag{2.17}$$

The procedure is initialized by using the identity matrix, which amounts to taking the first step in the direction of the negative gradient. The direction **-gG** is guaranteed to be a descent direction at each step, since the matrix **G** is positive definite. Since the full Newton step may take the search outside the range of validity of the quadratic approximation, line search should be used to find the minimum of the error function in the search direction. The weight vector is subsequently updated as follows

$$\mathbf{w}_{(j+1)} = \mathbf{w}_{(j)} + \alpha_{(j)}\mathbf{G}_{(j)}\mathbf{g}_{(j)} \tag{2.18}$$

where $\alpha_{(j)}$ is determined by line search at each iteration.

---

[9] This may be a limitation if the Hessian matrix is far from diagonal.

[10] For an m-dimensional quadratic form, the sequence of matrices **G**(j) is guaranteed to find the true Hessian after m iterations, as well as the exact minimum of the quadratic form after m steps, provided that the line minimizations are performed exactly.

Unlike conjugate gradient methods, quasi-Newton methods have the advantage that the line search does not have to be particularly accurate, as it is not critical to the performance of the algorithm. A disadvantage of the quasi-Newton method is that it needs to store and update a matrix of size m x m. This can lead to prohibitive memory requirements in large networks (more than a few thousand weights). Under these circumstances, conjugate gradient methods would be the better option, as they require only O(m) storage.

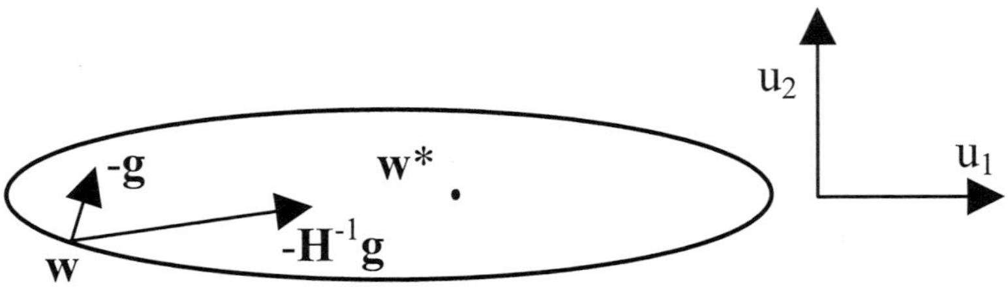

Figure 2.5. The Newton direction for a quadratic error surface. The Newton direction $-\mathbf{H}^{-1}\mathbf{g}$ points to the minimum of the error surface, whereas the local negative gradient does not necessarily.

2.4. LEVENBERG-MARQUARDT ALGORITHM

Optimization algorithms, such as those based on conjugate gradients or Newton directions are general-purpose methods designed to work with a wide range of error functions. In contrast, the Levenberg-Marquardt algorithm is specifically designed to minimize sum-of-squares error functions, of the form

$$E = \tfrac{1}{2}\Sigma_k(e_k)^2 = \tfrac{1}{2}||\mathbf{e}||^2 \qquad (2.19)$$

where e_k is the error in the k'th exemplar or pattern and e is a vector with elements e_k. If the difference between the previous weight vector and the new weight vector is small, the error vector can be expanded to first order by means of a Taylor series

$$\mathbf{e}_{(j+1)} = \mathbf{e}_{(j)} + \partial e_k/\partial w_i(\mathbf{w}_{(j+1)} - \mathbf{w}_{(j)}) \qquad (2.20)$$

As a consequence, the error function can be expressed as

$$E = \tfrac{1}{2}||\ \mathbf{e}_{(j)} + \partial e_k/\partial w_i(\mathbf{w}_{(j+1)} - \mathbf{w}_{(j)})||^2 \qquad (2.21)$$

Minimizing the error function with respect to the new weight vector, gives

$$\mathbf{w}_{(j+1)} = \mathbf{w}_{(j)} - (\mathbf{Z}^T\mathbf{Z})^{-1}\mathbf{Z}^T\mathbf{e}_{(j)} \qquad (2.22)$$

where

$$(\mathbf{Z})_{ki} \equiv \partial e_k/\partial w_i \qquad (2.23)$$

Since the Hessian for the sum-of-squares error function is

$$(\mathbf{H})_{ij} = \partial^2 E/\partial w_i \partial w_j = \Sigma\{(\partial e_k/\partial w_i)(\partial e_k/\partial w_i) + e_k \partial^2 e_k/\partial w_i \partial w_j\} \qquad (2.24)$$

Neglecting the second term, the Hessian can be written as

$$\mathbf{H} = \mathbf{Z}^T\mathbf{Z} \tag{2.25}$$

Updating of the weights therefore involves the inverse Hessian or an approximation thereof for nonlinear networks. The Hessian is relatively easy to compute, since it is based on first order derivatives with respect to the network weights that are easily accommodated by back propagation. Although the updating formula could be applied iteratively to minimize the error function, this may result in a large step size, which would invalidate the linear approximation on which the formula is based.

In the Levenberg-Marquardt algorithm, the error function is minimized, while the step size is kept small in order to ensure the validity of the linear approximation. This is accomplished by use of a modified error function of the form

$$E = \tfrac{1}{2}\|\, \mathbf{e}_{(j)} + \partial e_k/\partial w_i(\mathbf{w}_{(j+1)} - \mathbf{w}_{(j)})\|^2 + \lambda\|\mathbf{w}_{(j+1)} - \mathbf{w}_{(j)}\|^2 \tag{2.26}$$

where λ is a parameter governing the step size. Minimizing the modified error with respect to $\mathbf{w}_{(j+1)}$, gives

$$\mathbf{w}_{(j+1)} = \mathbf{w}_{(j)} - (\mathbf{Z}^T\mathbf{Z} + \lambda\mathbf{I})^{-1}\mathbf{Z}^T\mathbf{e}_{(j)} \tag{2.27}$$

Very large values of λ amount to standard gradient descent, while very small values of λ amount to the Newton method.

2.5. STOCHASTIC METHODS

Supervised training of neural networks can be seen as an unconstrained nonlinear continuous optimization problem in which the search space is defined by the weight space, while the objective function is defined by the error function. Since the response surface of the error function can be extremely rugged and have many local minima, search methods that cannot escape from local minima will find it difficult to minimize the error function.

The simplest probabilistic method restarts the training procedure when the algorithm gets stuck or fails to converge, while more advanced approaches, such as simulated annealing, depend on probability measures to determine the direction of the search, and therefore whether ascent or descent should take place in a local minimum. Other stochastic methods, such as random recombination or evolutionary algorithms, make use of probabilistic interpolation to determine new intermediate starting points for the algorithm. Some of these are described in more detail below.

2.5.1. Simulated annealing

Simulated annealing (Kirkpatrick et al., 1983) is a stochastic global minimization method that has been applied in many fields, including the training of neural networks (Kahng, 1992; Boese and Kahng, 1993) It starts at an initial point, takes a step, and evaluates the error function once. Any downhill movement is accepted, while uphill movements may also be accepted in order to escape from local minima. As minimization proceeds, the step length decreases and the probability of accepting uphill movements decreases as well. The search converges to a local (or sometimes global) minimum.

The development of simulated annealing methods was motivated by the behaviour of mechanical systems with large degrees of freedom, which is too large to be searched exhaustively. Physically, any such system will tend towards a state of minimum energy. Mathematical models of such systems likewise tend to minimize the energy of the system. For example, atoms in a molten metal tend to arrange themselves in a state of minimum energy upon cooling of the metal. These states tend to be local energy minima, rather than global minima and may be decreased upon reheating and slow cooling (annealing). Simulated annealing may therefore be an effective method for finding global or near-global minima of large systems.

By starting at an initial point in the search space, the corresponding initial value of the objective function (F_0) is evaluated. Next a search direction is generated randomly, e.g. by randomly generating a set of p coordinates (if the search space is p-dimensional), and a step of magnitude Δr is taken in that direction. After the step has been taken, the new value of the objective function (F_1) is evaluated and accepted with a probability p, given by

$$p = 1, \text{ if } \Delta F = F_1 - F_0 \leq 0 \text{ and} \tag{2.28}$$

$$p = \exp(-\beta\Delta F), \text{ if } \Delta F = F_1 - F_0 > 0 \text{ } (\beta \text{ is a positive parameter}) \tag{2.29}$$

As a result, beneficial steps are accepted unconditionally, while detrimental steps are accepted only occasionally. A random number ρ is generated from a uniform distribution [0,1] and compared with the value, $\exp(-\beta\Delta F)$. If $p < \exp(-\beta\Delta F)$, then the change is accepted, otherwise it is rejected, and a new search direction is generated.

The larger the change ΔF, the smaller its chance of being accepted. Reasonable values for β should be such that the inequalities $0.5 < \exp(-\beta\Delta F) < 0.9$ hold (Bohachevsky et al., 1986). Reasonable values for the function step size Δr should be such that local minima can be escaped from within two or three iterations.

The expression for the conditional acceptance of the probability p is motivated by the physical process in statistical mechanics, where the probability that a system will move from an energy state E_1 to energy state E_2, where $E_2 - E_1 = \Delta E > 0$ is $\exp(-\Delta E/kT)$. In this case k is the Boltzmann constant and T the absolute temperature of the system, i.e. $\beta = 1/kT$. The lower the temperature, the less likely the system is to move from a lower to a higher energy state. Therefore, optimization is guided by a temperature schedule, $T_1, T_2, T_3, \ldots T_M$ ($T_j \geq 0$)

The algorithm can be summarized as follows.

(i) Select an arbitrary starting point $\mathbf{x}_0$.

(ii) Set $F_0 = F(\mathbf{x}_0)$, if $|F_0 - F_m| < \varepsilon$, stop.

(iii) Select a random direction by generating p independent standard normal deviates $y_1, y_2, \ldots y_p$ and compute the components of u, where $u_i = y_i/(y_1^2 + y_2^2 + \ldots y_p^2)^{1/2}$, for i = 1, 2, ... p.

(iv) Set $\mathbf{x}^* = \mathbf{x}_0 + u\Delta\mathbf{r}$.

(v) Set $F_1 = F(\mathbf{x}^*)$ and $\Delta F = F_1 - F_0$.

(vi) If $F_1 \leq F_0$, set $\mathbf{x}_0 = \mathbf{x}^*$ and $F_0 = F_1$. If $|F_0 - F_m| < \varepsilon$, stop, otherwise go to step iii).

(vii) If $F_1 > F_0$, set $p = \exp(-\beta\Delta F)$ and generate a uniform variate $v \in [0,1]$. If $v \geq p$, go to iii). If $v < p$, set $\mathbf{x}_0 = \mathbf{x}^*$ and $F_0 = F_1$ and go to step iii).

2.5.2. Genetic algorithms

The shortcomings of gradient descent methods can be overcome by making use of evolutionary algorithms (Fogel et al., 1990; Bäck et al., 1997), which enable versatile exploration of complex search spaces, based on the Darwinian theory of natural selection.

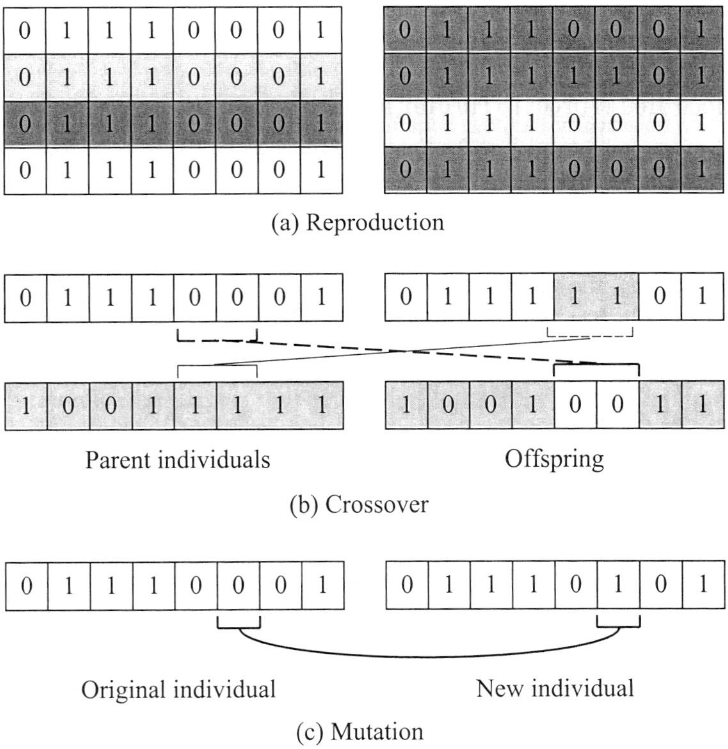

Figure 2.6. Schematic representation of (a) reproduction, (b) crossover and (c) mutation. Reproduction or selection of individuals is stochastic and favours fit individuals (dark coloured) over unfit individuals (lighter coloured).

It differs from conventional search strategies that attempt to improve a single solution to a problem, in that they operate on a population of individual solutions, each of which is rated in terms of some relevant optimization criterion. This fitness value is used as a basis for selection (reproduction), recombination (crossover) and mutation of individuals that are

2.6 REGULARIZATION AND PRUNING OF NEURAL NETWORK MODELS

In general, the methods discussed previously, are useful only when the network architecture has been chosen correctly. Networks that are too small will not be able to represent relationships well, while overly large networks will be prone to overfitting and poor generalization. Regularization is sometimes used to alleviate this problem by promoting smoother mappings via addition of a penalty term to the objective function (Chauvin, 1989; Hanson and Pratt; 1989; Weigend et al., 1991). Unlike pruning (Reed, 1993) and constructive algorithms, it cannot alter the topology of the network.

The trade-off between the reliability of the training data and the goodness of the model can be realized by minimizing the total risk, expressed as

$$R(\mathbf{w}) = E_s(\mathbf{w}) + \lambda E_c(\mathbf{w})$$

The first term $E_s(\mathbf{w})$, is the standard performance measure of the network, which depends on the model itself and the training data, e.g. $E(\mathbf{w}) = \Sigma_j\{d_j-f(\mathbf{x}_j,\mathbf{w})\}^2$, where d_j is the desired or target value of the jth pattern and $f(\mathbf{x}_j,\mathbf{w})$ is the network output corresponding to the jth target. The second term is the complexity penalty, which depends on the neural network model only. λ is referred to as a (complexity) regularization parameter, which represents the relative importance of the complexity penalty term over the standard performance measure. Training takes place with the objective to minimize the total risk, requiring the derivatives of the complexity penalty term to be computed efficiently.

An example of such a complexity penalty term $E_c(\mathbf{w})$ is the kth order smoothing integral

$$E_c(\mathbf{w},k) = \tfrac{1}{2}\int||\partial^k/\partial\mathbf{x}^k\{F(\mathbf{x},\mathbf{w})\}||^2\rho(\mathbf{x})d\mathbf{x} \quad (2.30)$$

where $F(\mathbf{x},\mathbf{w})$ is the input-output mapping performed by the model and $\rho(\mathbf{x})$ is some weighting function that determines the region over the input space over which $F(\mathbf{x},\mathbf{w})$ is required to be smooth. The larger the value of k, the smoother the function $F(\mathbf{x},\mathbf{w})$ will be. Complexity regularization for multilayer perceptrons include weight decay, weight elimination and approximate smoothers.

2.6.1 Weight decay

Consider a neural network with a q-dimensional weight vector $\mathbf{w}$, containing all the weights of the network. Given the training input vectors $\{x_j\}$ with dimension m and associated output vectors $\{d_j\}$, with dimension p, the objective is to find the value of $\mathbf{w}$, so that the network outputs $f(\mathbf{x}_j,\mathbf{w})$ are as close to the target values $\{d_j\}$ as possible. In the weight decay approach (Hinton, 1989), the complexity penalty term is defined as the squared norm of the parameters of the model (weight vectors of the neural network), i.e.

$$E_c(\mathbf{w}) = ||\mathbf{w}||^2 = \Sigma_{j\in\mathbf{w}} w_j^2 \quad (2.31)$$

Where $\mathbf{w}$ refers to all the parameters of the model. For 1st order learning rules, the updating weight value is

$$\Delta w_j = -\eta\partial E_s/\partial w_j - \eta\lambda w_j \quad (2.32)$$

Bartlett (1997) has shown that the magnitudes of the weights are more important than the number of the weights as far as generalization is concerned. Quadratic weight decay adjusts weights proportional to their magnitudes, so that large weights are decreased more than smaller weights. In this way all the weights are retained, albeit with small values.

The constant weight decay method (Ishikawa, 1996) is another simple approach with a complexity penalty term given by

$$E_c(\mathbf{w}) = \Sigma_{j\in\mathbf{w}}|w_j| \tag{2.33}$$

For 1st order learning rules, the weights are updated as follows

$$\Delta w_j = -\eta\partial E_s/\partial w_j - \eta\lambda \mathrm{sign}(w_j) \tag{2.34}$$

This results in constant decay, rather than decay proportional to the magnitude of the weight. This is less effective than the quadratic approach, since it can result in the removal of small weights, while large weights will remain relatively large. As a result smoothing and generalization may not be improved substantially.

2.6.2. Removal of weights

In this case, the complexity penalty term is defined as (Weigend et al., 1991)

$$E_c(\mathbf{w}) = \Sigma_{j\in\mathbf{w}}\, (w_j/w_0)^2/(1+w_j/w_0)^2 \tag{2.35}$$

where w_0 is some preassigned parameter. As indicated in Figure 2.8, the complexity penalty term varies between zero, when the magnitudes of the weights are very small and 1, when the magnitudes of the weights are very large. Also, it can be seen that the complexity penalty term includes weight decay as a special case, i.e. for large values of w_0.

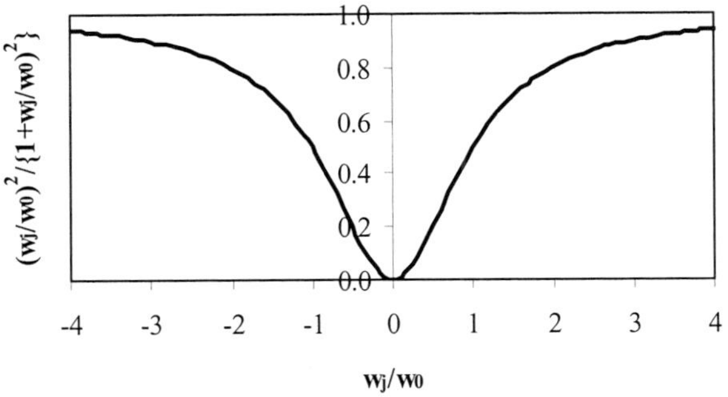

Figure 2.8. Complexity penalty term as a function of the magnitude of the jth weight.

2.6.3 Approximate smoother

Moody and Rögnvaldsson (1997) have proposed the following complexity penalty term for a multilayer perceptron with a single hidden layer and a single node in the output layer

$$E_c(\mathbf{w}) = \sum_{j=1}^{H} w_{0j}^2 \|\mathbf{w}_j\|^r \qquad (2.35)$$

where w_{0j} are the weights in the output layer, and w_j is the weight vector for the jth neuron in then hidden layer. For a global smoother, the power r is defined by $r = 2k-1$ and for a local smoother $r = 2k$, where k is the order of differentiation of $F(\mathbf{x},\mathbf{w})$ with respect to $\mathbf{x}$. The approximate smoother is a more sophisticated approach to regularization, because it distinguishes between the weights in different layers and it also takes the interaction between weights in different layers into account. The penalty for these advantages is higher computational complexity.

2.7 PRUNING ALGORITHMS FOR NEURAL NETWORKS

2.7.1 Hessian-based network pruning

With Hessian-based pruning, the 2nd order derivatives of the error surface are used to strike a balance between the performance of the network on the training data set and the complexity of the network. The standard performance measure of the network, $E_s(\mathbf{w})$, is approximated by a Taylor series expansion about the operating point

$$E_s(\mathbf{w} + \Delta\mathbf{w}) = E_s(\mathbf{w}) + \mathbf{g}^T(\mathbf{w})\Delta\mathbf{w} + \tfrac{1}{2}\Delta\mathbf{w}^T\mathbf{H}(\mathbf{w})\Delta\mathbf{w} + O(\|\mathbf{w}\|^3) \qquad (2.36)$$

$\Delta\mathbf{w}$ is a perturbation applied to the operating point $\mathbf{w}$, while $\mathbf{g}(\mathbf{w})$ is the gradient vector evaluated at the operating point. $\mathbf{H}(\mathbf{w})$ denotes the Hessian matrix. The idea is to identify the weights that will least affect the performance measure of the network if they are eliminated.

By assuming that training of the neural network has converged to a (local or global) minimum on the error surface, i.e. the gradient vector $\mathbf{g}(\mathbf{w})$ is zero and that the error surface around this minimum is nearly quadratic, i.e. the higher order terms in equation (2.36) can be neglected, equation (2.36) reduces approximately to

$$E_s(\mathbf{w} + \Delta\mathbf{w}) \approx E_s(\mathbf{w}) + \tfrac{1}{2}\Delta\mathbf{w}^T\mathbf{H}(\mathbf{w})\Delta\mathbf{w} \qquad (2.37)$$

or

$$E_s(\mathbf{w}) = E_s(\mathbf{w} + \Delta\mathbf{w}) - E_s(\mathbf{w}) \approx \tfrac{1}{2}\Delta\mathbf{w}^T\mathbf{H}(\mathbf{w})\Delta\mathbf{w} \qquad (2.38)$$

2.7.2 Optimal brain damage and optimal brain surgeon algorithms

In the optimal brain damage (OBD) algorithm (LeCun et al., 1990), the Hessian matrix in equations (2.36) and (2.37) is assumed to be diagonal, but no such assumption is made in the optimal brain surgeon (OBS) algorithm (Hassibi et al., 1992), so that the former can be considered as a special case of the latter. In the OBS algorithm, the quadratic form $\tfrac{1}{2}\Delta\mathbf{w}^T\mathbf{H}(\mathbf{w})\Delta\mathbf{w}$ is minimized with respect to an incremental change in the weight vector $\mathbf{w}$, subject to the constraint that $\mathbf{1}_j^T\Delta\mathbf{w} + w_j = 0$[11], after which the result is minimized with respect to the index j.

2.8. CONSTRUCTIVE ALGORITHMS FOR NEURAL NETWORKS

Since the architecture of a neural network can have a profound influence on the performance of the network, various techniques have been developed to optimize the architecture of the network, often in conjunction with the training of the network itself. An exhaustive search of all possible architectures can be computationally infeasible, even for small networks. This problem is ameliorated to some extent by making use of regularization, or by growing algorithms that start with a small network, adding nodes and connections incrementally and using the weights of the trained network as an initial estimate of the weights of the incrementally expanded network (plus the initialized weights of new nodes). In pruning algorithms, the opposite approach is followed, in that a relatively large network is gradually pruned to fit the data best.

2.8.1 State space search

The problem of constructing neural network models can be viewed as a search problem with a state space, initial state, a goal state (termination of the search) and a search strategy. The target function of a regression problem can be defined with regard to a function space, **F**, which is selected by the user and is related to the error criterion being minimized. To implement a function in **F** with a neural network, the number of hidden nodes in the network needs to be specified, as well as the directed connectivity graph (**C**) specifying the interconnectedness between nodes in the network, the functional forms of the nodes and the parameters (weights) of the entire network.

The collection of functions that can be implemented b a class of neural networks forms the state space **S**. With some simplification[12], each state $s \in \mathbf{S}$, may be represented by a 2-tuple, $s = \mathbf{C} = (\mathbf{V},\mathbf{E})$, where **V** and **E** are the respective sets of vertices and directed edges representing the topology of the network.

a) Initial state and goal state

Starting with some initial state[13] is straightforward and the search can be terminated in a number of ways, e.g. when the generalization of the network begins to deteriorate. The ability of the neural network to generalize can be based on its performance on separate validation sets (Parker and Tummala, 1992), cross-validation (Moody, 1994; Stone, 1974), bootstrapping (Efron and Tibshirani, 1993; Weigend and LeBaron, 1994) or information criteria, such as Akaike's information criterion (Akaike, 1974), Bayesian information criteria (Schwartz, 1978), general cross-validation (Craven and Wahba, 1979), minimum description length (Rissanen, 1975, 1989), predicted square error (Barron, 1984), generalized prediction error (Moody, 1991), Schwartz's information criterion (Schwartz, 1978), etc.

---

[11] It can be shown that the elimination of the j'th weight is equivalent to the condition $w_j + \Delta w_j = 0$ or $\mathbf{1}_j^T\Delta\mathbf{w} + w_j = 0$.

[12] E.g. by ignoring the effect of the functional forms of the hidden nodes and assuming the weights of the network can be uniquely determined from the connectivity of the neural network.

[13] This means a small network in general, often one without any hidden nodes.

b) Search strategy

The search strategy is a pivotal part of the constructive algorithm and it determines how movements from one state to another occur in the state space or equivalently, how the connectivity graph **C** evolves during the search. In movement from one state $s_1 = (V_1, E_1)$ to another $s_2 = (V_2, E_2)$, the following properties are usually satisfied, viz. $V_1 \subseteq V_2$ and $E_1 \subset E_2$. This means that nodes and connections in the current state are usually preserved in the next state, while the next state also contains some new connections or nodes. Note that the state transition mapping $\Delta: s_j \rightarrow s_{j+1}$, mapping the current state to the next is determined by the problem, not the search strategy.[14] The search strategy simply defines the order in which possible next states are visited. In principle, regression problems are unconstrained, as one may randomly move from one state to another.

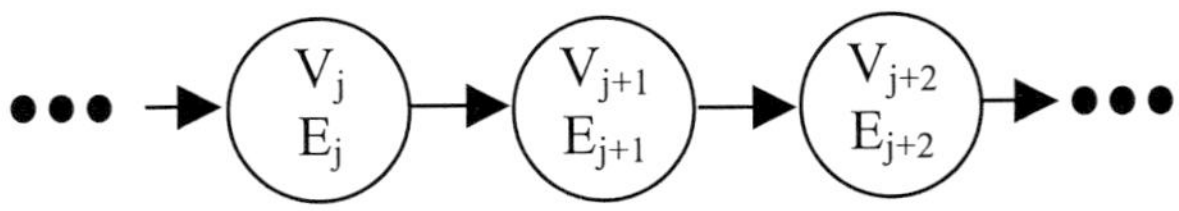

Figure 2.8. Typical state traversal with single-valued state transitions, $|V_{j+1}| = |V_j| + 1$.

The state transition Δ may be single- or multi-valued. In the former case there is only one state to be explored with each transition and the state traversal is reduced to a simple chain of states, as indicated in Figure 2.8. The disadvantage of single-state transitions is that they may lack the flexibility to find adequate solutions. With only one possible next state, the network may be stuck when a mismatch occurs between the problem and the architecture constructed by the algorithm.

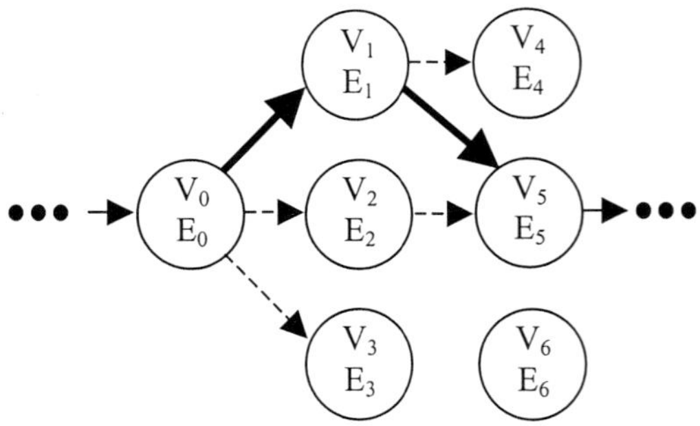

Figure 2.9. Typical state traversal with multi-valued state transitions, $|V_{j+1}| \neq |V_j| + 1$.

Multi-valued state transitions are not prone to these problems, as they involve multiple candidate solutions that may be adapted to the specific problem, as indicated in Figure 2.9. Nonetheless, the search method still has to define the order in which the candidate solutions are evaluated. Moreover, the computational cost incurred by multi-valued state transitions can be a major drawback. Other important issues regarding the search strategy are the definition

[14] For example, in chess problem, the possible board configurations of the next state are determined by the rules of the game, not the search strategy.

of new connectivity graphs, as well as the question of universal approximation and convergence.

Generation of the next state from the current state can be accomplished by simply requiring that all the nodes be in the same (single) hidden layer. This gives a regular structure with constant fan-in for the hidden nodes, which also simplifies hardware implementation. However, if the hidden nodes have only simple functional forms, they are limited in the types of relationships they can represent. For example, radial basis function nodes can only form local bumps in the input space, while sigmoidal nodes can only form hyperplanes, giving rise to networks with different properties (Hartman and Keeler, 1991; Renals, 1988). This may be problematic when the nodes are added one at a time in a greedy manner. Algorithms constructing multiple hidden layers or nodes with more complex activation functions are less limited, but may give difficulties by fitting some of the noise in the data.

It is assumed that the class of functions implemented in the search strategy is sufficiently broad to contain at least an adequate approximation to the target function f. The issue of universal approximation is of fundamental concern in the development of neural networks and also appears to be a prerequisite for the convergence[15] of the network sequence produced by the search algorithm.

c) Generalized search spaces

As discussed so far, each state in the state space corresponds with only one network and the constructive algorithm keeps only one network at any given time, even in the case of multi-valued transitions, where multiple networks are examined prior to actual state transitions. However, some algorithms allow for a wider simultaneous exploration of the state space by maintaining a pool of networks. This approach can be formalised by defining a generalized state space S_G from S, with each S_G being a set containing all the states of S explored at the time. An example of this approach is the use of genetic algorithms for the construction of neural networks, as indicated in Figure 2.10. Since the input and output layers are typically fixed, only the hidden layers need to be encoded by a particular chromosome. Owing to the flexibility of these representations, the parameters (weights), the activation functions and number of nodes in the hidden layer(s) of the neural network (Stork et al., 1990; Liu and Yao, 1996; Sebald and Chellapilla, 1998), as well as the learning rules (Mjolsness et al., 1989; Baxter, 1992) can be evolved simultaneously. In the latter case, the metadevelopment of neural networks is still in its infancy, but regarded as important where a premium is placed on the ability of neural networks to cope with complex dynamic environments.

Population-based learning systems are another example of generalized state space representations. In this case the objective is to design as good as possible a neural network under a given time constraint, by use of a constructive algorithm, such as cascade correlation learning. A direct approach would entail the sequential construction of several neural networks, until the available time has run out. In the population based approach, the time is divided into a number of different quanta. Estimates of network performance are obtained by extrapolation based on the training error, the rate of change, the number of epochs trained, etc. The most promising network is subsequently selected and grown further for a certain time quantum.

[15] A sequence $\{f_k\}$converges to f, if $\lim_{k\to\infty}||f-f_k|| = 0$. This norm must be based on the whole input space and not just the training data set.

The cycle is repeated until the total time allowed is expended. This appears to give superior results when the available time is long, but otherwise the system tends to spend too much time maintaining the population, with resultant inferior performance.

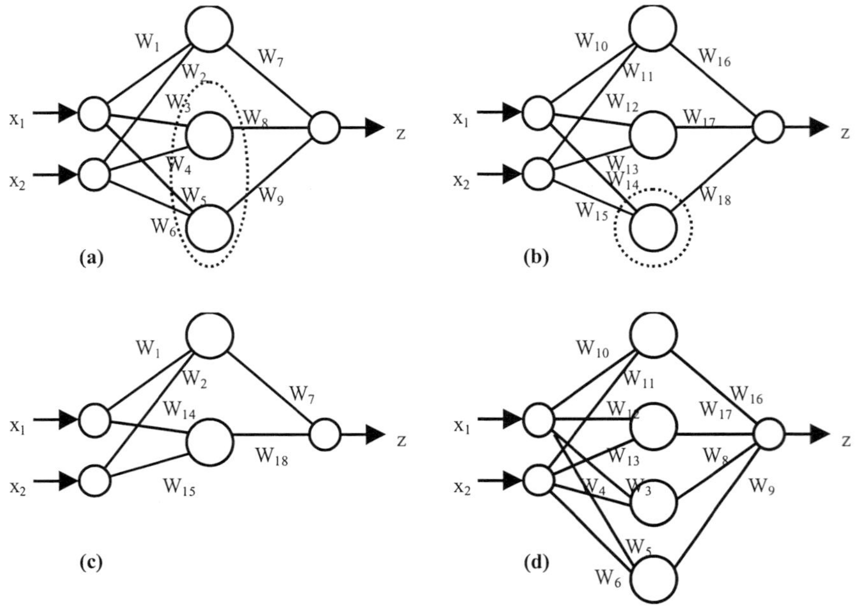

Figure 2.10. Crossover operation in a genetic algorithm, where parents (a) and (b) combine to generate offspring (c) and (d).

2.8.2. Training algorithms

Constructive procedures for the development of neural network models can be classified as indicated in Figure 2.11 (Kwok and Yeung, 1999). According to this classification scheme, constructive procedures are either based on single- or multi-valued transitions. Procedures based on single-valued transitions can be subdivided further into three categories, viz. those where the whole network is trained at every iteration, those where only the new nodes are trained, and those where training is marked by a memorization process.

Procedures where only new nodes are trained can deal with either single- or multiple hidden layers. Likewise, the category based on multi-value transitions can also be differentiated into those where the whole network is trained at each cycle, as opposed to those where only the new hidden layer is trained.

a) Dynamic node creation (DNC)

The state transition mapping is single-valued in dynamic node creation algorithms and their variants (Moody, 1994; Chung and Lee, 1995; Azimi-Sadjadi et al., 1993; Setiono and Hui, 1995; Ash, 1989; Bartlett, 1994; Hirose et al., 1991; Khorasani and Weng, 1994; Zhang, 1994). Sigmoidal nodes are added one at a time to the same hidden layer and the whole network has to be retrained after the addition of each new hidden node. Although these networks are simple and display robust convergence, the computational cost becomes exorbitant for large complex problems.

b) Projection pursuit regression (PPR)

Projection pursuit regression (described in more detail in 3.9.4) is closely related to neural networks, as can be seen from the similarity between these models and multilayer perceptrons. A set of parameters ($\mathbf{a}_m$) defines the projection of the input vectors onto a set of planes (m = 1, 2, ... M). As is the case with multilayer perceptrons, these projections are transformed by nonlinear activations functions (ϕ_m). In turn, the activation functions are linearly combined to give the output **y**. However, unlike a multilayer perceptron, the nonlinear transformation or activation functions are neither all the same, nor prescribed a priori, but are fitted to the data. In addition, the parameters in the projection pursuit regression model are optimized cyclically in groups, unlike the simultaneous optimization of the weights of a multilayer perceptron. In short, projection pursuit regression can be seen as a generalization of multilayer perceptrons, i.e. with more flexible activation functions. As such, they are also universal approximators (Diaconis and Shahshahani, 1984; Jones and Sibson, 1987).

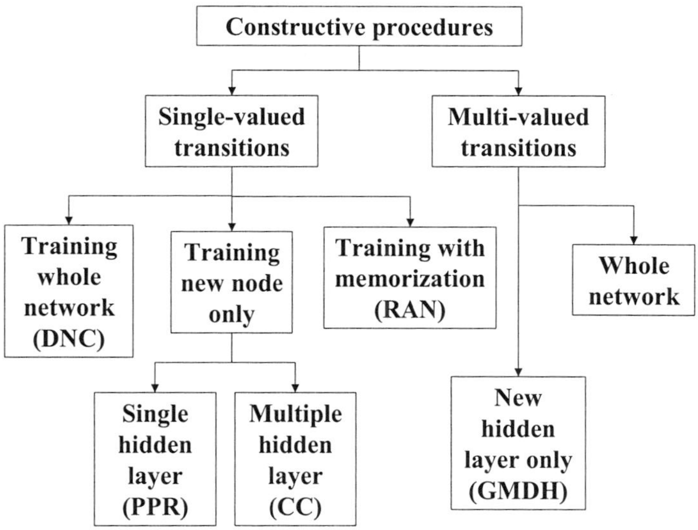

Figure 2.11. Classification of constructive procedures for neural networks.

The activation functions of projection pursuit regression networks are nonparametric and are obtained with a variable span or supersmoother. It is based on a symmetric k-nearest neighbour linear least squares fitting algorithm. The optimal value of k (the span of the supersmoother) is selected for each input point by means of a local cross-validation procedure. Supersmoothers suffer from several disadvantages, viz. unstable approximations when derivatives are calculated, piecewise interpolation when activation values are calculated and the use of cumbersome regression tables. Automatic smoothing spline projections uses smoothing splines to alleviate some of these drawbacks. Although the projections are also nonparametric, they can give accurate derivative calculations and smooth interpolation. However, they still require the use of a smoother matrix and in addition, the generalized cross-validation statistic used to select the degree of smoothness, tends to undersmooth (Roosen and Hastie, 1994).

Instead of using nonparametric hidden nodes, some algorithms use parametric hidden nodes. For example in projection pursuit learning (PPL) neural networks (Hwang et al., 1994) each hidden node is represented as a linear combination of Hermite functions

$$g(\mathbf{x}) = \sum_{r=1}^{R} c_r h_r(\mathbf{x}) \tag{2.30}$$

where $h_r(x)$ are orthonormal Hermite functions, c_r are constants and R is a user-defined parameter (the order).

Parametric nodes expedite smooth interpolation and do not require large regression tables for fast and accurate calculation of the derivatives. Apart from Hermite functions, other parametric forms may also be used, e.g. B-splines in multidimensional additive spline approximation, normalized Legendre polynomial expansion in exploratory projection pursuit (Friedman and Tukey, 1974; Friedman, 1987), the expansion of basis functions, pi-sigma networks in ridge polynomial networks (Shin and Gosh, 1992, 1995), etc. Moreover, as Laghi and Lizzani (1997) have indicated, regression dces not have to be restricted to continuous variables.

In all these algorithms where parametric hidden nodes are used, the number or complexity of the nodes is controlled by some parameter. This can be the order parameter in Hermite functions, the number of hidden nodes in fixed neural network architectures, etc. The setting of these parameters can sometimes have a considerable effect on the performance of the neural network (Hwang et al., 1994). Only new nodes are trained as they are added to the network, i.e. only the smoother function and the projection vector need to be retrained.

c) Cascade correlation method

Unlike the algorithms discussed above, the cascade correlation algorithm constructs neural networks with multiple hidden layers. The cascade correlation method of Fahlman and Lebiere (1990) starts with an input and output layer only. The weights of the m output nodes are consequently trained over all the output classes via the single layer perceptron mode. Once the network has converged in this mode, a single hidden node is adjoined to the network, as indicated in Figure 2.11.

The hidden node is connected to all the input nodes and trained with a gradient descent method over all the exemplars, until the output of the hidden node correlates maximally with the residual error of the output nodes. The weights of the newly added hidden node are subsequently saved. The weights of the hidden node are then reinitialized and trained several more times, after which the weights associated with the maximum correlation with the residual error of the output nodes are retained. These weights are set, and not changed again during training. In the next step a new node is added to the hidden layer, as indicated in Figure 2.11. This node is connected to all the input nodes, as well as all other nodes in the hidden layer. The weights of this new node are again trained in a multistart mode to maximize the correlation between the output of the new hidden node and the output nodes of the network Once this correlation has been maximized, the weights of the *output* nodes only are trained by means of a simple back propagation algorithm. The weights of the hidden nodes are not trained at this stage. If the output error (SSE) of the network is acceptably small, training is stopped, otherwise another node is added to the hidden layer, and the training procedure is repeated.

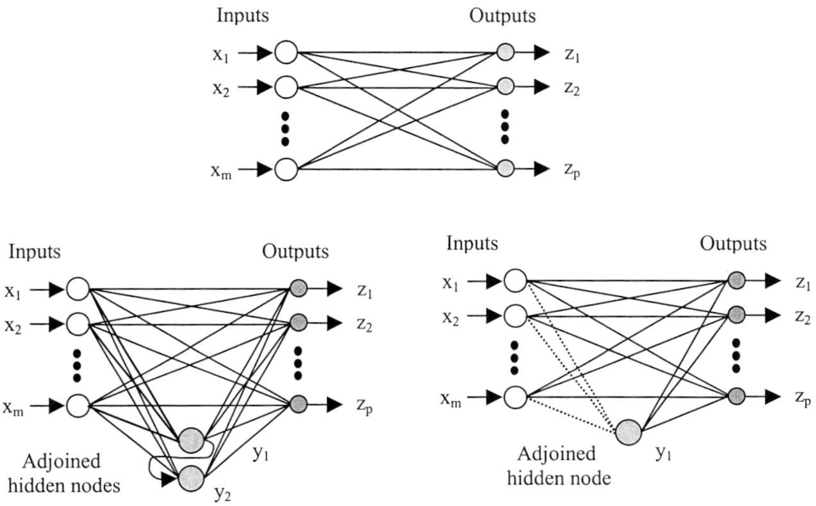

Figure 2.11. Training of cascade correlation neural networks.

The advantage of the cascade correlation algorithm is that training initially involves a single layer of output nodes to account for the linear relationships in the data. Nonlinear relationships are accounted for by incremental addition of nodes in the hidden layer, which need not be larger than is necessary. Moreover, training in the early stages is fast, owing to the small size of the hidden layer and neural networks trained by the cascade correlation method perform well on difficult nonlinear problems.

The disadvantage is that when many nodes are added to the hidden layer, the network becomes progressively slower, owing to the increased sequential path length through the network.

d) Resource allocating neural networks (RAN)

Like dynamic node creation algorithms, resource allocating algorithms also add hidden nodes one at a time to the same hidden layer of the neural network. However, in contrast with other networks, memorization of the training data is used to reduce the computational cost of training. Methods like Parzen windows (Duda and Hart, 1973) and k-nearest neighbours also use memorization, but produce networks with potential growth linearly related to the size of the training data set, thus placing large demands on space and time in the computation of network outputs. In contrast, resource allocating algorithms attempt to strike a balance in that memorization is used only on novel or difficult patterns.[16] Novel patterns are detected if they

[16] Since individual patterns are memorized, training takes place on a pattern-by-pattern basis, unlike other algorithms, where it can also take place in an epochal mode.

are further away from the hidden nodes[17] than dictated by a resolution parameter and the difference between the target output of the pattern and the network output is larger than the desired accuracy. Therefore, if the network performs well on a particular training pattern or if the pattern is already close to a stored vector, the weights of the network is adjusted. Otherwise, if the pattern is novel, no training takes place, since the network memorizes the pattern (input) and the associated output.

Resource allocating neural networks can be used for sequential learning, but noisy data may lead to a large number of hidden nodes. In addition, the resolution parameter δ specifies the smallest permissible distance between nodes, so that features smaller than that will get averaged out. This makes it very difficult to model functions that vary on a fine scale. These problems are alleviated by making use of growing cell structures (Fritzke, 1994b). In addition, a disadvantage of these types of networks is that their convergence properties are unknown.

e) Group method of data handling

With group method of data handling (GMDH) methods, the state transitions are multi-valued. Although each hidden node has a fixed number of incoming connections, the sources of these connections are not fixed and may be derived from any possible combination of input and existing hidden nodes in the neural network. A number of candidate networks will therefore be available when a new hidden node has to be added. Searching of the candidates can be done in different ways, such as with simulated annealing (Tenorio and Lee, 1990) or other ad hoc approaches (Parker and Tummala, 1992). The resultant architecture is typically not a regular structure, but may have many intralayer connections, as indicated in Figure 2.13.

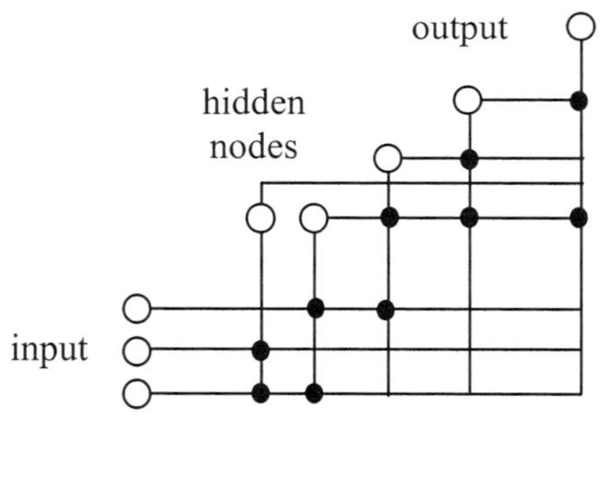

Figure 2.13. Structure of a GMDH neural network.

Only the new hidden node is trained[18] and it is treated as an interim output node during training, so that there is only one layer of weights (feeding into the node) to train. By treating the hidden node as an interim output node and making use of the squared error criterion,

[17] These algorithms typically use radial basis function nodes, which respond to small, localized regions in the input space.
[18] Not many constructive algorithms with multi-valued state transitions train the whole network after the addition of each hidden node, probably owing to the high computational cost involved.

training of the weights reduces to a linear problem, which can be solved by simply computing the pseudo-inverse.

Although any functional form can be used, further reduction in the computational load of this class of network is achieved by use of polynomial nodes. With p incoming connections to the hidden node[19], the activation functions may be of the form

$$g(x_1, x_2, \ldots x_p) = a + \sum_{i=1}^{p} b_i x_1 + \sum_{i=1}^{p} c_i x_i^2 + \sum_{i,j=1, i \neq j}^{p} d_{ij} x_i x_j \qquad (2.31)$$

where a, b_i, c_i and d_{ij} are connection weights and x_i and x_j are the corresponding incoming activations. By using polynomial activation functions, GMDH networks are universal approximators, even with low order polynomials. However, the convergence properties of these networks are not well-known, owing to their ad hoc nature.

In summary, it should be borne in mind that although regularization, pruning and constructive algorithms are designed for the same purpose, they should not be considered as competing alternatives. Instead, these methods complement each other and are sometimes also used concurrently. Moreover, the problem of model selection and the estimation of generalization occur in a much wider context within statistics and it is anticipated that the close relationship between many statistical methodologies and neural networks will lead to further mutual advances in both fields.

---

[19] p is usually set to a small value, such as 2.

Chapter 3
Latent Variable Methods

3.1. BASICS OF LATENT STRUCTURE ANALYSIS

A latent variable is simply a variable that cannot be measured directly, but instead has to be inferred from manifest variables that can be measured directly. More formally, consider a data set x, with k variables, $x_1, x_2, \ldots x_k$ measured on some process. Moreover, assume that this process is actually driven by a set of $p \leq k$ latent variables (**z**) that cannot be observed, but whose influence is reflected in the measured variables (**x**). The relationship between **x** and **z** can be modelled as follows

$$\mathbf{x} = \mathbf{zP} + \mathbf{e} \tag{3.1}$$

where $\mathbf{x} \in \Re^{1xk}$, $\mathbf{z} \in \Re^{1xp}$, $\mathbf{P} \in \Re^{pxk}$ and $\mathbf{e} \in \Re^{1xk}$. The last term **e** is considered to be a random error vector, representing sampling errors, measurement errors and other unknown process disturbances. Since **z** cannot be observed and **P** is unknown, **z** cannot be identified from equation (3.1). In fact, it can be shown that **x** can be modelled also by replacing **z** and **P** with $\mathbf{z}^* = \mathbf{zC}$ and $\mathbf{P}^* = \mathbf{C}^{-1}\mathbf{P}$ respectively, where **C** is any non-singular p x p matrix (Burnham et al., 1999). The measured data are therefore more commonly expressed as

$$\mathbf{x} = \mathbf{tP} + \mathbf{e} \tag{3.2}$$

where **t** is some transform **zC** of the actual latent variables **z**. In effect, the transformation of **z** to **t** is simply a transformation of basis, i.e. the observations in t-space lie in the same vector space as those in **z**, albeit with a different basis. Since the actual latent variables are not as important as the space they generate, any basis **t** will suffice. Consequently, for a given set of n observations, the model can be expressed in matrix form as

$$\mathbf{X} = \mathbf{TP} + \mathbf{E} \tag{3.3}$$

where $\mathbf{X} \in \Re^{nxk}$, $\mathbf{T} \in \Re^{nxp}$, $\mathbf{P} \in \Re^{pxk}$ and $\mathbf{E} \in \Re^{nxk}$

This approach is often varied by making a distinction between two blocks of data, i.e. by considering two spaces $\mathbf{X} \in \Re^{nxk}$ and $\mathbf{Y} \in \Re^{nxm}$.

In general, these spaces can be expressed as models (3.3-3.4),

$$\mathbf{X} = \mathbf{TP} + \mathbf{E} \tag{3.3}$$

$$\mathbf{Y} = \mathbf{UQ} + \mathbf{F} \tag{3.4}$$

where **E** and **F** represent random errors.

This can be contrasted with the standard multivariate regression model, where **X** and **Y** are related as follows

$$\mathbf{Y} = \mathbf{XB} + \mathbf{F} \tag{3.5}$$

where **B** is a (k x m) matrix of regression coefficients that have to be estimated. This model does not account for any latent structure in either **X** or **Y**. No measurement errors in **X** are considered, in fact the observations are usually considered to be fixed constant points.

In reduced rank regression models (Izenman, 1975; Tso, 1981; Davies and Tso, 1982), an additional constraint is imposed on the model, in that the rank of the regression vector **B** is considered to be reduced, i.e. **B** is an a x m matrix, with a < m. In this way the latent structure in **Y** is considered, but not the errors or latent structure in **X**.

Errors in **X** are accounted for in errors-in-variables models, i.e. in terms of the observed values **X** and **Y** and the true values $\mathbf{X}_t$ and $\mathbf{Y}_t$.

$$\mathbf{Y}_t = \mathbf{X}_t\mathbf{B} \tag{3.6}$$

$$\mathbf{Y} = \mathbf{Y}_t + \mathbf{F} \tag{3.7}$$

$$\mathbf{X} = \mathbf{X}_t + \mathbf{E} \tag{3.8}$$

Although error-in-variables models accommodate measurement errors in **X** and **Y**, they do not take any latent structure into account. In latent variable multiple regression models, the **X** and **Y** spaces are assumed to have a common underlying latent structure (Burnham et al., 1999), so that in addition to equation (3.3),

$$\mathbf{Y} = \mathbf{TQ} + \mathbf{F} \tag{3.9}$$

where **F** again represents random errors.

This could be equivalent to equation 3.3, by formulating a single space model

$$[\mathbf{XY}] = \mathbf{T}[\mathbf{PQ}] + [\mathbf{EF}] \tag{3.10}$$

Likewise, in principal component regression models, the latent structure of **X** is considered, but not that of **Y**. In partial least squares models, both the latent structure in **X** and **Y** are considered (Geladi and Kowalski, 1986). Both these approaches have been used extensively in chemical engineering.

3.2. PRINCIPAL COMPONENT ANALYSIS

Principal component analysis linearly transforms a set of variables into a substantially smaller set of uncorrelated variables containing most of the information of the original set of variables. A small set of uncorrelated variables is easier to understand and work with than a large set of correlated variables.

For example, a large set of variables may be measured on a plant, such as flow rates, flow concentrations, temperatures and pressures. If these variables are correlated, and especially if they are highly correlated, they can be transformed to a small set of uncorrelated variables. The derived variables that maximize the variance accounted for in the original set of variables are referred to as *principal components*.

Apart from being a useful technique to reduce the dimensionality of the variable space, principal component analysis can also be used to visualize multivariate data sets, so that outlying or atypical observations can be detected.

In contrast to direct methods in which multivariate observations are transformed in such a way so as to enable the observations to be depicted graphically in a direct form, geometrical configurations of multivariate observations can also be constructed, whereupon the 'best' 2-D subspace to view the data can be found, and the data can be projected into this 2-D subspace for visualization and interpretation. Of all the techniques used to explore multivariate data, principal component analysis is probably one of the most popular, so much so, that it is often erroneously seen as the only technique for doing so.

3.2.1. Mathematical perspective

In general it can be supposed that a (m x 1) vector $\mathbf{x}^T = (x_1, x_2, \ldots x_m)$ of variables is observed on each of n independent samples or units, which gives rise to an (n x m) matrix of observations with elements x_{ij}, (i = 1, 2, ... n and j = 1, 2, ... m) Moreover, the m values observed on the i'th unit or sample can be denoted by $\mathbf{x}_i$. The mean of the j'th variable in the sample is then $x_{avg,j} = 1/n\Sigma_{i=1}^{n} x_{ij}$. The *sample mean vector* is $\mathbf{x}_{avg}^T = (x_{avg,1}, x_{avg,2}, \ldots x_{avg,m})$. The variance of the j'th variable is given by $s_{jj} = [1/(n-1)]\Sigma_{i=1}^{n}(x_{ij}-x_{avg,j})^2$, while the covariance between the j'th and k'th variables is given by $s_{jk} = [1/(n-1)]\Sigma_{i=1}^{n}(x_{ij}-x_{avg,j})(x_{ik}-x_{avg,k})$. These variances and covariances are represented by the *sample covariance matrix* **S**, which has s_{jk} as the (j,k)'th element, that is $\mathbf{S} = [1/(n-1)]\Sigma_{i=1}^{n}(\mathbf{x}_i - \mathbf{x}_{avg})(\mathbf{x}_i - \mathbf{x}_{avg})^T$ or in terms of matrix notation

$$\mathbf{S} = \text{cov}(\mathbf{X}) = \mathbf{X}^T\mathbf{X}/(n-1) \tag{3.11}$$

where the columns of **X** have been mean centred. If the columns of **X** have been autoscaled to zero mean and unit variance, **S** is also the correlation matrix (**R**) of **X**.

With principal component analysis, the matrix **X** is decomposed into the sum of the outer products of two vectors, viz.

$$\mathbf{X} = \mathbf{T}\mathbf{P}^T = \mathbf{t}_1\mathbf{p}_1^T + \mathbf{t}_2\mathbf{p}_2^T + \mathbf{t}_3\mathbf{p}_3^T + \ldots + \mathbf{t}_k\mathbf{p}_k^T + \ldots \mathbf{E} \tag{3.12}$$

The index k is less than or equal to the smallest dimension of the data matrix, i.e. k ≤ min(n,m). The $\mathbf{p}_j$ vectors are the eigenvectors of the covariance matrix **S** (equation 3.11) and the t_j are the score vectors representing the relationships between the samples, that is

$$\mathbf{S}\mathbf{p}_j = \lambda_j\mathbf{p}_j \tag{3.13}$$

This implies that the derivation of principal components is based on the successive projection of lines through the m-dimensional space, as indicated in Figure 3.1, so that the variance of the projections of the n points in the space onto these lines is maximal at each instance. It can be shown that a line in a space can be represented by a vector **p** of unit length, which means that per definition $\mathbf{p}^T\mathbf{p} = \Sigma_{j=1}^{m}p_j^2 = 1$. It is thus algebraically possible to define the principal components to be the linear combination $\mathbf{p}_i = \mathbf{p}_i^T\mathbf{X}$ of the original variables that maximizes the value of $\mathbf{p}_i^T\mathbf{S}\mathbf{p}_i$, subject to the constraint that $\mathbf{p}_i^T\mathbf{p}_i = 1$.

The *loading* of the k'th original variable x_k on the j'th principal component p_j is defined by $p_{jk}(\lambda_j)^{1/2}$, where $\mathbf{p}^T_j = (p_{j1}, p_{j2}, \ldots p_{jm})$. The *score* of the i'th individual or sample point on the j'th principal component p_j is defined as $t_{ij} = \mathbf{p}^T_j\mathbf{x}_i = p_{j1}x_{i1} + p_{j2}x_{i2} + \ldots + p_{jm}x_{im}$ (j = 1, 2, ... m).

Note that because the $(\mathbf{t}_i,\mathbf{p}_i)$ pairs are extracted in descending order according to the associated eigenvalues λ_i, the first pair captures the largest amount of variation in the data that can possibly be captured by a linear factor, while each subsequent pair captures the largest amount of the variation of the data not captured by the preceding pairs (that is after subtracting $\mathbf{t}_i\mathbf{p}_i^T$ from $\mathbf{X}$).

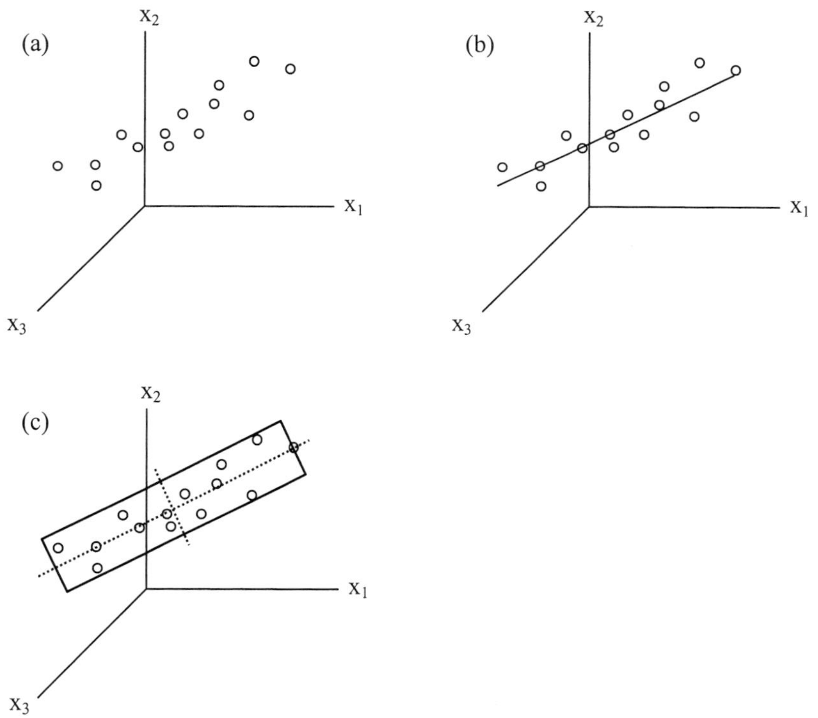

Figure 3.1. Geometric representation of the steps in principal component analysis, showing (a) the data points in the observation space, (b) the first principal component, (c) the plane defined by the first two principal components

3.2.2. Statistics associated with principal component analysis models

The scores from principal component analysis models have a zero mean and a variance equal to the eigenvalue associated with the score vector $(\mathbf{t}_j)$. Assuming that the scores are normally distributed[20], confidence limits can be calculated on the scores by use of the standard normal distribution or Student's t-distribution.

The limits on the scores of the j'th principal component can be expressed as

$$t_{j,\alpha} = \pm(\lambda_j)^{1/2} t_{n-1,\alpha/2} \tag{3.14}$$

where n is the number of observations and $t_{n-1,\alpha/2}$ corresponds with the probability point on the single-sided t-distribution with n-1 degrees of freedom and area $\alpha/2$.

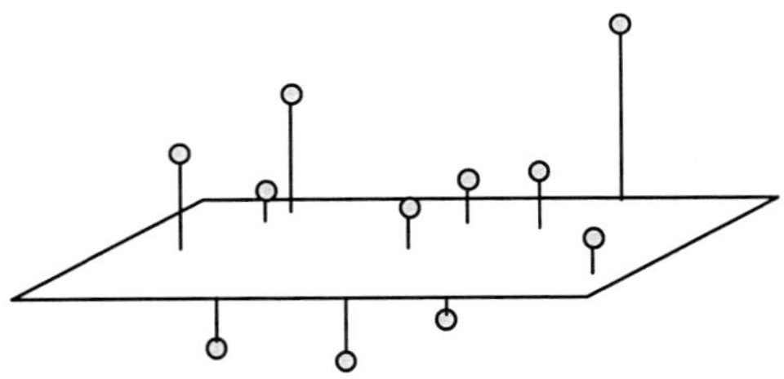

Figure 3.2. Residuals between observations and their projections onto a two-dimensional plane in the data space.

The calculation of lack of fit statistics for principal component analysis models is also possible, and the Q-statistic that is often used is simply the sum of squares of each row of the residual matrix **E** (Wise et al., 1990). That is, for the i'th observation $\mathbf{x}_i$ in **X**,

$$Q_i = \mathbf{e}_i\mathbf{e}_i^T = \mathbf{x}_i(\mathbf{I} - \mathbf{P}_k\mathbf{P}_k^T)\mathbf{x}_i^T \tag{3.15}$$

$\mathbf{P}_k$ is the matrix of k loading vectors retained in the principal component analysis model (with column in the matrix representing a loading vector). It is a measure of the residual between an observation and its projection onto the k principal components retained in the model. Geometrically this can be interpreted as the distance of the i'th observation off the hyperplane defined by the k principal components, as indicated in Figure 3.2, for a k = 2.

The confidence limits for the overall residual Q can be calculated, based on all the eigenvalues λ_i of the covariance matrix of **X**, i.e.

$$Q_\alpha = \Lambda_1[1 + c_\alpha(2\Lambda_2\theta^2)^{1/2}/\Lambda_1 + \Lambda_2\theta(\theta-1)/\Lambda_1^2]^{1/\theta} \tag{3.16}$$

where $\Lambda_i = \Sigma_{k+1}{}^m\lambda_j{}^i$ (for i = 1, 2, 3) and $\theta = 1 - 2\Lambda_1\Lambda_3/(3\Lambda_2^2)$. c_α is the standard normal deviate corresponding to the upper (1-α) percentile, while m is the total number of principal components. The residual Q is more likely to have a normal distribution than the principal component scores, since Q is a measure of the non-deterministic behaviour of the system. Q values lying outside the confidence limits indicate data lying unusually far from the hyperplane defined by the k principal components.

---

[20] This will only be the case when the underlying physical phenomena being observed are related to Gaussian processes, since the principal components are weighted linear combinations of the variables.

The variation of each observation *within* the principal component analysis model is indicated by Hotelling's T^2 statistic, defined as

$$T_i^2 = \mathbf{t}_i\lambda^{-1}\mathbf{t}_i^T = \mathbf{x}_i\mathbf{P}\lambda^{-1}\mathbf{P}^T\mathbf{x}_i^T \tag{3.17}$$

where t_i refers to the i'th row of $\mathbf{T}_k$, the matrix of k score vectors from the principal component analysis model. λ^{-1} is the diagonal matrix containing the inverse of the eigenvalues associated with the k eigenvectors (principal components) of the model.

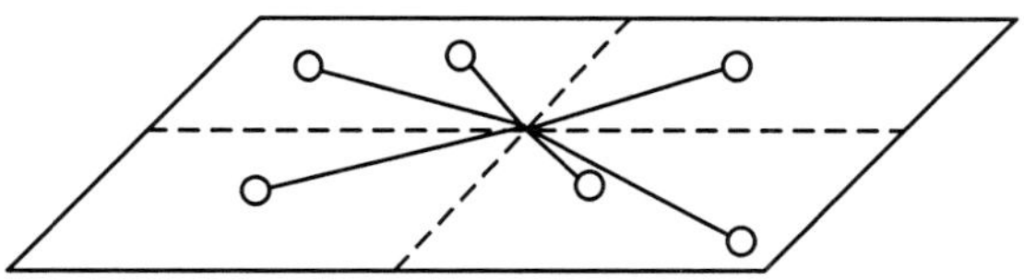

Figure 3.3. Residuals between observations and their projections onto a two-dimensional plane in the data space.

The confidence limits for T^2 can be calculated by means of the F-distribution as follows

$$T^2_{k,m,\alpha} = F_{k,m-k,\alpha}k(m-1)/(m-k) \tag{3.18}$$

T^2 is a measure of the distance from the multivariate mean to the projection of the observation onto the principal component hyperplane, as indicated by Figure 3.3.

3.2.3. Practical considerations regarding principal component analysis

One of the first decisions to be taken before the application of principal component analysis, concerns the data on which the analysis is to be performed. As was explained above, the first principal component is the linear combination $\mathbf{p}_1 = \mathbf{t}_1^T\mathbf{X}$ of the original variables and has the maximum variance among all such possible combinations. The presence of considerable differences among the variables (e.g. variables representing vastly different entities, or having different units or scales of measurement, etc.) result in *variables* with large variances, corresponding to axes in the geometrical representation along which individuals or sample points have a wide dispersion, while other variables will correspond to axes with smaller scatter. This will yield principal components simply corresponding to the original variables with the largest variance and little will be learnt from the data.

It is therefore often necessary to ensure that the measured variables are comparable in their magnitude of variance and their units of measurements. If required, this can be achieved by *standardizing* the variables prior to principal component analysis. This is accomplished by considering each variable separately, whereupon the mean value of the variable is subtracted from each observation of the variable, and the results are divided by the standard deviation (s) of the variables. By doing so, the sample covariance matrix **S** is converted to the sample correlation matrix **R**, and as a result the principal components are obtained from the eigenvalues of matrix **R**. These principal components are different from the ones that are obtainable from S, and cannot be used to derive those obtainable from **S**. Since the results of principal component analysis are greatly affected by standardization of the variables, this decision should be carefully considered.

The next issue concerns the adequacy of any number of components as a representation of the original data. Since variance forms the basis on which the principal components are derived, a measure of the adequacy of the projection of the n sample points into the subspace defined by the first r principal components might be related to the variance of the projected points.

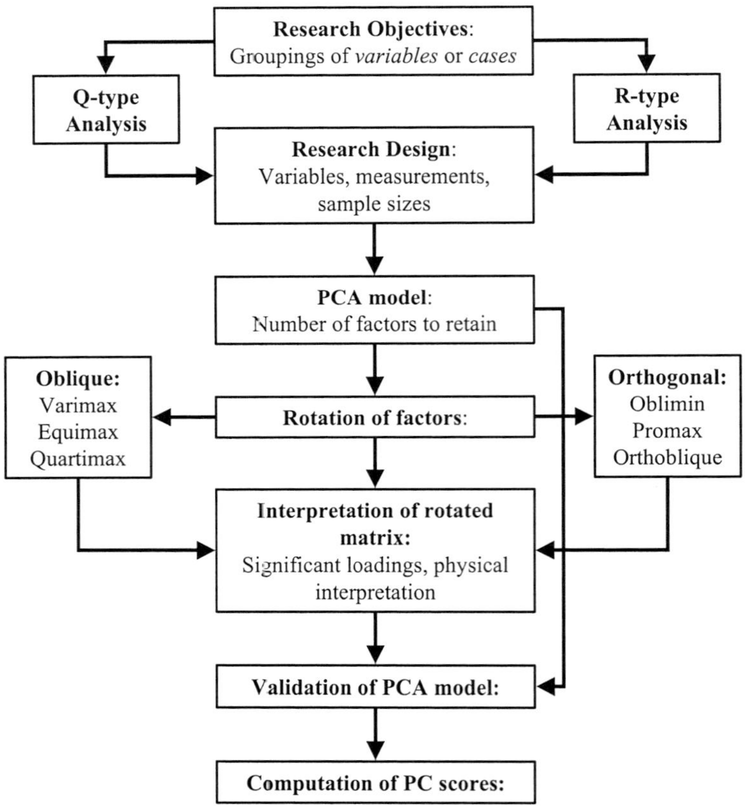

Figure 3.4. Application of principal component analysis.

Moreover, each eigenvalue λ_i of the covariance matrix **S** gives the variance of the points with respect to the corresponding component. If the centroid of the points is at the origin of the co-ordinate axes (e.g. after standardization of the data), then the total variance of the points S_r^2 in the subspace defined by the first r principal components is equal to the sum of the variances of the points when projected onto each component as an axis, i.e.

$$S_r^2 = \lambda_1 + \lambda_2 + \ldots + \lambda_r = \Sigma_{i=1}^{r} \lambda_i \quad (3.19)$$

The total variance in the original system is simply the sum of the variances of each original variable x_i, which is the sum of the diagonal elements of the covariance matrix S[21]. It can be shown that trace(**S**) = $\Sigma_{i=1}^{r} \lambda_i$, since principal component analysis merely results in a rotation of axes (the frame of reference for the sample points), so that the total variance of the x_i has to be the same as the total variance of the Y_i, so that $S_p^2 = \Sigma_{i=1}^{r} \lambda_i$. As a consequence, the proportion of the total variance explained by the first r principal components is given by

$$P_r = S_r^2/S_p^2 = \Sigma_{i=1}^{r} \lambda_i/\Sigma_{i=1}^{p} \lambda_i = \Sigma_{i=1}^{r} \lambda_i/\text{trace}(\mathbf{S}) \quad (3.20)$$

The ratio P_r thus serves as a measure of the adequacy of the principal component projection into the subspace defined by the first r principal components, of the actual configuration of the data in p dimensions. Adequacy is strongly determined by the context. For simple visualization of the data, 60% or more may be acceptable. For models based on the principal component scores, rather than the observations on the original variables, more stringent criteria are usually required, so that only those components essentially containing noise are left out. Therefore, the choice of r as far as the above criterion is concerned cannot be objective, and depends on other (prior) knowledge of the data or the system being investigated.

3.2.4. Interpretation of principal components

Ideally, any column of the loadings matrix should have mostly small values, as close to zero as possible, any row of the loadings matrix should have only a few entries far from zero and any two columns of the matrix should show a different pattern of low and high loadings. When this is not the case, the principal components may be rotated to better approach these ideal conditions, as indicated in Figure 3.5.

For example, consider the case where three variables can be represented by two principal components without undue loss of information. Also, assume that the first variable has a high loading on the 1st principal component and a low loading on the 2nd principal component, while the 2nd and 3rd variables have reasonably high loadings on both principal components.

This will make the 1st principal component difficult to interpret in terms of the original variables, since all the variables are highly loaded onto this component. By rotating the principal component axes, variables two and three are loaded highly on the rotated principal component (RPC_2), while the first variable is highly loaded on the first rotated principal component (RPC_1). This makes interpretation of the (rotated) principal components easier. However, this is done by sacrificing some of the explanatory power of the components themselves, since they do not account for the maximum possible amount of variation in the variables anymore.

Principal component axes can be rotated either orthogonally or obliquely. In the first, the original orientation between the principal components is preserved, so that they are still perpendicular after rotation. In the second instance, no such restrictions are imposed and the axes can be rotated independently, so that they may not necessarily be perpendicular after rotation.

---

[21]Mathematically denoted by trace(**S**), i.e. the sum of the elements on the diagonal of matrix **S**.

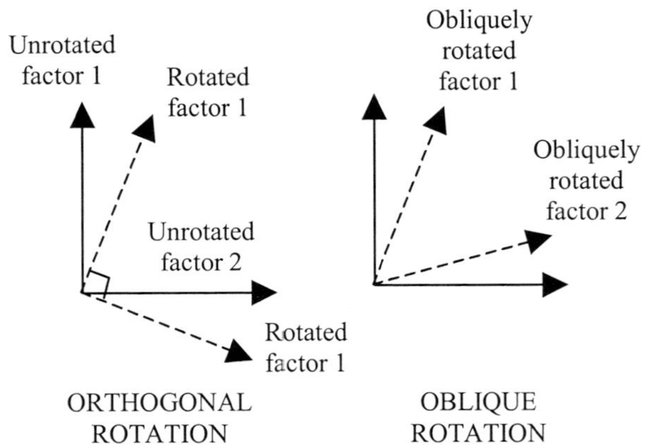

Figure 3.5. Rotation of principal components.

The three most popular (and readily available) methods of orthogonal rotation are the *varimax*, *quartimax* and *equimax* methods. Of these three, the *varimax* procedure is used most often by far to rotate principal components. With the *varimax* procedure, the objective is to enlarge the variation of the squared loadings for a given principal component. In the *quartimax* approach, the objective is to accomplish, for a given variable, one and only one major loading on a given principal component. In practice this may not be feasible, but even so, the method attempts to get as close as possible to this ideal. The undesirable result is a tendency to generate a general principal component onto which most of the variables are highly loaded. The *equimax* method attempts to obtain a simple structure with respect to both the rows and columns of the loading matrix, but in practice this may again not always be feasible.

3.2.5. Simple examples of the application of principal component analysis

a) *Example 1*

Consider the two variables $\mathbf{x'}_1$ and $\mathbf{x'}_2$, shown in Table 3.1[22]. As can be seen from the scatter plot in Figure 3.6, the two variables are highly correlated.

Table 3.1. Observations of variables x_1 and x_2.

$\mathbf{x'}_1$	0.401	0.597	0.640	0.137	0.832	0.540	0.406	0.473	0.096	0.077
$\mathbf{x'}_2$	0.131	0.228	0.434	0.146	0.812	0.465	-0.054	0.200	0.119	-0.103

The data matrix therefore consists of two column vectors, each with ten rows (observations) that is $\mathbf{X'} = [\mathbf{x'}_1,\mathbf{x'}_2]^T \in \Re^{10x2}$.

The variables are first scaled to zero mean and unit variance, by subtracting the average of each column from the entries in the column, and dividing the result by the standard deviation

[22] The variables are printed in bold face to indicate that they are (column) vectors (of observations).

of each column. In general the j'th observation of the i'th variable is scaled $x_{ij} = (x'_{ij} - x'_{i,avg})/s'_i$ to give the observations in indicated in Table 3.2.

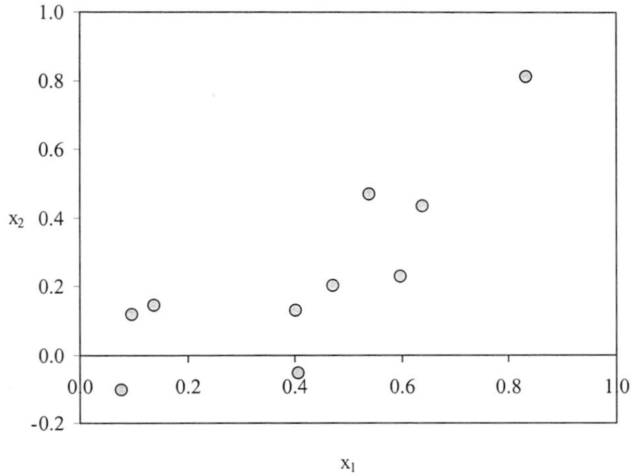

Figure 3.6. Scatter plot of variables x'_1 and x'_2 in Table 3.1.

Table 3.2. Scaled observations of variables x_1 and x_2 based on the mean and standard deviation vectors of the variables, $[x'_{1,avg}, x'_{2,avg}] = [0.420, 0.238]$ and $[s'_1, s'_2] = [0.252, 0.270]$.

$\mathbf{x_1}$	-0.076	0.703	0.874	-1.124	1.638	0.478	-0.057	0.212	-1.287	-1.362
$\mathbf{x_2}$	-0.397	-0.036	0.726	-0.339	2.126	0.841	-1.080	-0.139	-0.441	-1.261

A scatter plot of the scaled (and centred) data is shown in Figure 3.7. The covariance matrix of the variables is consequently calculated, that is $\text{cov}(\mathbf{X}) = \mathbf{X}^T\mathbf{X}/(n-1)$, or $\text{cov}(x_1, x_2) = \Sigma_{j=1}{}^n x_1x_2/(n-1)$. This gives

$$\text{cov}(\mathbf{X}) = \mathbf{S} = \mathbf{R} = \begin{bmatrix} 1.00 & 0.803 \\ 0.803 & 1.00 \end{bmatrix}$$

This covariance matrix is subsequently used to extract the principal components from the data based on spectral decomposition of the covariance matrix S, $(S - \lambda_j I)p_j = 0$, that is solving[23] for λ_j and p_j, $(j = 1, 2)$ given that the principal components (p_j) are orthonormal, i.e. $p_j{}^T p_k = 1$ $(j = k)$ and $p_j{}^T p_k = 0$ $(j \neq k)$.

Since $(\mathbf{S} - \lambda_j\mathbf{I})\mathbf{p}_j = \mathbf{0}$, $\det(\mathbf{S} - \lambda_j\mathbf{I}) = |\mathbf{S} - \lambda_j\mathbf{I}| = 0$.

More specifically,

---

[23] Note that several iterative methods are available for the spectral decomposition of larger matrices, for example Jacobi's diagonalization method (Golub and van Loan, 1996; Parlett, 1980), the power method (Golub and van Loan, 1996), the method of Lanczos (Parlett, 1980), Arnoldi's method (Arnoldi, 1951; Saad, 1992) and Davidson's method (Davidson, 1975; Crouzeix et al., 1992; Morgan and Scott, 1993).

$$\det(S-\lambda I)=\begin{vmatrix}1-\lambda & 0.803\\ 0.803 & 1-\lambda\end{vmatrix}=0$$

That is $(1-\lambda)(1-\lambda) - 0.803^2 = 0$, or

$0.355 - 2\lambda + \lambda^2 = 0$, which gives two latent roots

$\lambda_1 = 1.803$ and $\lambda_2 = 0.197$

Note that the sum of the two roots is 2, as expected, since the variables have each been scaled to unit variance.

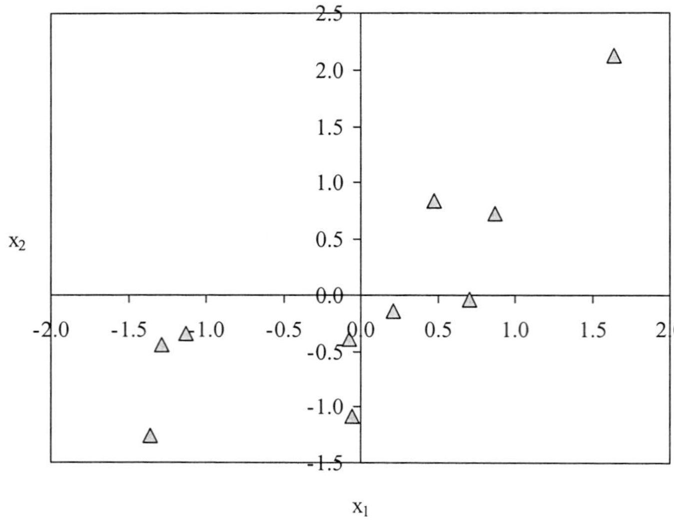

Figure 3.7. Scatter plot of scaled variables x_1 and x_2 in Table 3.2.

The eigenvectors associated with these two eigenvalues are found by solving

$$(\mathbf{S}-\lambda_j\mathbf{I})\mathbf{p}_j = \mathbf{0}$$

The first eigenvector $\mathbf{p}_1$ with vector components p_{11} and p_{12} can be found from

$$\begin{bmatrix}-0.803 & 0.803\\ 0.803 & -0.803\end{bmatrix}\begin{bmatrix}p_{11}\\ p_{12}\end{bmatrix}=\begin{bmatrix}0\\ 0\end{bmatrix}$$

This gives $p_{11} = p_{12}$, and with the constraint that $\mathbf{p}_1^T\mathbf{p}_1 = 1$, $p_{11} = 0.707$ and $p_{12} = 0.707$, or

$p_1 = 0.707x_1 + 0.707x_2$

Likewise, the second eigenvector with vector components p_{21} and p_{22} can be found from

$$\begin{bmatrix} 0.803 & 0.803 \\ 0.803 & 0.803 \end{bmatrix} \begin{bmatrix} p_{21} \\ p_{22} \end{bmatrix} = \begin{bmatrix} 0 \\ 0 \end{bmatrix}$$

This gives $p_{21} = -p_{22}$, and with the constraint that $\mathbf{p}_2^T\mathbf{p}_2 = 1$, $p_{21} = 0.707$ and $p_{22} = -0.707$, or

$$p_2 = 0.707x_1 - 0.707x_2$$

$$\mathbf{P} = \begin{bmatrix} 0.707 & 0.707 \\ 0.707 & -0.707 \end{bmatrix}$$

The first eigenvector explains 100x1.803/(1.803+0.197) = 90.1% of the variance of the data.

Principal scores

The principal component scores are obtained by projecting each variable onto the principal component, i.e.

$$\mathbf{t}_i = \mathbf{X}\mathbf{p}_i^T$$

The scores are shown in Table 3.3.

Table 3.3. Scores of the observations in Table 3.2, after projection on to the principal component axes.

$\mathbf{t}_1$	-0.180	0.230	0.580	-0.520	1.370	0.480	-0.430	0.020	-0.610	-0.950
$\mathbf{t}_2$	-0.120	-0.270	-0.060	0.290	0.170	0.130	-0.370	-0.130	0.310	0.040

Put in a different way, the matrix **X** can therefore be represented by the principal components and their loadings

$$\mathbf{X} = \mathbf{t}_1\mathbf{p}_1^T + \mathbf{t}_2\mathbf{p}_2^T$$

or

$$\begin{bmatrix} x_{1,1} & x_{2,1} \\ x_{1,2} & x_{2,2} \\ x_{1,3} & x_{2,3} \\ x_{1,4} & x_{2,4} \\ x_{1,5} & x_{2,5} \\ x_{1,6} & x_{2,6} \\ x_{1,7} & x_{2,7} \\ x_{1,8} & x_{2,8} \\ x_{1,9} & x_{2,9} \\ x_{1,10} & x_{2,10} \end{bmatrix} = \begin{bmatrix} t_1 \\ t_2 \\ t_3 \\ t_4 \\ t_5 \\ t_6 \\ t_7 \\ t_8 \\ t_9 \\ t_{10} \end{bmatrix} \begin{bmatrix} p_{11} & p_{12} \end{bmatrix} + \begin{bmatrix} t_1 \\ t_2 \\ t_3 \\ t_4 \\ t_5 \\ t_6 \\ t_7 \\ t_8 \\ t_9 \\ t_{10} \end{bmatrix} \begin{bmatrix} p_{21} & p_{22} \end{bmatrix}$$

or since the first principal component accounts for more than 90% of the variance in the data,

$$\mathbf{X} \approx \mathbf{X}_{rec} = \mathbf{t}_1\mathbf{p}_1^T$$

The reconstructed data $\mathbf{X}_{rec}$ are shown in Table 3.4 and Figure 3.8.

Table 3.4. Scores of the observations in Table 3.2, after projection on to the principal component axes.

$\mathbf{x}^{rec}_1$	-0.120	0.160	0.400	-0.360	0.940	0.330	-0.290	0.010	-0.420	-0.650
$\mathbf{x}^{rec}_2$	-0.130	0.170	0.420	-0.370	0.990	0.350	-0.310	0.020	-0.440	-0.680

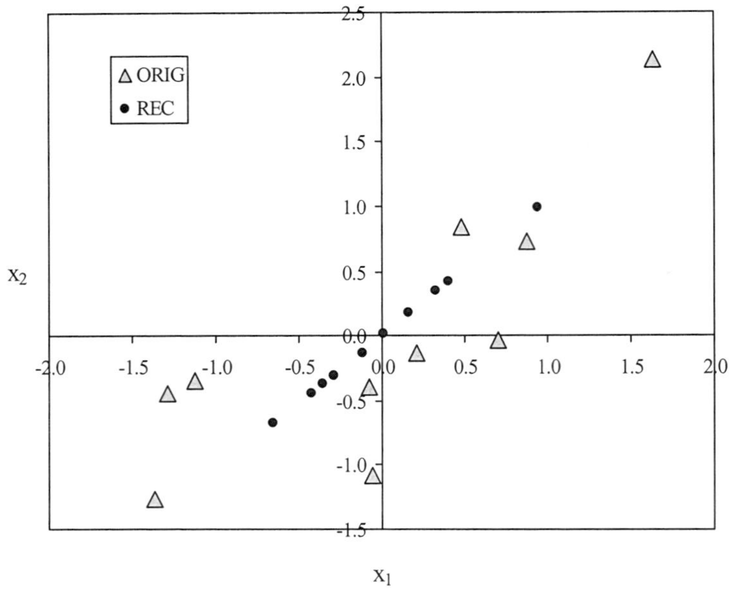

Figure 3.8. Reconstructed observations based on first principal component, i.e. $\mathbf{X}_{rec} = \mathbf{t}_1\mathbf{p}_1^T$.

b) *Example 2*

The daily average composition of the feed to a base metal plant is shown below, for 19 consecutive days. Some of the operators on the plant suspect that the feed conditions have changed over the last ten days of operation. This hypothesis can be evaluated by making use of principal component analysis to portray the feed graphically.

Table 3.5. Daily variation of feed composition.

Day	%Cu	%Pb	%Zn
1	0.44	4.82	3.21
2	0.45	4.80	3.20
3	0.47	4.74	3.18
4	0.46	4.65	3.12
5	0.47	4.59	3.11
6	0.48	4.44	3.08

7	0.47	4.41	3.08
8	0.47	4.38	3.07
9	0.47	4.40	3.05
10	0.47	4.40	3.04
11	0.48	4.47	3.06
12	0.47	4.45	3.04
13	0.46	4.49	3.05
14	0.46	4.54	3.05
15	0.47	4.51	3.07
16	0.44	4.68	3.05
17	0.44	4.58	3.10
18	0.43	4.58	3.11
19	0.41	4.56	3.10

Table 3.6. Covariance (**S**) and correlation matrices (**R**) of the data in Table 3.5.

Covariance Matrix (**S**)

	%Cu	%Pb	%Zn
%Cu	0.00035		
%Pb	-0.00112	0.01804	
%Zn	-0.00030	0.00595	0.00280

Correlation matrix (**R**)

	%Cu	%Pb	%Zn
%Cu	1		
%Pb	-0.444	1	
%Zn	-0.298	0.837	1

Step 1: Calculation of covariance/correlation matrix

The covariance and correlation matrix of the feed is shown on the left, that is $\mathbf{S} = \mathbf{X}^T\mathbf{X}/(m-1)$, where $\mathbf{X} = [\mathbf{x}_1, \mathbf{x}_2, \mathbf{x}_3]$ denotes the data matrix in Table 3.6. The 19 x 3 matrix consists of 19 observations on 3 variables $\mathbf{x}_1$ (%Cu), $\mathbf{x}_2$ (%Pb) and $\mathbf{x}_3$ (%Zn). Since we want to weight the %Cu, %Pb and %Zn equally, we need to decompose the *correlation* matrix, rather than the covariance matrix, into eigenvalue-eigenvector pairs.

Step 2: Eigenvalue-eigenvector decomposition of correlation matrix

It can be shown that[24], $\mathbf{Rp} = \lambda\mathbf{p}$, or

$$\begin{bmatrix} 1 & -0.444 & -0.298 \\ -0.444 & 1 & 0.837 \\ -0.298 & 0.837 & 1 \end{bmatrix} * \begin{bmatrix} 0.888 & -0.148 & -0.435 \\ 0.198 & -0.731 & 0.653 \\ 0.415 & 0.666 & 0.620 \end{bmatrix} = \begin{bmatrix} 0.762 \\ 0.147 \\ 2.091 \end{bmatrix} * \begin{bmatrix} 0.888 & -0.148 & -0.435 \\ 0.198 & -0.731 & 0.653 \\ 0.415 & 0.666 & 0.620 \end{bmatrix}$$

[24] Matlab's command [D,V] = eig(R).

The first principal component ($\mathbf{p}_1$) is associated with the largest eigenvalue, $\lambda_1 = 2.091$, the second principal component ($\mathbf{p}_2$) is associated with the second largest eigenvalue, $\lambda_2 = 0.762$, and the last principal component ($\mathbf{p}_3$) is associated with the smallest eigenvalue, $\lambda_3 = 0.147$.

Therefore

$$\mathbf{p}_1 = -0.435\mathbf{x}_1+0.653\mathbf{x}_2+0.620\mathbf{x}_3$$

$$\mathbf{p}_2 = 0.888\mathbf{x}_1+0.198\mathbf{x}_2+0.415\mathbf{x}_3$$

$$\mathbf{p}_3 = -0.148\mathbf{x}_1-0.731\mathbf{x}_2+0.666\mathbf{x}_3$$

The total variance explained by the three principal components $\mathbf{p}_1$, $\mathbf{p}_2$ and $\mathbf{p}_3$ is

$$\lambda_1 + \lambda_2 + \lambda_3 = 2.091 + 0.762 + 0.147 = 3$$

which is the same as the variance explained by $\mathbf{x}_1$, $\mathbf{x}_2$ and $\mathbf{x}_3$. The variance explained by the first two principal components is 2.853, which is 95.1% of the total variance. The first principal component explains 69.7% of the total variance. It is easy to show that the eigenvectors are orthonormal, i.e. $\mathbf{p}_1^T\mathbf{p}_1 = \mathbf{p}_2^T\mathbf{p}_2 = \mathbf{p}_3^T\mathbf{p}_3 = 1$

Step 3: Calculation of principal component scores

The principal component scores are calculated from $\mathbf{Xp}_i = \mathbf{t}_i$. The values of the scores are summarized in Table 3.7.

Table 3.7. Scores of the observations shown in Table 3.5.

$\mathbf{t}_1$	$\mathbf{t}_2$	$\mathbf{t}_3$
0.0880	-0.1171	0.2511
0.0865	-0.1059	0.2304
0.0811	-0.0827	0.1768
0.0358	-0.0529	0.0831
0.0224	-0.0141	0.0310
-0.0076	0.0685	-0.0910
-0.0213	0.0974	-0.1081
-0.0321	0.1042	-0.1324
-0.0357	0.0747	-0.1322
-0.0455	0.0725	-0.1378
-0.0116	0.0364	-0.0813
-0.0325	0.0363	-0.1011
-0.0275	0.0156	-0.0676
-0.0129	-0.0202	-0.0325
-0.0097	0.0091	-0.0451
-0.0088	-0.1189	0.0579
-0.0121	-0.0081	0.0297
-0.0138	-0.0026	0.0427
-0.0426	0.0079	0.0263

Step 4: Plot the scores of the first two principal components, i.e. t_1 versus t_2.

From the scatter diagram in Figure 3.9, it certainly appears as if the plant feed had been more consistent over the last 10 days (empty circles), compared to the first 9 days (solid diamonds).

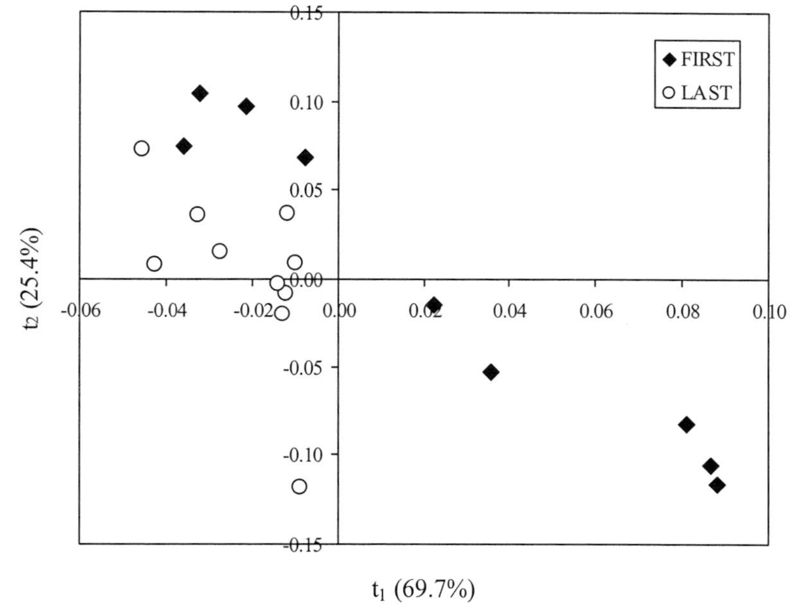

Figure 3.9. Score plot of the data in Table 3.7. Plant feed composition in the first 9 days and last ten days is indicated by empty circles and solid diamonds respectively.

3.3. NONLINEAR APPROACHES TO LATENT VARIABLE EXTRACTION

Principal component analysis is a specific approach to the general problem of latent variable mapping, which can be formalised as follows (Chang and Ghosh, 2001). Given a set of observations, $\mathbf{x} \in \Re^{nxm}$ drawn from a random vector X, the problem is to find the mappings $G:\Re^m \rightarrow \Re^p$, and $F:\Re^p \rightarrow \Re^m$, for all j =1, 2, … n. That is $G(\mathbf{x}_j) = \mathbf{y}_j \in \Re^{nxp}$ and $F(\mathbf{y}_j) = \mathbf{x}_{j'} \approx \mathbf{x}_j \in \Re^{nxm}$.

Mappings G and F can be derived by optimizing one of several possible criteria, e.g. minimum square error (MSE) or maximum likelihood. In principal component analysis, both G and F are linear and the empirical reconstruction MSE is minimized. The forward mapping G is computed via eigenvalue decomposition of the sample correlation (covariance) matrix and the derivation of G automatically leads to the corresponding reverse mapping F, as discussed previously.

Latent variable models, such as independent component analysis and factor analysis, first compute F, from which G can be obtained trivially via pseudo-inverses. In the case of nonlinear mappings G is usually approximated by some projection operator, since inverse mappings may not be easy to find otherwise. Neural networks are often used as this projection operator and these approaches are considered in more detail below.

3.4. PRINCIPAL COMPONENT ANALYSIS WITH NEURAL NETWORKS

Although various configurations of neural networks can be used for the extraction of principal components from data, autoassociative neural networks are the most widely used. Autoassociative neural networks are multilayer perceptron neural networks, which are trained with exemplars in which the inputs are identical to the targets. These networks have at least one hidden layer with dimension $p < m$, the dimensionality of the training vectors. The hidden layer acts as a bottleneck, which forces the network to represent the inputs with fewer parameters than m, allowing the outputs to be optimally reconstructed from these features. Unless the mapped variables have the same intrinsic dimension as the number of nodes in the bottleneck layer of the neural network, the identity mapping from inputs to outputs is generally not attainable. For example, consider an autoassociative neural network with a single hidden layer and units with linear activation functions, i.e. $\mathbf{g}(\mathbf{x}) = \mathbf{x}$. For simplicity, but without loss of generality, also consider the bias nodes to be absent. The output is then a linear function of the input

$$\mathbf{y} = \mathbf{W}^*\mathbf{W}^T\mathbf{x} \tag{3.21}$$

where $\mathbf{W}^* = [w^*_{ij}] \in \Re^{mxp}$ and $\mathbf{W} = [w_{ij}] \in \Re^{pxm}$ are the weight matrices of the encoding and decoding layers respectively. The input-output mapping $\mathbf{W}_o = \mathbf{W}^*\mathbf{W}^T$ is the product of two matrices with rank less than m, so that rank$(\mathbf{W}_o) \leq m$, $\mathbf{W}_o \neq \mathbf{I}$ and $\mathbf{y} \neq \mathbf{x}$. The quality of the match between the observed inputs and the reconstructed inputs is typically measured by a cost function such as the mean square error

$$C = \tfrac{1}{2}\Sigma_{j=1}^{n}\|\mathbf{x}_j-\mathbf{y}_j\|^2 \tag{3.22}$$

Instead of using single hidden layer autoassociative neural networks, the use of such networks with multiple hidden layers has been proposed by Kramer (1991, 1992). These networks are similar to single hidden layer autoassociative neural networks, except that they usually have three hidden layers, as shown in Figure 3.10 (Oja, 1996).

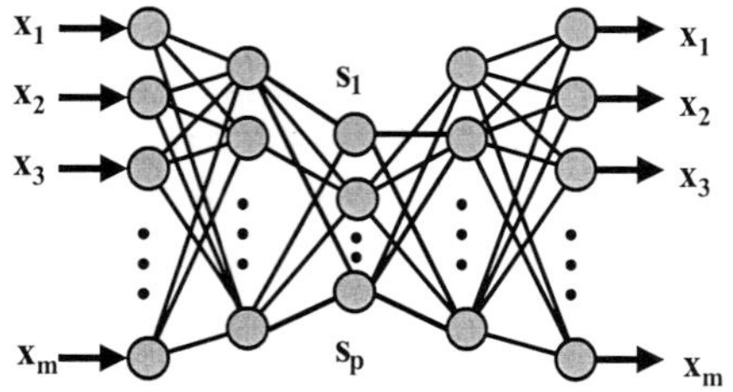

Figure 3.10. An autoassociative neural network with three hidden layers.

For an analysis of autoassociative neural networks, let $\mathbf{X}$ be an n x m matrix containing n observations on m variables. The i'th row vector of $\mathbf{X}$ can be denoted by the m x 1 vector $\mathbf{x}_i$, while the j'th column can be denoted by the n x 1 vector $\mathbf{x}_{(j)}$. Although the superficial dimension of $\mathbf{X}$ is m, it is assumed that there are dependencies among the columns of $\mathbf{X}$, and that the intrinsic dimension of $\mathbf{X}$ is p, which can be appreciably smaller than m. Given these circumstances, it may be desirable to extract a new set of variables or features from **X**, which contain essentially the same information as **X**, but the smaller intrinsic dimension (p).

This is accomplished by an input-output mapping formed by the network, $\mathbf{g}:\Re^m \rightarrow \Re^p$, where $\mathbf{S} = \mathbf{g}(\mathbf{x}) = \sigma(\mathbf{w}_2\sigma(\mathbf{w}_1\mathbf{x}))$. The values of these features are referred to as scores (s_i) and they are denoted by the n x p matrix S. The original vectors can be reconstructed by a mapping of the scores in the p-dimensional subspace to the original m-dimensional space by minimizing the squared error on the training data set

$$J(\mathbf{g}) = E\{||\mathbf{x}_i - \mathbf{g}(\mathbf{x}_i)||^2 \qquad (3.23)$$

$$J(\mathbf{g}) = E\{||\mathbf{x} - \sigma(\mathbf{w}_3\sigma(\mathbf{w}_2\sigma(\mathbf{w}_i\mathbf{x})))\ ||^2 \qquad (3.24)$$

The observations and scores are presumed to be related as follows

$$\mathbf{x}_i = \mathbf{f}(\mathbf{s}_i) + \mathbf{e}_i = [\mathbf{f}_1(\mathbf{s}_i), \mathbf{f}_2(\mathbf{s}_i), \ldots \mathbf{f}_m(\mathbf{s}_i)]^T + [\mathbf{e}_{i1}, \mathbf{e}_{i2}, \ldots \mathbf{e}_{im}]^T \qquad (3.25)$$

In equation (3.25), $\mathbf{e}_i$ is a vector of noise and $\mathbf{f}:\Re^p \rightarrow \Re^m$ is a continuous function (usually nonlinear) reconstructing the outputs from the scores, also referred to as a (globally) parameterized k-surface in $\Re^m$. The nonlinear feature extraction problem therefore concerns the estimation of **S** and **g** (Malthouse, 1998), while the reconstruction of the original data from the scores concerns estimation of the demapping function **f**.

In linear principal component analysis, **S** and **g** are estimated by finding the k-dimensional hyperplane that has smaller mean square deviations from the inputs than any other hyperplane, as discussed in more detail in chapter 3. Many nonlinear generalizations of principal component analysis have been proposed and in order to facilitate discussion of some of these approaches, some concepts need to be reviewed.

Networks with a single hidden node in the bottleneck layer are similar to principal component methods, and both give the principal component solution when $\mathbf{s}_f$ and **g** are linear. Moreover, the neural network defines a function $\mathbf{g}:\Re^1\rightarrow \Re^m$, which is a curve in $\Re^m$. The network also defines a function $\mathbf{s}_f$: $\Re^m\rightarrow \Re^1$. The functions g and s_f are both continuous in the case of the autoassociative neural network. As Malthouse (1998) has shown, this has some practical significance, in that autoassociative neural networks cannot model curves or surfaces that intersect themselves.

In contrast with simultaneous feature extraction, sequential feature extraction entails the use of a series of autoassociative neural networks with one bottleneck node each, which are trained sequentially, as indicated diagrammatically in Figure 3.11. With this method, the primary factor of the residual matrix **e**, is the second factor of the original data matrix **X**. The series of autoassociative neural networks are therefore trained sequentially. Only one factor is extracted at each stage, and the residuals from the i'th stage become the input of the (i+1)'th stage. That is

$$\mathbf{s}_{i+1} = \mathbf{g}(\mathbf{X}_{i+1})$$

$$\mathbf{X}_{i+1} = \mathbf{X}_i - \mathbf{X}_i'' = \mathbf{X}_i - \mathbf{f}(\mathbf{s}_i) \qquad (3.26)$$

Where $\mathbf{s}_{i+1}$ is the (i+1)'th score vector of the residuals resulting from the mapping and reconstruction of the i'th residual matrix of the original data set.

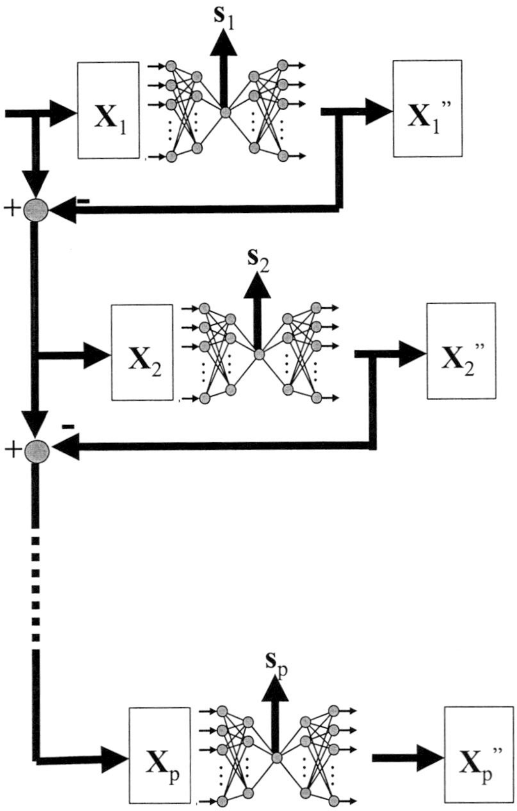

Figure 3.11. Sequential extraction of features with a series of autoassociative neural networks.

Sequential feature extraction from the data is more tedious than simultaneous extraction of the scores, but affords certain advantages over the latter approach. Training of the neural network can be expedited by rescaling of the residual matrices between stages (Kramer, 1991), which can alleviate potential problems caused by progressively smaller residual values. More importantly, the serial extraction of features forces the neural network to assign a single feature to each latent variable in the data. This generally does not happen with the simultaneous extraction of features, where the neural network tends to distribute the latent variables across all the nodes in the bottleneck layer.

3.5. EXAMPLE 2: FEATURE EXTRACTION FROM DIGITISED IMAGES OF INDUSTRIAL FLOTATION FROTHS WITH AUTOASSOCIATIVE NEURAL NETWORKS

Froth flotation is a process of major economic importance, mainly as a result of its efficiency in the treatment of large ore tonnage and its relative ease of implementation in industry. The process is complex, and in practice the control of industrial flotation plants is often based on the visual appearance of the froth phase, and depends on the experience and ability of a human operator. A standard control cycle comprises the fixing of initial set points, a settling period for transient dynamics to subside, a period of measurement and evaluation, and a final estimate of appropriate set points. Some studies have indicated that operators often tend to make the periods for settling and measurement too small. Apart from these aspects, the inexperience or inability of the operator can have a further significant impact on the control of the plant. Consequently, optimal control is not usually maintained, especially where incipient erratic behaviour in the plant is difficult to detect.

Figure 3.12. Machine vision system on an industrial flotation plant.

These factors have led to investigation of the use of machine vision systems as an integral part of the monitoring and control of flotation plants (Moolman et al., 1994, 1995a, 1995b, Aldrich et al., 1995, Niemi et al., 1997). The most important aspect of the machine vision system concerns the pattern recognition problem, where the digitised images have to be related to the performance of the flotation cell, as well as the storage of machine vision data, since the cost associated with the storage of the images themselves could be exorbitant.

These grey-scale or colour images can be viewed as large-dimensional data samples, in which each pixel represents a variable. Each captured image would therefore represent a sample vector of dimension n x n, where n is the resolution of the image ($\mathbf{x} = x_1, x_2, \ldots x_{nxn}$). With typical values of n = 256, this would result in a 65536-dimensional vector, where the dimensionality of the vector exceeds the number of observations by one or two orders of magnitude.

With higher resolution images (n = 1024) or colour images the dimensionality of the sample vector can be much higher. Clearly these multivariate observations contain a large amount of redundant information and cannot be used direct. As illustrated below, an autoassociative neural network can be used to extract features from these images, while ensuring that as little information as possible is lost in the process. Depending on the compression ratio, these features can both be used to store and reconstruct the original images, or as low-dimensional sets of features which can implemented directly in multivariate statistical process control schemes.

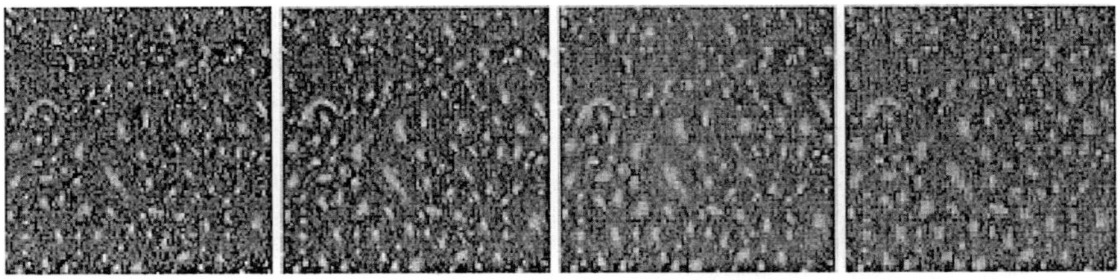

(a) (b) (c) (d)

Figure 3.13. The effect of compression ratio on the quality of the reconstructed image, based on a training set derived from a 4 x 4 window, with (a) original image, (b) compression ratio 4:1, (c) compression ratio 8:1 and (d) compression ratio 16:1 (Moolman et al., 1995c).

Figure 3.12 shows a machine vision system consisting of a videocamera in a protective casing mounted on top of a flotation cell on a South African platinum plant. The camera is connected to a personal computer equipped with a frame grabber, which monitors one of the cells in a bank of primary roughers.

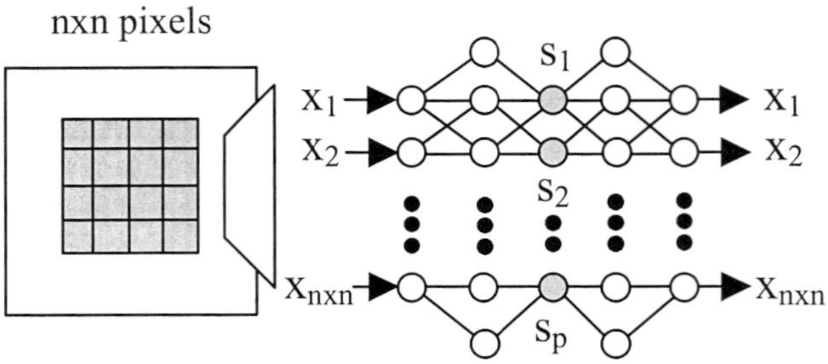

Figure 3.14. Extraction of features from rasterized froth images.

In Figure 3.13(a) a typical image of a surface froth captured on a platinum flotation plant is shown. The compression of this image, as outlined below, is shown in Figure 3.13(b)-3.13(d) for various compression ratios. Note that with a compression ratio of 16:1, the reconstructed image still shows a good resemblance to the original.

Relatively large images of homogeneous structures (such as flotation froths) need not be sampled in their entirety[25], and in order to present the image data to an autoassociative neural network, the images can be rasterized by use of a moving window of size n x n, which orders the pixels in vectors of size n x n, as indicated in Figure 3.14. From these $x \in \Re^{nxn}$ vectors, appropriate training and validation data sets can be compiled, which can be used to train autoassociative neural networks. Once these neural networks are trained, the layers succeeding the bottleneck layer of each network are removed. The layers preceding and including the bottleneck layers of the neural networks could subsequently be used to extract features from the reduced images. These features are then stored over a longer period, instead of the actual images. For visualization of the process in terms of changes in the froth structures of the flo-

[25] More formally, the images are assumed to be ergodic.

tation cells, rasterization need not be implemented, and the images can be reduced directly. Results obtained with the use of an autoassociative neural network with three unipolar sigmoidal hidden layers are shown in Figure 3.15(a). The first three nonlinear principal components were extracted successively by an autoassociative neural network with a 12:1:12 unipolar sigmoidal hidden layer configuration. The neural network was trained with the Levenberg-Marquardt algorithm. The asterisks indicate two-phase froths, while the stars indicate the same froths loaded with quartz. The first three features (s_1, s_2 and s_3) extracted with the autoassociative neural network collectively explained 97.1% and 95.6% of the variance of the data pertaining to the two- and three-phase froths respectively.

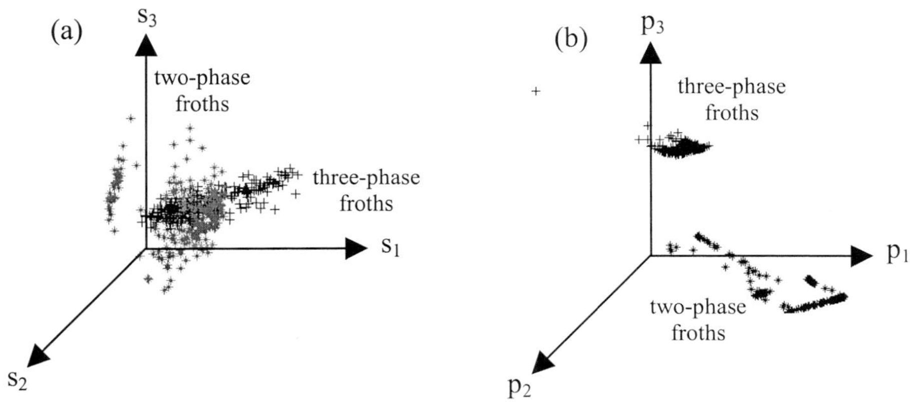

Figure 3.15. Visualization of flotation data with (a) an autoassociative neural network with three hidden layers, consisting of a 12:1:12 configuration of unipolar sigmoidal nodes, and (b) the same data portrayed in terms of the first three principal components.

The same data are shown in Figure 3.15(b), where the features were extracted by means of linear principal component analysis. In this case the variance explained by the first three principal components (p_1, p_2 and p_3) was similar to that of the neural network, viz. 97.8% and 96.9% of data pertaining to the two- and three-phase froths respectively. However, as is clear from a comparison of Figure 3.15(a) and 3.15(b), the different groups are more sharply defined by die features extracted by the neural network.

3.6. ALTERNATIVE APPROACHES TO NONLINEAR PRINCIPAL COMPONENT ANALYSIS

3.6.1. Principal curves and surfaces

The method of principal curves (Hastie and Stuetzle, 1989) has a clear geometric interpretation in the input space. The structure of the data is captured by the iterative estimation of a curve or a surface to which the data are mapped. The curve or surface has the property that each point on the curve or surface is the average of all data points projected onto it. Principal curves are a direct generalization of principal components, where the curves are straight lines.

A one-dimensional curve in p-dimensional space is a vector $\mathbf{f}(\gamma)$ of p functions of a single variable γ. These functions are called the coordinate functions. By definition, if these coordinate functions are smooth then the curve is smooth. The curve is invariant under any monotone transformation on γ, with appropriate modification of the coordinate functions. The curve is parameterized by the variable γ and it also provides an ordering along the curve, that is, γ is the arc length along the curve, starting from either side. If $\mathbf{x} \in \Re^p$ is a continuous random vector with a known distribution h, then the curve f is called a principal curve of h, if

$$E\,[\mathbf{x} \mid \gamma_f(\mathbf{x}) = \gamma] = \mathbf{f}(\gamma) \tag{3.27}$$

where γ_f is defined as a projection index of $\Re^p \rightarrow \Re^1$, and

$$\gamma_f(\mathrm{x}) = \sup_\gamma\{\gamma \| \mathbf{x} - \mathbf{f}(\gamma)\| = \inf_\gamma \| \mathbf{x} - \mathbf{f}(\mu)\|\} \tag{3.28}$$

Thus, the projection index $\gamma_f(\mathrm{x})$ of $\mathbf{x}$ is the value of γ for which $f(\gamma)$ is closest to $\mathbf{x}$. If several such values exist, the largest one is used.

Principal curves can be parameterized by arc length, that is $g(s_2) - g(s_1)$ is the length of the curve between the points s_1 and s_2 on the curve. The arc length L of such a curve is

$$L = \int^{s2}_{s1}\{\Sigma_{i=1}^{P}(dg_i/ds)^2\}^{0.5}ds = s_2 - s_1 \tag{3.29}$$

The curve f is parameterized by arc length, if, and only if it has the property of unit-speed, defined by

$$\{\Sigma_{i=1}^{P}(df_i/ds)^2\}^{0.5} = 1 \tag{3.30}$$

Every smooth curve can be parameterized by arc length and this parameterization is unique with regard to choice of origin and sign flips. Also, any vector with unit length has the unit-speed property.

The evaluation of the projection index $\mathbf{s}_f(\mathbf{x})$ at some point $\mathbf{x}$, gives the arc length along the curve $\mathbf{f}$ from some fixed origin to the point that is closest to $\mathbf{x}$ (Malthouse, 1998). Chang and Ghosh (1998) have proposed the following algorithm for the computation of principal curves.

a) *Initialization*

The first principal curve $g^{(0)}$ is initialized, for example as the first principal component extracted from the data set, or otherwise as small random values. Although both these approaches work well, they yield different results, since principal curves are not uniquely defined.

b) *Projection*

At the (j+1)'th iteration, given the current curve $g^{(j)}(\lambda_i)$, new projection indices are generated, that is

$$\lambda^{(j+1)} = \lambda_f^{(j)}:$$

i) Let $d_{i,k}$ denote the (Euclidean) distance from x_i to the closest point on the curve parameterized by $\lambda_k \in [\lambda_k^{(j)}, \lambda_{k+1}^{(j)}]$ on each line segment joining $g^{(j)}(\lambda_k^{(j)})$ and $g^{(j)}(\lambda_{k+1}^{(j)})$

ii) Set $\lambda_i^{(j+1)} = \lambda_k^*$ if $d_{i,k} = \min_k d_{i,k}$

iii) Find the interpolated $g^{(j)}$, corresponding to $\lambda_i^{(j+1)}$

iv) Replace $\lambda_i^{(j+1)}$ by the arc length from $g^{(j)}(\lambda_i^{(j)})$ to $g^{(j)}(\lambda_i^{(j+1)})$

c) Expectation

At the (j+1)'th iteration, given the updated projection indices $\lambda_i^{(j+1)}$, compute $f^{(j+1)}(\lambda_i^{(j+1)})$ as the solution to the following minimization problem for each dimension

$$D^2(f) = \Sigma_{i=1}^{N}[x_i - f(j)(\lambda_i^{(j+1)})]^2 + s\int(\partial^2 f^{(j)}/\partial\lambda^2)d\lambda \tag{3.31}$$

In equation 3.31, s is a regularization parameter, which controls the smoothness of the curve (the larger s, the smoother the curve). It can also be viewed as a neighbourhood parameter. It is well-known that the solution to equation 3.31 is a cubic spline (Chang and Ghosh, 1998), which is orders of magnitude faster than other methods for finding principal curves. It should be noted that the formulation represented by equation 3.31 is biased towards the centre of curvature.

3.6.2. Local principal component analysis

With local principal component analysis, the data space $\Re^m$ is partitioned into Q disjoint regions $\{R^{(1)}, R^{(2)}, \dots R^{(Q)}\}$. The local covariance matrices (Σ) of each of these regions are subsequently calculated

$$\Sigma^{(j)} = E\{[\mathbf{x}-E(\mathbf{x})][\mathbf{x}-E(\mathbf{x})]^T | \mathbf{x} \in R^{(j)}\}; j = 1, 2, \dots Q \tag{3.32}$$

Next, the eigenvectors of these covariance matrices are calculated, i.e. $e_k^{(j)}$, k = 1, 2, ... m and relabeled so that their corresponding eigenvectors are in descending order, $\lambda_1^{(j)} > \lambda_2^{(j)} > \dots > \lambda_m^{(j)}$. Finally, a target dimension p is chosen and the p dominant eigenvectors are retained.

Vector quantization provides a convenient means of partitioning the data space. The trained quantizer consists of Q reference vectors, $\mathbf{r}^{(j)}$, j = 1, 2, ... Q, corresponding to the regions $R^{(j)}$. The definition of these regions satisfy Lloyd's optimality conditions (Kambhatla and Leen, 1997).

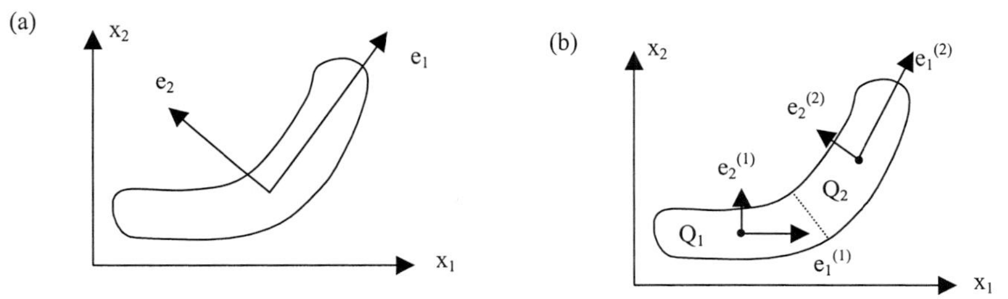

Figure 3.16. (a) Principal component analysis and (b) localized principal component analysis.

Mathematically, $R^{(j)} = \{\mathbf{x} \mid d_E(\mathbf{x},\mathbf{r}^{(j)}) < d_E(\mathbf{x},\mathbf{r}^{(i)})$, for all $i \neq j$. That is each region $R^{(j)}$ corresponds to all $\mathbf{x} \in \Re^m$ that are closer to $\mathbf{r}^{(j)}$ than to any other reference vector $\mathbf{r}^{(i)}$ based on Euclidean distance as metric, where $d_E(\mathbf{a},\mathbf{b})$ denotes the Euclidean distance between points **a** and **b**. Moreover, each reference vector $\mathbf{r}^{(j)}$ is located at the centroid of the corresponding region $R^{(j)}$, the centroid being the mean $\mathbf{r}^{(j)} = E[\, \mathbf{x} \mid \mathbf{x} \in R^{(j)}]$.

The idea is illustrated diagrammatically in Figure 3.16 In Figure 3.16(a), the dominant eigenvector $\mathbf{e}_1$ captures a relatively small percentage of the variance in the data, owing to the nonlinear distribution of the data. When localized eigenvectors are used, most of the variance can be captured by $\mathbf{e}_1^{(1)}$ and $\mathbf{e}_1^{(2)}$, as indicated in Figure 3.16(b). If a manifold does not represent the data or if the dependencies in the data vary nonlinearly, it may be advantageous to model the data locally. In this case, crossvalidation or some cost function penalizing complexity can be used to select the appropriate dimension for each region.

3.6.3. Kernel principal component analysis

Kernel principal component analysis (Schölkopf et al., 1998) can be viewed as a nonlinear generalization of principal component analysis in that it performs principal component analysis in feature spaces of arbitrarily high (possibly infinite) dimension, as indicated diagrammatically in Figure 3.17. Standard principal component analysis is recovered by use of the kernel $k(\mathbf{x}_1,\mathbf{x}_2) = (\mathbf{x}_1 \cdot \mathbf{x}_2)$.

Like classical principal component analysis, kernel principal component analysis has the advantage that no nonlinear optimization is required, since the calculations are essentially based on linear algebra. The disadvantage of kernel principal component analysis is that unlike neural network approaches, it is not well suited to the processing of very large data sets (resulting in large kernel matrices). Also, it may be more difficult to interpret in the input space (compared to principal curves, for example). A few examples will highlight the use of these nonlinear methods.

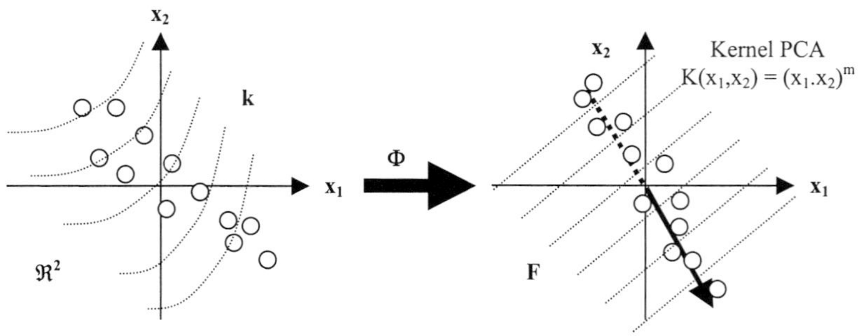

Figure 3.17. Basic concepts in kernel principal component analysis. Linear principal component analysis is performed in some high-dimensional feature space F. Since F is nonlinearly related to the input space (via Φ) the contour lines of the constant projections onto the principal eigenvector (bold arrow) are nonlinear in the input space (a preimage of the eigenvector in input space is not shown, since it may not even exist).

3.7. EXAMPLE 1: LOW-DIMENSIONAL RECONSTRUCTION OF DATA WITH NONLINEAR PRINCIPAL COMPONENT METHODS

In the first simple example, observations on two variables x_1 and x_2 were considered, with a piecewise linear distribution as indicated in Figure 3.18. These data could be reconstructed from a single linear principal component, as indicated in Figure 3.18 by the solid gray line. Clearly, a linear representation of the data was not particularly accurate, as only 50.9% of the variance of the data can be captured.

The reconstruction of the data by use of an autoassociative neural network with three hidden layers, the first and the third of which contained 3 hyperbolic tangent nodes and the second of which contained a single hyperbolic tangent node is shown by a solid black curve. As can be seen from Figure 3.18, the neural network was capable of forming a reasonably accurate representation of the data by capturing 90.2% of the variance in the data. (The data shown were not used to train the neural network.) More hidden nodes led to slight improvement, e.g. six nodes in each of the first and third layers improved the captured variance to 93.2%.

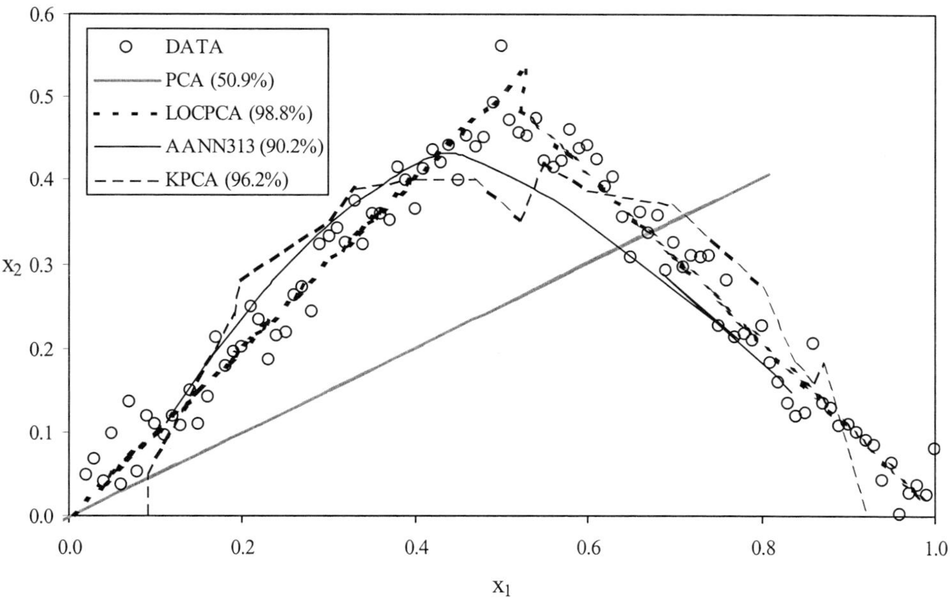

Figure 3.18. Mapping of data by use of (non)linear principal component analysis. With principal component analysis the data are mapped by means of linear principal component analysis, with R^2 = 0.509, with AANN313 with an autoassociative neural network with 3 hyperbolic tangent hidden layers, with $R^2 = 0.902$, with local principal component analysis with local principal components after partitioning with k-means cluster analysis (k=2), with R^2 = 0.988, and with kernel principal component analysis with nonlinear principal components based on 10 Rosenblatt kernels, with R^2 = 0.962.

With the reconstruction of the data from a local principal component model, the data were first partitioned into two regions by use of k-means cluster analysis. The two regions $Q^{(1)}$ and

$Q^{(2)}$ were subsequently modelled separately by linear principal components and the dominant eigenvector of each region was used to reconstruct the data in the particular region. This resulted in a one-dimensional representation of the data that could capture approximately 98.8% of variance of the data. This is shown in the figure by the broken lines. Finally, Figure 3.18 also shows the reconstruction of the data from a single kernel-based principal component. The principal component was derived from ten Rosenblatt kernels and captured approximately 96.2% of the variance in the data.

3.8. PARTIAL LEAST SQUARES (PLS) MODELS

In the two-block predictive model, the matrix of explanatory variables **X** and the matrix of predicted or output variables **Y** are decomposed as $\mathbf{X} = \mathbf{TP}^T$ and $\mathbf{Y} = \mathbf{UQ}^T$, so that **X** is well represented by $\mathbf{TP}^T$, and the matrix **T** predicts **U** well by means of a linear model, $\mathbf{U} = \mathbf{TB}$, where **B** is a diagonal matrix. **Y** is therefore modelled linearly by $\mathbf{Y} = \mathbf{TQ}^T\mathbf{B}$. Although the decomposition of the **X**- and **Y**-blocks is similar to that of principal component regression, it addresses a different problem and is therefore statistically and numerically different from principal component analysis. However, the model tends to converge to the multiple regression model when the dimensionality of the PLS model tends towards the number of **X**-variables. The dimensionality of the model, i.e. the number of columns in **T** and the rows in $\mathbf{P}^T$ is usually determined by cross-validation.

After scaling (zero mean and unit variance) the matrices **X** and **Y** are decomposed as

$$\mathbf{X} = \mathbf{TP}^T + \mathbf{E} \tag{3.33}$$

$$\mathbf{Y} = \mathbf{UQ}^T + \mathbf{F} \tag{3.34}$$

Where **E** and **F** are the residual matrices. **U** is simultaneously related to **T** by the diagonal matrix **B**, yielding the residual matrix **G**. That is

$$\mathbf{U} = \mathbf{TB} + \mathbf{G} \tag{3.35}$$

In turn, this gives for **Y**

$$\mathbf{Y} = \mathbf{TBQ}^T + \mathbf{H} = \mathbf{TQ}^{*T} + \mathbf{F}^* \tag{3.36}$$

The diagonal matrix **B** represents the so-called inner relationship. If there is only one y variable, **Q** and **H** become column vectors.

PLS model parameters can be calculated in several ways (Marengo and Todeschini, 1991). The classic approach is the method known as non-iterative partial least squares (NIPALS). With NIPALS, scores **T** and **U** and loadings **P** and **Q**, as well as a weight vector **w** (required to maintain the orthogonality of the scores) and a vector **B** (relating the **X** and **Y** blocks) are calculated.

The PLS decomposition starts with the selection of one column[26] of **Y**, $\mathbf{y}_j$, as an initial estimate of $\mathbf{u}_1$. Starting in the **X**-data block

[26] Usually the column with the largest variance or in the case of univariate **y**, then $\mathbf{u}_1 = \mathbf{y}$.

$$\mathbf{w}_1 = \mathbf{X}^T\mathbf{u}_1/||\mathbf{X}^T\mathbf{u}_1|| \tag{3.37}$$

$$\mathbf{t}_1 = \mathbf{X}\mathbf{w}_1 \tag{3.38}$$

while in the **Y**-block

$$\mathbf{q}_1 = \mathbf{u}_1{}^T\mathbf{t}_1/||\mathbf{u}_1{}^T\mathbf{t}_1|| \tag{3.39}$$

$$\mathbf{u}_1 = \mathbf{Y}\mathbf{q}_1 \tag{3.40}$$

Convergence is checked by comparing $\mathbf{t}_1$ in equation 3.38 with the value from the previous iteration. If they are not equal to within some specified error criterion, the estimate of $\mathbf{u}_1$ in equation 3.37 is updated with that from equation 3.40. Equations 3.39 and 3.40 can be omitted if **Y** is univariate. Upon convergence the **X**-data block loadings are calculated and the scores rescaled and weighted accordingly.

$$\mathbf{p}_1 = \mathbf{X}^T\mathbf{t}_1/||\mathbf{X}^T\mathbf{t}_1|| \tag{3.41}$$

$$\mathbf{p}_{1,new} = \mathbf{p}_{1,old}/||\mathbf{p}_{1,old}|| \tag{3.42}$$

$$\mathbf{t}_{1,new} = \mathbf{t}_{1,old}/||\mathbf{p}_{1,old}|| \tag{3.43}$$

$$\mathbf{w}_{1,new} = \mathbf{w}_{1,old}/||\mathbf{p}_{1,old}|| \tag{3.44}$$

The regression coefficient for the inner relation is subsequently calculated from

$$\mathbf{b}_1 = \mathbf{u}_1{}^T\mathbf{t}_1/\ \mathbf{t}_1{}^T\mathbf{t}_1 \tag{3.45}$$

The **X**- and **Y**-block residuals are calculated by

$$\mathbf{E}_1 = \mathbf{X} - \mathbf{t}_1\mathbf{p}_1{}^T \tag{3.46}$$

$$\mathbf{F}_1 = \mathbf{Y} - b_1\mathbf{t}_1\mathbf{q}_1{}^T \tag{3.47}$$

The next latent variable is extracted by repetition of the entire procedure, replacing **X** and **Y** with the residual matrices $\mathbf{E}_1$ and $\mathbf{F}_1$ respectively, and incrementing subscripts by 1.

Partial least squares forms the matrix inverse defined by

$$\mathbf{X}^+ = \mathbf{W}(\mathbf{P}^T\mathbf{W})^{-1}(\mathbf{T}^T\mathbf{T})^{-1}\mathbf{T}^T \tag{3.48}$$

The scores and loadings calculated in PLS are not the same as those calculated in principal component analysis, but can be thought of as principal component analysis scores and loadings that have been rotated in order to improve the prediction of **Y**. Alternatively, it can be shown that the first PLS weight vector is given by the eigenvector of $\mathbf{X}^T\mathbf{Y}\mathbf{Y}^T\mathbf{X}$ associated with the largest eigenvalue, that is

$$\lambda\mathbf{w}_1 = \mathbf{X}^T\mathbf{Y}\mathbf{Y}^T\mathbf{X}\mathbf{w}_1 \tag{3.49}$$

Like principal component regression, if all latent variables are included, the PLS model converges to the multiple linear regression model (MLR). PLS is a flexible method that can handle high-dimensional data sets, even where the number of variables exceeds the number of observations and some versions of the algorithm can handle missing data. Moreover, PLS methods can fit data with fewer components than other methods. The drawback of PLS models are loadings that are difficult to interpret.

3.9. MULTIVARIATE STATISTICAL PROCESS CONTROL

Although modern instrumentation allows large volumes of process data to be accumulated, these data are of limited use without appropriate processing, especially as far as the discovery of abnormal events derived from the interaction between variables, etc., is concerned.

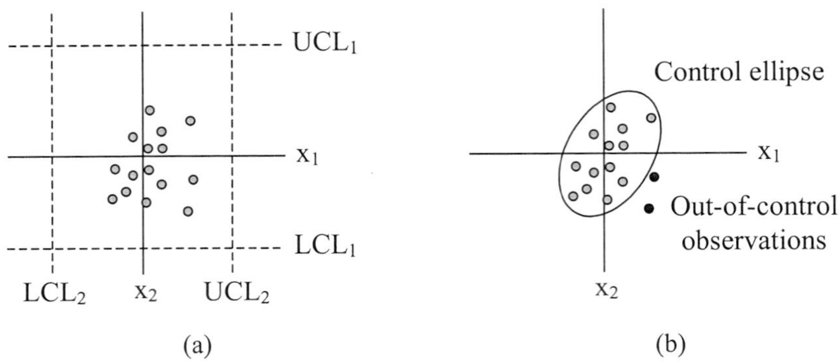

Figure 3.19(a) Two superimposed univariate process control charts, and (b) a control chart taking the correlation between the variables into account.

In the last few decades, several multivariate statistical methods have been developed to identify correlations between variables and to exploit process information for monitoring and control purposes (Wise et al., 1990; Kourti and MacGregor, 1995; Martin et al., 1996; Martin et al., 1999). The advantages of multivariate statistical process control over univariate control can be demonstrated by superimposing two univariate control charts on top of each other, as illustrated in Figure 3.19(a). The individual lower and upper control limits (LCL_1, UCL_1) and (LCL_2, UCL_2) of each of the variables x_1 and x_2 are indicated by broken lines. With correlated variables, the approach outlined in Figure 3.19(a) is not particularly accurate, since the redundancy of the variables is not exploited. The true control region for the two variables is shown in Figure 3.19(b). The correlation between the two variables is reflected by the aspect ratio of the principal axes of the control ellipse. Note that two of the observations that were considered to be in-control in Figure 3.19(a), are shown to be out-of-control in Figure 3.19(b).

Principal component analysis and related techniques are commonly used to construct charts like the one shown in Figure 3.19(b), as discussed by Kresta et al. (1991), Nomikos and MacGregor (1995), Amirthalingam and Lee (1997), Kosanovich et al. (1996), Kourti and MacGregor (1995), Jia et al. (1998), Jaeckle and MacGregor (1998), Seasholtz (1999) and Zhang et al. (1996, 1997, 1999), among many others. Principal component analysis effectively reduces the measurement space, which facilitates process failure and disturbance diagnosis through the capture of correlations between variables and characterizing normal operating conditions (NOC). The ability of principal component analysis to facilitate visualization of abnormal conditions is limited to processes that can be adequately represented by three PCs at most. Where more than two or three PCs are needed, either measures have to be adopted thast account for all the dimensions in the problem space, e.g. Hotelling's T^2 statistics or other methods of visualization have to be used, e.g. topographical methods. Control schemes very similar to those used in univariate statistical process control can also be used in multivariate statistical process control, as explained briefly below.

3.9.1. Multivariate Schewhart charts

Multivariate χ^2 and Hotelling's T^2 charts (Hotelling, 1947; Ryan, 1989; Jackson, 1991; MacGregor and Kourti, 1995) are used where a vector of m variables are tracked. Given an m x 1 vector of measurements y on m normally distributed variables with an in-control covariance matrix Σ, the idea is to test whether the vector μ of the means of these variables is at the desired target κ, by computing the statistic

$$\chi^2 = (\mathbf{y}-\kappa)^T\Sigma^{-1}(\mathrm{y}-\kappa) \tag{3.50}$$

which has a central χ^2 distribution with m degrees of freedom is $\mu = \kappa$. The χ^2 statistic in equation (3.50) represents the Mahalanobis distance of any point from the target κ. Multivariate Shewhart charts are constructed by plotting χ^2 versus time with an upper control limit given by $\chi^2_{\alpha,m}$, with α an appropriate level of significance (e.g. $\alpha = 0.05$ or 0.01). All points outside the ellipsoid described by $\chi^2 = \chi^2_{\alpha,m}$ are therefore detected as special events or out-of-control situations.

Since the in-control covariance matrix Σ is often not known a priori, it has to be estimated from a historic sample of n multivariate observations, as

$$\mathbf{S} = (n-1)^{-1}\Sigma_{j=1}{}^{n}(y_j-y_{avg})(y_j-y_{avg})^T \tag{3.51}$$

The corresponding Hotelling's T^2 statistic is subsequently calculated as

$$T^2 = (\mathbf{y}-\kappa)^T\mathbf{S}^{-1}(\mathbf{y}-\kappa) \tag{3.52}$$

The upper control limit of the chart is given by

$$T^2_{UCL} = (n-1)(n+1)mF_{\alpha,m,n-m}/[n(n-m)] \tag{3.53}$$

where $F_{\alpha,m,n-m}$ is the upper $100\alpha\%$ critical point of the F-distribution with m and n-m degrees of freedom. The lower control limit is zero.

Note that the definitions of χ^2 and T^2 have to be redefined if more than a single multivariate observation is considered at each time (Wierda, 1994). Although the simplicity of the Shewhart chart is an advantage, if an out-of-control condition is detected, the challenge is to determine which variables are responsible for the process upset. Another practical problem with Shewhart charts is their lack of robustness, since they are sensitive to multivariate outliers (observations with large T statistics).

3.9.2. Multivariate CUSUM charts

Shewhart control charts use single observations only to determine the state of control. In contrast, multivariate CUSUM and EWMA charts make use of sequences of previous observations. CUSUM quality control schemes can be viewed as a sequence of sequential probability ratio tests (Crosier, 1988). The concept of multivariate CUSUM charts can be explained by considering a sequence of random variables, y_1, y_2, ... y_m having independent multivariate normal distributions N(μ,Σ). In this case the hypothesis H_0: $\mu = \kappa$ is tested agains H_1: $\mu = \mu_1$. If the squared distance between the mean μ_1 and the target κ is defined as $d^2 = (\mu_1\text{-}\kappa)^T\Sigma^{-1}(\mu_1\text{-}\kappa)$, then the sequential probability ratio test rejects the null hypothesis, when

$$\Sigma_{j=1}^{n}\{d^{-1}(\mu_1-\kappa)\Sigma^{-1}(y_j-\kappa)-d/2\} > -\log(\alpha)/d \qquad (3.54)$$

with α the significance level of the test. Healy (1987) has proposed the use of a CUSUM plot of

$$Cj = \max\{0,Cj + d^{-1}(\mu_1-\kappa)\Sigma^{-1}(y_j-\kappa)-d/2\} \qquad (3.55)$$

to assess the shift in the mean from κ to μ_1.

A drawback of the CUSUM scheme as outlined above, is that in some applications there is no basis for the selection of a specific value μ_1 for the alternative hypothesis. This is not a serious problem in the univariate case, as the mean can only shift either up or down, while the magnitude of the shift is not particularly important (a shift of one standard deviation is often used).

Extensions to the univariate CUSUM have been proposed to generalize the test to H_1: $\mu \neq \kappa$. Crosier (1988) has proposed the calculation of T^2 at each observation, followed by computation of the CUSUM of the scalar distance T^2 as

$$C_j = \max\{0, C_{j-1} + T_j - k\} \qquad (3.56)$$

with initial condition $C_0 \geq 0$. Out-of-control events are signalled by $C_j > h$.

CUSUM charts use all the observations in-between special events, in contrast with Shewhart charts, where only the last observation is used. The advantage of this is that CUSUM charts require a smaller average run length for detection of small shifts in the process mean.

3.9.3. Multivariate EWMA charts

Multivariate multivariate exponentially weighted moving average (EWMA) charts, have properties very similar to CUSUM charts, and are based on the statistic (Lowry et al., 1992)

$$\mathbf{z}_j = \mathbf{R}(y_j- \kappa) + (\mathbf{I}\text{-}\mathbf{R})\mathbf{z}_{j-1} \qquad (3.57)$$

R is a diagonal matrix with elements r_1, r_2, ... r_m and $0 \leq r_j \leq 1$, for all j. Large values of r_j give more smoothing and better detection of small process deviations. An out-of-control signal is generated when

$$Q_j^2 = \mathbf{z}_j^T \Omega_j^{-1} \mathbf{z}_j > h \tag{3.58}$$

When all the elements of **R** are equal, the covariance matrix Ω_j is computed as

$$\Omega_j = r[1-(1-r)^{2j}]\Sigma/(2-r) \tag{3.59}$$

where Σ is the covariance matrix of all the **y** variables.

3.9.4. Multivariate statistical process control based on principal components

For a long time, the development of statistical quality control methods has focussed on the monitoring of product quality data only (**Y**), ignoring variables related to process conditions (**X**), where special events may be reflected as well. The latter variables are often measured considerably more frequently (and more accurately) than the process quality variables. In fact, the measurement of process quality may only occur after the product has been made, i.e. too late for control purposes.

The multivariate methods discussed above may also be applied to the process condition variables (**X**), but with large numbers of highly correlated variables, these methods are impractical and besides, they do not take the relationship between the **X** and **Y** data into account. In addition, these methods cannot deal with missing data arising from sensor failure or other causes. Therefore, the most practical approaches to multivariate SPC are based on statistical projection methods such as principal component analysis or PLS. These methods can handle large numbers of highly correlated and noisy process variables, as well as missing data.

a) *Principal component models*

The philosophy underpinning the development of SPC procedures based on projection methods is the same as that for univariate or multivariate Shewhart charts. Normal operating conditions (NOC) are defined by an appropriate set of historic process data. As discussed by Kresta et al. (1991), the selection of these data is critical in the successful application of the procedure. The multivariate control chart consists of the T^2 chart of the first k latent variables, as well as a squared prediction error (SPE) or Q-statistic (see equation 3.15), defined as

$$SPE = \Sigma_{j=1}^{n}(x_j - x_{rec,j})^2 \tag{3.60}$$

where x_j is the j'th new point mapped to the chart and $x_{rec,j}$ is the j'th new point reconstructed from the principal component analysis model, as discussed previously. The control limits for the T^2 chart are determined as before. The upper control limit of the SPE is based on the χ^2 approximation or Q-statistic, as discussed by Nomikos and MacGregor (1995). This plot will detect the occurrence of any novel events that will move the process away from the hyperplane or NOC volume defined by the model.

As an example, consider the following set of process data, consisting of five process variables, which have been scaled to zero mean and unit variance. The principal components of a set of 99 observations associated with normal operating conditions are shown in Table 3.8. At a 95% significance level the T^2 and Q confidence limits based on the first three principal components of the data are 8.26 and 0.528 respectively.

Table 3.8. Percent variance captured by principal component analysis model.

Index		% Variance explained	% Cumulative variance explained
1	3.250	65.0	65.0
2	0.945	18.9	83.9
3	0.661	13.2	97.1
4	0.137	2.75	99.9
5	0.0077	0.15	100

The scores of the first two principal components are shown in Figure 3.20. As can be seen from this figure, the normal operating region has a relatively wide dispersion to compared to the scores (DEV1 and DEV2), which represent plant operations under different conditions. The Hotelling's T^2 chart for the data is shown in Figure 3.21, with the solid line representing normal operating conditions and the broken lines representing the deviations from normal operation, as indicated on the chart. The horizontal broken line represents the 95% confidence limit for the T^2 values.

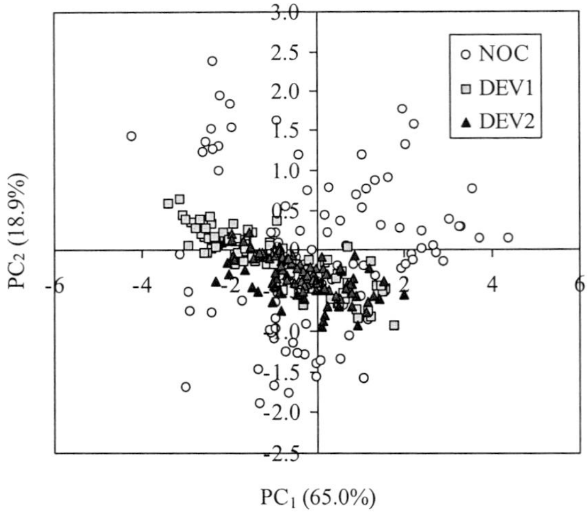

Figure 3.20. Principal component scores of the data associated with NOC and deviations DEV1 and DEV2.

Likewise, in Figure 3.22, the squared prediction errors (SPE) for the observations are shown, with the 95% confidence limit. On this chart, the difference between the classes DEV1 and DEV2 is not as obvious as in Figure 3.21 and in practice both charts would be used. To trace the deviations back to the original variables, the average squared prediction errors for each variable can be calculated, as shown in Figure 3.23. From this figure it is clear that x_2 and x_3 are the most and least important variables respectively.

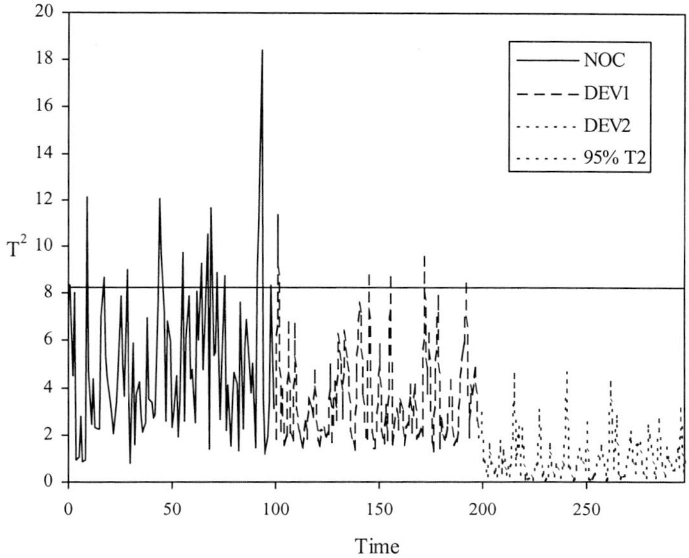

Figure 3.21. Hotelling's T^2 chart based on three latent variables for monitoring of a platinum group metals flotation plant.

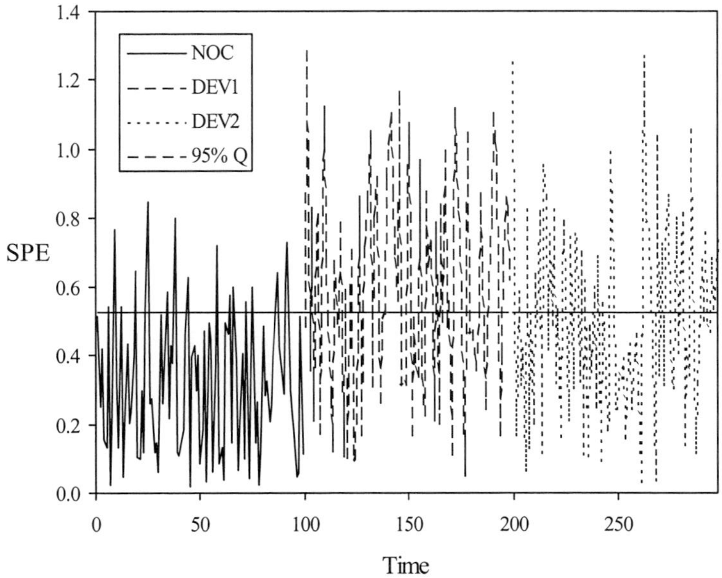

Figure 3.22. SPE chart for monitoring of a platinum group metals flotation plant.

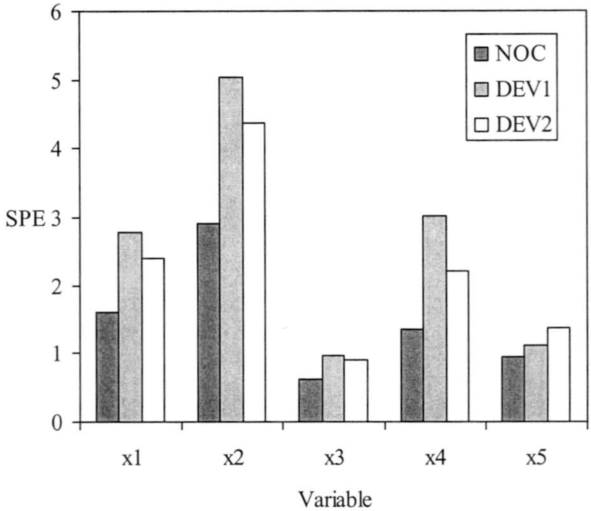

Figure 3.23. Fault diagnosis based on the contributions of variables related to each of the two deviations.

As with principal component analysis, given a process data matrix $\mathbf{X} \in \Re^{nxm}$ and a matrix of process quality data or key performance variables $\mathbf{Y} \in \Re^{nxp}$, partial least squares can be used to extract latent variables not only explaining the variation in the process data, but also in the performance variables.

b) Partial least squares models

Standard partial least squares methods are based on both the predictor matrix, $\mathbf{X} \in \Re^{nxm}$ and the response matrix $\mathbf{Y} \in \Re^{nxp}$, which are decomposed to a sum of rank one component matrixes. The partial least squares decomposition of **X** and **Y** gives

$$\mathbf{X} = \mathbf{T}\mathbf{P}^T = \mathbf{E} \tag{3.61}$$

$$\mathbf{Y} = \mathbf{U}\mathbf{Q}^T + \mathbf{F} \tag{3.62}$$

With conventional partial least squares methods, two statistical monitoring charts are derived, based on Hotelling's T^2 and Q statistics, like those used in principal component analysis charts, i.e.

$$T^2_{k(PLS)} = \Sigma_{i=1}^{n}(t_{ki}^2/\sigma_{T,i})^2 \tag{3.63}$$

$$Q^2_{k(PLS)} = \Sigma_{i=1}^{m}(e_{ki}/\sigma_{Q,i})^2 \tag{3.64}$$

where the t_{ki} denotes the value of the i'th t-score and $\sigma_{T,i}$ the standard deviation of the i'th score vector of the prediction residuals, e_{ki} represents the residual of the i'th predictor variable $\mathbf{x}_i$ and $\sigma_{Q,i}$ is the standard deviation of the i'th residual variable. The statistics in each of these univariate monitoring charts T_k^2 and Q_k follow a control χ^2 distribution with p and m degrees of freedom respectively, if the predictor variables are normally distributed. Under these circumstances, the confidence limits are easy to calculate. Variants on these algorithms have been proposed by Geladi and Kowlaski (1986) and Kruger et al. (2001), among other.

c) Multidimensional NOC volume, V_{NOC}

The volume spanned by what is considered to be the NOC in the multidimensional scores space, V_{NOC}, is a measure of the resolution afforded by principal component analysis (Wachs and Lewin, 1999). This volume can be computed as the product of the principal component standard deviations, or equivalently, as the product of the roots of the eigenvalues of the correlation matrix.

$$V_{NOC} = \Pi_{i=1}^{p} \sigma_i(\mathbf{A}) = = \Pi_{i=1}^{p} \{\lambda_i(\mathbf{A})\}^{1/2}$$

Since 3σ limits are usually imposed on the NOC, the V_{NOC} would become

$$V_{NOC} = \Pi_{i=1}^{p} 3\{\lambda_i(\mathbf{A})\}^{1/2}$$

Since the determinant of a correlation matrix equals the product of its eigenvalues, V_{NOC} can also be expressed as

$$V_{NOC} = 3^p|\mathbf{A}|^{1/2}$$

Increased resolution through a reduction in V_{NOC} can therefore be attained by minimizing the determinant of the correlation matrix, for example by making use of delay-adjusted principal component analysis (Wachs and Lewin, 1999). With delay-adjusted principal component analysis the variables from which the PCs are extracted are shifted in time to increase the correlation between the variables and hence to minimize the determinant of the associated correlation matrix. Alternatively, descriptive statistics can be computed using a vector of the last s principal component analysis scores (Wachs and Lewin, 1999), or by calculating accumulated scores, i.e. $s \to \infty$ (Zhang et al., 1996). This is essentially an extension of the standard univariate moving average technique, applied to the principal component analysis model.

3.9.5. Methodology for process monitoring

As discussed previously, most multivariate process control strategies are based on the projection of the process variables to their latent structures. The structure of the process data can be visualized by constructing a three-dimensional latent variable or topographic map of the data, as discussed previously. This is usually followed by delineation of a region in the mapping space that constitutes normal operation, e.g. based on the relevant statistics in the case of latent variable maps. Once this normal operating region (NOR) has been defined in the feature space of the process, plant operation is monitored through on-line mapping of measurements and visualization of the location or trajectory of the plant in this reduced variable space. The concept is illustrated in Figure 3.24. In this figure the process is monitored in a 3D-space, where the squared prediction error (SPE) on the z-coordinate is mapped against two representative plant features on the x- and y-coordinates.

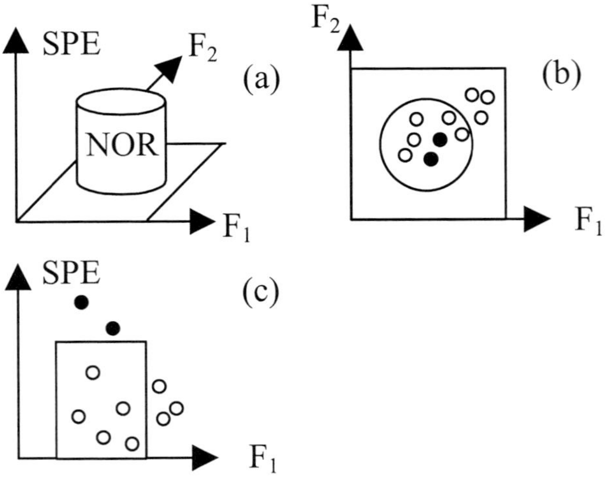

Figure 3.24. Typical multivariate maps used to diagnose process behaviour, showing (a) the cylindrical normal operating region (NOR) in terms of two process features (F_1 and F_2) and the squared prediction error of the process data reconstructed from these features, (b) a top view of the process map and (c) a side view of the process map.

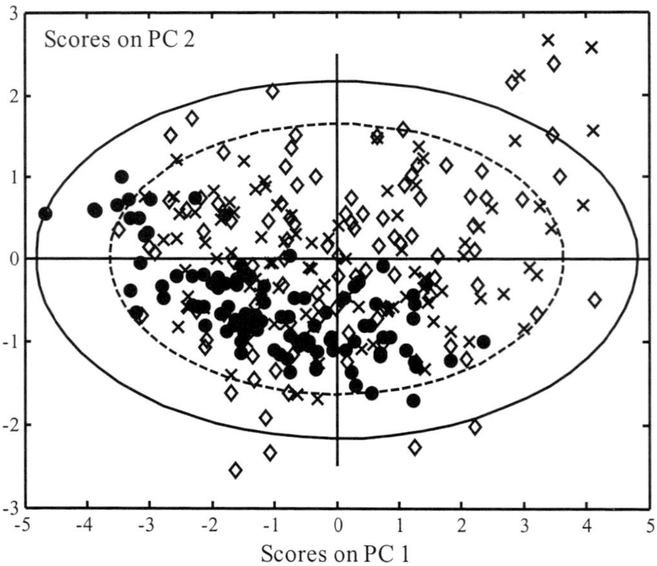

Figure 3.25. Score plot of the first two principal components of the data shown in Figure 3.19. The broken and solid lines indicate the 95% and 99% confidence intervals respectively. The solid circles indicate normal operating conditions (NOC), while the crosses indicate class DEV1 and the diamonds indicate class DEV2.

The squared prediction error is generated through reconstruction of the measurement vector from the feature vector and comparison of the reconstructed vector (x_r) with the original (x_m), i.e. SPE = $||x_m-x_r||^2$.

In the second case, the features lie outside the NOR, as indicated by the top view of the system in Figure 3.24(b), but which may not necessarily be visible from a side view, as indicated in Figure 3.24(c). This simply means that that process conditions themselves have moved outside the normal range of these features, as could happen when the feed has changed significantly, etc.

Figure 3.25 portray the data shown in Figure 3.19 in the score space of the first two principal components of the NOC observations, based on retention of the first three principal components, as summarized in Table 3.8. The 95% and 99% confidence ellipse are indicated by the broken and solid lines respectively. The classes NOC, DEV1 and DEV2 are indicated by the solid circles, crosses and diamonds respectively.

Chapter 4
Regression Models

In experimental work, it is often desirable to find out whether, and by how much, the level of one quantity changes with changes in the level of some other quantity. One of the best ways to accomplish this is by use of regression analysis. For example, it may be particularly important to estimate the dependence of yield and product quality on reaction conditions, such as temperature, pressure, concentration, time of reaction, etc. Apart from determining the best operating conditions, an empirical relation may also increase one's understanding of the chemistry and kinetics of the reaction.

In process systems, the relationship between two variables x and y is often not unique, as can be illustrated by means of a scatter diagram. In the scatter diagram in Figure 4.1, the relationship between mass% phlogopite and mass% apatite in a phosphate ore is shown. Clearly the relationship between the two constituents is not unique. However, in general it can be seen that the average observed mass% apatite decreases, with an increase in the average mass% apatite. The graph of the mean value of one variable, for given variables of the other variable, is referred to as a regression[27] curve, or simply regression.

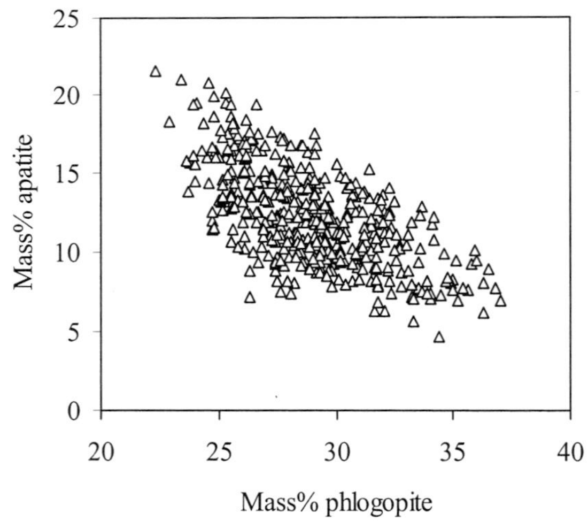

Figure 4.1. Relationship between mass% apatite and mass% phlogopite in a phosphate ore.

Although the method of fitting curves to data based on the least squares principle is not dependent on the probability distributions of the variables involved, much of the appeal of linear (regression) models is based on the fact that these models are easy to interpret if the

[27] The term regression does not make particular sense in the context of data analysis in general. Historically, the term is accredited to Galton (1885), who conducted studies on the relationship between the height of parents and children. He had observed that tall fathers tended to have tall children, although not as tall as the fathers, while short fathers tended to have short children, although not as short as the fathers. The children therefore tended to *regress* to the mean of the population.

probability distributions of the variables are normal. Regression analysis is one of the most widely used methods for the modelling and interpretation of data and as was mentioned previously, simple multilayer perceptron neural networks (no hidden layers, linear activation functions) are equivalent to linear regression or least squares models. It is therefore worthwhile to have a basic understanding of these types of models and related issues, as these can sometimes be obscured by the greater complexity of nonlinear models. Since many excellent texts on the methodology can be found elsewhere (e.g. Chatterjee and Price, 1977; Draper and Smith, 1981; Dillon and Goldstein, 1984; Frank, 1995), the basic concepts are reviewed only briefly below.

4.1. THEORETICAL BACKGROUND TO MODEL DEVELOPMENT

First consider the simplest case of a linear model of the form

$$\mathbf{y} = \mathbf{X}\beta + \mathbf{e} \tag{4.1}$$

where $\mathbf{y}$ is an n x 1 response vector, $\mathbf{X}$ is an n x m matrix of data (i.e. n observations on m variables), with rank m, β is a m x 1 vector of parameters, and e is an n x 1 random vector with independent, identically and normally distributed elements, i.e. $e_i \sim N(0, \sigma^2)$ for i = 1, 2, … n. This is equivalent to the model $E(\mathbf{y}) = \mathbf{X}\beta$, with

$$\mathbf{y} = \begin{bmatrix} Y_1 \\ Y_2 \\ \vdots \\ Y_n \end{bmatrix}, \ \mathbf{X} = \begin{bmatrix} 1 & x_{11} & x_{21} & \cdots & x_{m1} \\ 1 & x_{12} & x_{22} & \cdots & x_{m2} \\ \vdots & \vdots & \vdots & \cdots & \vdots \\ 1 & x_{1n} & x_{2n} & \cdots & x_{mn} \end{bmatrix} \text{ and } \beta = \begin{bmatrix} \beta_o \\ \beta_1 \\ \vdots \\ \beta_m \end{bmatrix} \tag{4.2}$$

4.1.1. Estimation of model parameters

If $E(\mathbf{y}) = \mathbf{X}\beta$ and $E[(\mathbf{y} - \mathbf{X}\beta)(\mathbf{y} - \mathbf{X}\beta)^T] = \mathbf{V}$, and $\mathbf{V}$ is known, it is possible to obtain the least squares estimate of β, without having to make further assumptions with regard to the distribution of $\mathbf{y}$. This is accomplished by minimizing

$$\mathbf{S} = (\mathbf{y} - \mathbf{X}\beta)^T\mathbf{V}^{-1}(\mathbf{y} - \mathbf{X}\beta) \tag{4.3}$$

That is

$$\partial\mathbf{S}/\partial\beta = -2\mathbf{X}^T\mathbf{V}^{-1}(\mathbf{y} - \mathbf{X}\beta) = \mathbf{0} \tag{4.4}$$

or

$$\mathbf{b} = (\mathbf{X}^T\mathbf{V}^{-1}\mathbf{X})^{-1}\mathbf{X}^T\mathbf{V}^{-1}\mathbf{y} \tag{4.5}$$

provided that the matrices can be inverted. If the elements of $\mathbf{y}$ are independent, and have a common variance, then

$$\mathbf{b} = (\mathbf{X}^T\mathbf{X})^{-1}\mathbf{X}^T\mathbf{y} \tag{4.6}$$

The fitted regression model is then

$$\mathbf{y} = \mathbf{Xb} \tag{4.7}$$

4.1.2. Assumptions of regression models with best linear unbiased estimators (BLUE)

The statistical properties of ordinary least squares models are based on several assumptions, as follows.

i) The expected value of the residual vector is zero, i.e.

$$E(\mathbf{e}) = \mathbf{0} \tag{4.8}$$

ii) The assumptions that there are no serial correlation between residuals and that the residuals exhibit constant variance (i.e. they are homoscedastic), i.e.

$$E(\mathbf{ee}^T) = \sigma^2\mathbf{I} \tag{4.9}$$

iii) Moreover, the covariance between the independent variables and the residual terms is zero. This assumption is automatically satisfied when the explanatory variables are non-stochastic, so that the n x m matrix consists of fixed numbers (the usual case)

$$\text{cov}(\mathbf{e},\mathbf{X}) = \mathbf{0} \tag{4.10}$$

iv) The rank of the data matrix **X** is m, the number of columns in **X**, and less than n, the number of observations, i.e.

$$\text{Rank}(\mathbf{X}) = m, \text{ where } m < n \tag{4.11}$$

This means that the population regression function should include only those variables that are not linear functions of some of the variables in the model (that is no multicollinearity).

4.2. REGRESSION AND CORRELATION

Regression and correlation are closely related, but conceptually very different. Correlation analysis measures the strength or degree of the linear association between two variables, while regression estimates the mean value of the dependent variable, based on the known values of one or more predictor variables. In regression analysis a distinction is therefore made between dependent and independent variables. Moreover, dependent variables are assumed to have distributions, while the independent or explanatory variables are usually assumed to be fixed. In correlation analysis the variables are not treated asymmetrically. No distinction is made between dependent and independent variables, and both variables are assumed to be stochastic.

4.2.1. Multiple correlation coefficient, R^2 and adjusted R^2

If $\mathbf{y} = \mathbf{X}\beta + \mathbf{e}$ and the elements of **e** are independently distributed with $E(e_i) = 0$, and $\text{var}(e_i) = \sigma^2$ for $i = 1, 2, \dots n$, then the least squares criterion is

$$S_{est} = \sum_{i=1}^{n} e_i^2 = \mathbf{e}^T\mathbf{e} = (\mathbf{Y}\text{-}\mathbf{X}\beta)^T(\mathbf{Y}\text{-}\mathbf{X}\beta) \tag{4.12}$$

and the minimum value of S for the model is

$$S = (\mathbf{y}-\mathbf{Xb})^T(\mathbf{y}-\mathbf{X}\beta) = \mathbf{y}^T\mathbf{y} - \mathbf{b}^T\mathbf{X}^T\mathbf{y} \tag{4.13}$$

and can be used as a measure of the fit of the model.

This value S_{est} is often compared with the fit of the simplest or minimal model $E(y_i) = \mu$ for all i. In the general form the minimal model can be expressed as

$$E(\mathbf{y}) = \mathbf{X}\beta, \text{ with } \beta = \mu \text{ and } \mathbf{X} = \mathbf{1} \tag{4.14}$$

As a result $\mathbf{X}^T\mathbf{X} = n$, $\mathbf{X}^T\mathbf{y} = \varepsilon y_i$ and $\mathbf{b} = \hat{u} = y_{avg}$. The least squares criterion for the minimal model is therefore

$$S_o = \mathbf{y}^T\mathbf{y} - n\text{-}\ y_{avg}^2 = \Sigma_{i=1}^n (y_i - y_{avg})^2 \tag{4.15}$$

This is proportional to the variance of the observations, and is regarded as the worst possible or pessimal value of S. Other models can consequently be assessed with regard to the minimal model

$$S_o - S_{est} = \mathbf{b}^T\mathbf{X}^T\mathbf{y} - n - y_{avg}^2 \tag{4.16}$$

This difference is the improvement in fit attributed to the model $E(\mathbf{y}) = \mathbf{X}\beta$

$$R^2 = (S_o\text{-}S_{est})/S_o = (\mathbf{b}^T\mathbf{X}^T\mathbf{y}\text{-}ny_{avg}^2)/(\mathbf{y}^T\mathbf{y}\text{-}ny_{avg}^2) \tag{4.17}$$

This can be viewed as the proportion of the total variation in the data explained by the model. The square root of R^2 is referred to as the multiple correlation coefficient.

4.2.2. Adjustment of R^2

Just as the correlation coefficient of the sample is expected to fluctuate around the value of the correlation in the population, the sample R^2 value is expected to fluctuate around the R^2-value of the population. However, the R^2-value can never be negative, and therefore fluctuation is always in the positive direction, and tends to inflate the sample R^2-value. The smaller the sample, the larger these fluctuations, and the more the sample R^2-value tends to be overestimated. It is for this reason that the R^2-value is routinely adjusted, to account for this inflation. One simple way of doing so is as follows.

$$R^2_{adj} = 1 - (1 - R^2)(n - 1)/(n - m - 1) \tag{4.18}$$

In the above adjustment, n represents the number of samples and m the number of independent variables. R^2_{adj} spuriously becoming negative should be reported as zero.

When the number of samples is few ($n < 60$), and the number of independent variables is large ($m > 20$ approximately), the above equation may prove to be inadequate, with adjusted values being off by up to 0.10. Under these circumstances the adjusted value R^2_{adj} is sometimes further adjusted, for example, by

$$R^2_s = [(n - m - 3)R^2_{adj} + R^2_{adj}]/[(n - 2k - 2)\, R^2_{adj} + m] \quad (4.19)$$

4.2.3. Analysis of residuals

Residuals should be plotted in various ways to detect possible anomalies. The following graphs are useful for general diagnostic purposes, as indicated in Figure 4.2. In Figure 4.2(a), the dependence of the residuals on the predicted values of the dependent variable indicates incorrect numerical calculations or the omission of a β_0 term from the model. The variance that is not constant in Figure 4.2(b) requires transformation of the variables or the use of weighted least squares or both. Likewise, a systematic pattern in the residuals plotted against any of the explanatory variables in Figure 4.2(c) suggests an inadequate model in need of more terms. Figure 4.2(d) depicts the ideal situation, where the residuals are uniformly distributed against the dependent variable. These types of graphs are by no means the only way to analyse residuals and often more than one method would be used, including Q-Q plots and histograms.

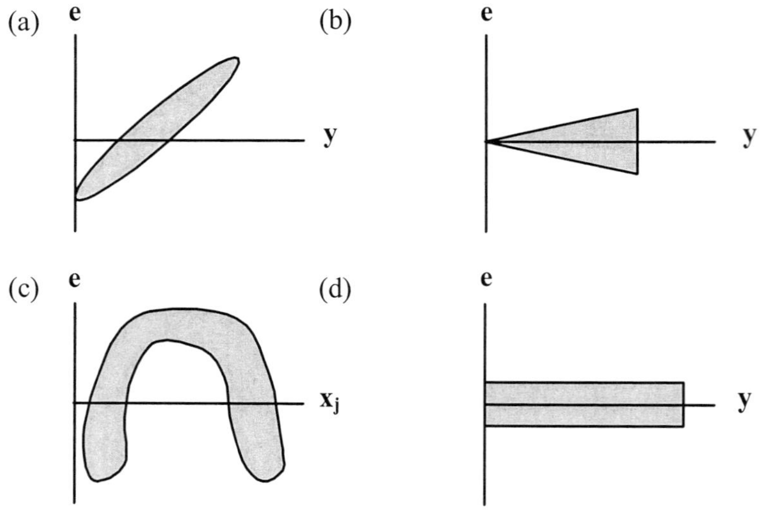

Figure 4.2. Residual diagnostics of regression models.

4.2.4. Confidence intervals of individual model coefficients

The confidence limits around the unstandardized regression coefficients can be calculated as follows (Myers and Montgomery, 1995). For the j'th regression coefficient

$$\begin{aligned} CL(\beta_j) &= b_j \pm t_{\alpha/2,n-m} SE(b_j) \\ &= b_j \pm t_{\alpha/2,n-m}(s^2 C_{jj})^{1/2} \end{aligned} \quad (4.20)$$

where C_{jj} is the jj'th element of the $(\mathbf{X}^T\mathbf{X})^{-1}$ matrix.

The (1-α) confidence interval around around the j'th regression coefficient in the model is the product of the standard error[28] of the coefficient and the two-tailed value of t for the desired confidence level (based on n-m degrees of freedom, where n is the sample size and m the number of parameters in the model). If the confidence interval contains the value zero, the null hypothesis that the coefficient is zero, has to be retained.

4.2.5. Joint confidence regions on model coefficients

It can be shown that $(\mathbf{b}-\beta)^T\mathbf{X}^T\mathbf{X}(\mathbf{b}-\beta)/pMSE$ has an F-distribution with numerator m and denominator n-m degrees of freedom (Myers and Montgomery, 1995), which implies that

$$P[(\mathbf{b}-\beta)^T\mathbf{X}^T\mathbf{X}(\mathbf{b}-\beta)/pMSE \leq F_{\alpha,m,n-m}] = 1 - \alpha \qquad (4.21)$$

This means that a 100(1-α)% point confidence region (an ellipsoidal region) for all the parameters β is

$$(\mathbf{b}-\beta)^T\mathbf{X}^T\mathbf{X}(\mathbf{b}-\beta)/pMSE \leq F_{\alpha,m,n-m} \qquad (4.22)$$

4.2.6. Confidence interval on the mean response

The 100(1-α)% confidence interval on the mean response at the point x_0 is

$$y_{avg}(\mathbf{x}_0) \pm t_{\alpha/2,n-m}(s^2\mathbf{x}_0^T(\mathbf{X}^T\mathbf{X})^{-1}\mathbf{x}_0)^{½} \qquad (4.23)$$

4.2.7. Confidence interval on individual predicted responses

The 100(1-α)% confidence interval on the mean response at the point x_0 is

$$y_{avg}(\mathbf{x}_0) \pm t_{\alpha/2,n-m}[s^2(1+\mathbf{x}_0^T(\mathbf{X}^T\mathbf{X})^{-1}\mathbf{x}_0)]^{½} \qquad (4.24)$$

4.2.8. Prediction intervals for neural networks

In principle, the bias-variance trade-off in neural network model selection can be dealt with on the basis of standard asymptotic theory for nonlinear regression models (Hwang and Ding, 1997). Consider a neural network (or a nonlinear regression model in general) defined by

$$y = f(\mathbf{X},\mathbf{w}) + \varepsilon \qquad (4.25)$$

where $y \in \Re^{nx1}$ is the output of the model, $\mathbf{X} \in \Re^{nxm}$ the input data, $\mathbf{w}$ the parameter (weight) vector and ε a normally distributed error term with a zero mean and variance σ^2, i.e. $\varepsilon \sim N(\mathbf{0}, \sigma^2\mathbf{I})$. The weight vector $\mathbf{w}^\#$ is chosen to minimize the residual sum of squares (RSS) = $(y-\mathbf{y}_{est})^T(\mathbf{y}-\mathbf{y}_{est})$, where $\mathbf{y}_{est}$ is the vector of predicted or estimated values of the output variable and y the vector of actual values. Close to the true parameter values, denoted by $\mathbf{w}^*$, f can be approximated by a Taylor series

$$\mathbf{y}_{est} = f(\mathbf{w}^\#)f(\mathbf{w}^*) + J(\mathbf{w}^\#-\mathbf{w}^*) \qquad (4.26)$$

[28] The standard error is related to the standard deviation of the population, that is the standard deviation of the mean of n observations is $\sigma_m = \sigma\surd n$.

where $\mathbf{J}$ is the Jacobian matrix with ij'th element $\partial f(x_i)/\partial w_j$, evaluated at the true parameter vector $\mathbf{w}^*$ and with x_i the i'th observation (row) of $\mathbf{X}$. With this, the residual sum of squares can be approximated as (De Veaux et al., 1998)

$$RSS \approx (\mathbf{y} - f(\mathbf{w}^*) + \mathbf{Jw}^* - \mathbf{Jw}^{\#})^T(\mathbf{y} - f(\mathbf{w}^*) + \mathbf{Jw}^* - \mathbf{Jw}^{\#})$$

$$RSS \approx (\mathbf{k} - \mathbf{Jw}^{\#})^T(\mathbf{k} - \mathbf{Jw}^{\#}) \tag{4.27}$$

where $\mathbf{k} = \mathbf{y} - f(\mathbf{w}^*) + \mathbf{Jw}^* = \varepsilon + \mathbf{Jw}^*$

If the derivatives with respect to $\mathbf{w}\#$ are set to zero, then

$$\mathbf{w}^{\#} \approx (\mathbf{J}^T\mathbf{J})^{-1}\mathbf{J}^T\mathbf{k} \tag{4.28}$$

so that the asymptotic variance of the parameters (weights) becomes

$$\text{var}(\mathbf{w}^{\#}) \approx \sigma^2(\mathbf{J}^T\mathbf{J})^{-1} \tag{4.29}$$

The value predicted by the neural network model with weights $\mathbf{w}^{\#}$ at a new point $\mathbf{x}_0$, can subsequently be approximated by

$$y_{0,est} = f(\mathbf{x}_0, \mathbf{w}^{\#}) \approx f(\mathbf{x}_0, \mathbf{w}^*) + \mathbf{g}_0^T(\mathbf{w}^{\#}-\mathbf{w}^*) \tag{4.30}$$

where $\mathbf{g}_0$ is a vector with i'th element $\partial f(x_0)/\partial w_j$, evaluated at the true parameter vector $\mathbf{w}^*$. An asymptotic estimate of the variance of the prediction at $\mathbf{x}_0$ is therefore

$$\text{var}(y_{est}|\mathbf{x}_0) \approx \sigma^2\mathbf{g}_0^T(\mathbf{J}^T\mathbf{J})^{-1}\mathbf{g}_0 \tag{4.31}$$

$\mathbf{J}$ and $\mathbf{g}_0$ can be estimated by evaluating the partial derivatives at $\mathbf{w}^{\#}$. As De Veaux et al. (1998) has shown, an asymptotically unbiased estimate of σ^2 is

$$s^2 = RSS/(n-m) \tag{4.32}$$

where n is the number of observations on m variables. Since $s[1+\mathbf{g}_0^T(\mathbf{J}^T\mathbf{J})^{-1}\mathbf{g}_0]^{½}$ has an approximately t_{n-m} distribution, the prediction intervals of half-width can be expressed as

$$t_{n-m,\alpha/2}s[1+\mathbf{g}_0^T(\mathbf{J}^T\mathbf{J})^{-1}\mathbf{g}_0]^{½} \tag{4.33}$$

Hwang and Ding (1997) have shown that although the individual weights of neural networks are not identifiable, equation (4.35) is still valid when the networks are trained to convergence. Studies have shown that when neural networks are trained on relatively small data sets, the prediction intervals tend to be unreliable, while stopping training early to avoid overfitting can lead to estimation of prediction intervals that are too wide. Although other possibly better methods have been proposed for the computation of confidence intervals in nonlinear regression models (Donaldson and Schnabel, 1987), the application of these methods to neural networks is not evident. De Veaux et al. (1998) have proposed an alternative approach to the estimation of prediction intervals based on weight decay to fit the weights of neural networks. Simulations have shown that this approach may alleviate some of the problems encountered with the above approach.

4.3. MULTICOLLINEARITY

When the explanatory variables are correlated, it is difficult, if not impossible, to disentangle the unique effects of each predictor on the response variable. In the extreme case of perfect collinearity (singularity), the regression coefficients become indeterminate. In the case of less than perfect collinearity, the regression coefficients can be estimated, but their standard errors tend to become large, as the degree of collinearity increases. Moreover, if collinearity is severe, the parameter estimates and their standard errors become increasingly sensitive to the slightest change in the data points.

4.3.1. Historic approaches to the detection of multicollinearity

Many approaches have been proposed to detect multicollinearity and although some of them are still quite popular, they suffer from certain weaknesses.

a) Methods based on the correlation coefficient matrix

One of the most obvious methods is based on an examination of the correlation coefficient matrix of the explanatory variables (**R**), or the inverse of this matrix ($\mathbf{R}^{-1}$). The diagonal elements of $\mathbf{R}^{-1} = [1/(n-1)](\mathbf{X}^T\mathbf{X})^{-1}$ are often called variance inflation factors, denoted by VIF_i, i.e. $VIF_i = (1-R_i^2)$, where R_i^2 is the coefficient of multiple correlation of x_i regressed on all the other explanatory variables. It is important to realize that a high simple correlation between two variables is a sufficient, but not a necessary condition for multicollinearity. Multicollinearity can be a major problem, even when the correlations between variables appear to be low. The same conclusions are valid for the inverse of the correlation matrix.

b) Characteristics of regression coefficients

Provided that the model specification is correct, multicollinearity can be indicated by the following.

- Some of the regression coefficients have signs opposite of what would have been expected.
- Several simple correlations between the response and explanatory variables are high, but the corresponding model parameters for these effects are not statistically significant (i.e. $\beta_j = 0$ cannot be rejected).
- The R^2-value for the model is high, but the partial correlation coefficients are low.

Although these effects can be helpful in the identification of multicollinearity, they are neither necessary, nor sufficient conditions for multicollinearity. In fact, collinearity is often cited as an explanation for one or more of the above effects, without any substantiation.

c) Eigenstructure of the crossproduct or correlation matrices

The eigenvalues and eigenvectors of the crossproduct ($\mathbf{X}^T\mathbf{X}$) or correlation matrix (**R**) can also be used to assess multicollinearity. Denoting the p eigenvalues of **R** by λ_i (i = 1, 2, ... p), where $\lambda_1 > \lambda_2 > \ldots > \lambda_{p}$., then in an orthogonal system

$$\Sigma_{i=1}^{p}\lambda_i^{*} = \Sigma_{i=1}^{p}\lambda_i^{*-1} = p \qquad (4.34)$$

where λ_i* (i = 1, 2, ... p) correspond to the p eigenvalues of **R*** = **I**. For a sample-based correlation matrix **R** with eigenvalues λ_i (i = 1, 2, ... p), it is thus possible to compare p with $\Sigma_{i=1}^{p}\lambda_i^{-1}$.

Large values of $\Sigma_{i=1}^{p}\lambda_i^{-1}$ indicate sever multicollinearity, but small eigenvalues should not be taken as prima facie evidence of collinearity. The drawback of this approach is that there are few guidelines as to the interpretation of the ratio of the smallest to the largest eigenvalues of a matrix.

4.3.2. Recent approaches to the detection of multicollinearity

The methods described above can neither be used for reliable detection of multicollinearity, nor can they be used to assess the potentially deleterious effects thereof. More recent strategies are based on the singular value decomposition (SVD) of the data matrix **X**.

a) Condition indices

The condition index of an n x n matrix **A**, is defined as $||\mathbf{A}||.||\mathbf{A}^{-1}||$, where $||\mathbf{A}||$ is the spectral norm of **A**. The magnitude of the condition index provides a bound for the relative change in the length of the solution vector of the system of linear equations, caused by perturbations in the elements of the matrix. Large values of the condition index ($||\mathbf{A}||.||\mathbf{A}^{-1}|| > 30$) are indicative of ill-conditioned matrices.

It can be shown that

$$||\mathbf{A}|| = \delta_{max} \text{ and } ||\mathbf{A}^{-1}|| = 1/\delta_{min} \qquad (4.35)$$

where δ_{min} and δ_{max} are the minimal and maximal singular values[29] of the matrix **A**.

The condition index can therefore be calculated from the ratio $\delta_{max}/\delta_{min}$. The number of near-dependences among the columns of the data matrix **X** can be assessed by calculating a condition index[30] for each variable, i.e.

$$\delta_{max}/\delta_k \text{, for } k = 1, 2, \dots p \qquad (4.36)$$

We can therefore conclude that there are as many near-dependences, as there are high condition indices.

b) Decomposition of regression coefficient variance

Although the condition indices can indicate the presence of near-dependences among the explanatory variables, it does not give any information with regard to the adverse effects that correlation among the variables may have on the variance of the regression coefficients. This

---

[29] Note that the squares of the singular values of **A** are the eigenvalues of $\mathbf{A}^T\mathbf{A}$.
[30] The maximum value of δ_{max}/δ_k is of course the condition index of the matrix **X**.

can be accomplished by decomposing the estimated variance of each regression coefficient into a sum of terms, each of which is associated with an eigenvalue of the data matrix **X**.

The method is based on the singular value decomposition of the covariance matrix of the regression coefficients. The covariance matrix of **b** is $\sigma^2(\mathbf{X}^T\mathbf{X})^{-1}$. Since **X** can be decomposed as $\mathbf{X} = \mathbf{PLQ}^T$, the covariance matrix of **b** can be rewritten in terms of the singular values **D**, and the eigenvectors **Q** as follows

$$\text{cov}(\mathbf{b}) = \sigma^2(\mathbf{X}^T\mathbf{X})^{-1} = \sigma^2\{(\mathbf{PLQ}^T)^T(\mathbf{PLQ}^T)^{-1} = \sigma^2(\mathbf{QLP}^T\mathbf{PLQ}^T)^{-1} = \sigma^2\mathbf{QL}^{-2}\mathbf{Q}^T \quad (4.37)$$

For the k-th component of the **b** vector

$$\text{var}(b_k) = \sigma^2\Sigma_j q_{kj}^2/\delta_j^2 \quad (4.38)$$

In equation (4.40) the δ_j's are the singular values of **X**, and q_{kj} denotes the $(k,j)^{th}$ element of the matrix **Q**. If

$$\phi_{kj} = q_{kj}^2/\delta_j^2 \text{ and } \phi_k = \Sigma_{j=1}^{P}\phi_{kj}, \text{ for } k=1,2, \ldots p \quad (4.39)$$

then the proportion of the variance of the k'th regression coefficient b_k, associated with the j'th component of its decomposition, is

$$\Phi_{kj} = \phi_{kj}/\phi_k \quad (4.40)$$

Since δ_j^2 appears in the denominator, components associated with near dependences will tend to be large relative to other components. As a result, an unusually *large proportion of the variance* (typically > 0.50) of two or more coefficients[31] concentrated in components associated with the same *small singular value*, suggests that the problem can be attributed to the corresponding near dependence of the two columns (variables). The uncovered near dependence of variables can be examined by regressing one of the variates involved on the others.

As was shown above, singular value decomposition of the matrix **X** forms the basis of the condition indices and the variance decomposition proportions. The singular value decomposition, $\mathbf{X} = \mathbf{PLQ}^T$ holds regardless of whether **X** is scaled or centred. However, for the purpose of diagnosing collinearity, it is recommended that **X** is scaled to have unit column lengths. Also, if the regression model being fitted contains an intercept term, then **X** should be uncentred with a column of ones.

4.3.3. Remedies for multicollinearity

It may seem to be a simple remedy to identify pairs of correlated explanatory variables and then dropping one of these variables. Although this approach can improve the computational accuracy of the estimates, the remedy may be worse than the disease (Dillon and Goldstein, 1984). For example, if the true regression model is

[31] At least two columns (variables) of **X** have to be involved.

$$y = \beta_0 + \beta_1 x_1 + \beta_2 x_2 + e \tag{4.41}$$

and the correlation between x_1 and x_2 is large, if x_2 is dropped an new model can be fitted

$$y = \alpha_0 + \alpha_1 x_1 + e \tag{4.42}$$

then it can be shown that

$$E(a_1) = \beta_1 + b_{21}\beta_2 \tag{4.43}$$

where b_{21} is the slope coefficient in the regression of x_2 on x_1. Thus a_1 is a biased estimate of β_1 as long as $b_{21} \neq 0$.

4.3.4. Examples

Consider the following process in a digestive tank, where the pulp viscosity depends on several process variables, viz. pH, temperature, SO_2 and a composite process variable CRM. As indicated in Tables 4.1 and 4.2, the model is highly significant and can explain approximately 73% of the variation in the viscosity in the tank. The model coefficients are summarized in Table 4.3. In the same table, the tolerance factors and variance inflation factors (VIF) associated with each parameter are given. As can be seen, the variance inflation factors are relatively low, while the tolerance values are relativey high.

In Table 4.4 the condition indices are given and it can be seen that the last two indices are relatively high. Although the second last condition index exceeds the criterion of 40, it is not associated with a pair of high variance proportions. Further examination of the variance proportions associated with the highest condition index suggests that the coefficients of the variables TEMP and the model constant are confounded. Collectively they may be important to the model, but is it difficult to separate their individual contributions.

Table 4.1. Summary of model of pulp viscosity in a digestive tank.

Model Summary[a,b]

Model	Variables		R	R Square	Adjusted R Square	Std. Error of the Estimate
	Entered	Removed				
1	pH, CRM, SO2, TEMP[c,d]	.	.854	.730	.728	9.7454

a. Dependent Variable: VISCOS

b. Method: Enter

c. Independent Variables: (Constant), pH, CRM, SO2, TEMP

d. All requested variables entered.

Table 4.2. Significance of the model (ANOVA) of pulp viscosity in a digestive tank.

ANOVA[a]

Model		Sum of Squares	df	Mean Square	F	Sig.
1	Regression	115622	4	28905.5	304.354	.000[b]
	Residual	42738.0	450	94.973		
	Total	158360	454			

a. Dependent Variable: VISCOS

b. Independent Variables: (Constant), pH, CRM, SO2, TEMP

Table 4.3. Model coefficients and collinearity statistics.

Coefficients[a]

Model		Unstandardized Coefficients		Standardized Coefficients	t	Sig.	Collinearity Statistics	
		B	Std. Error	Beta			Tolerance	VIF
1	(Constant)	42.691	53.937		.791	.429		
	SO2	3.754	1.227	.087	3.059	.002	.738	1.354
	CRM	2.626	.094	.771	27.925	.000	.786	1.273
	TEMP	-1.025	.374	-.084	-2.744	.006	.639	1.564
	pH	37.279	6.444	.161	5.785	.000	.777	1.287

a. Dependent Variable: VISCOS

Table 4.4. Collinearity diagnostics of digester model.

Collinearity Diagnostics[a]

Model	Dimension	Eigenvalue	Condition Index	Variance Proportions				
				(Constant)	SO2	CRM	TEMP	pH
1	1	4.931	1.000	.00	.00	.00	.00	.00
	2	6.425E-02	8.760	.00	.00	.79	.00	.00
	3	4.402E-03	33.468	.00	.59	.00	.00	.03
	4	4.712E-04	102.295	.03	.18	.07	.03	.96
	5	3.876E-05	356.687	.97	.23	.13	.96	.01

a. Dependent Variable: VISCOS

4.3.5. Multicollinearity and neural networks

As implicit, weakly parameterized or distributed models, neural networks are difficult to interpret. With the redundancy associated with relatively large numbers of weights (model parameters), the values of the individual weights in neural networks are unimportant. Neural networks are thus not susceptible to the problem of multicollinearity, at the expense of interpretability. However, multicollinearity should be taken into account when sensitivity analyses are performed with neural networks.

The effect of multicollinearity on the training of multilayer perceptron neural networks can be demonstrated by considering a simple network with a single hidden layer and one output node (Qin, 1996). If the input variables are collinear, many sets of weights can minimize the error function of the neural network. Apart from having a potentially adverse effect on the training of the network, multicollinearity also affects the reliability of the neural network model.

This can be illustrated by means of an example, where a hypothetical process relationship between a single input variable (x) and an output variable (z) is given by

$$z = ax + \varepsilon$$

where ε denotes normal Gaussian noise. If linear neural network models without hidden layers are trained on two collinear inputs (x and y, with $y = x$) in order to predict z (which does not depend on y) by means of the generalized delta backpropagation algorithm, the resultant model is of the form

$$z = bx + cy$$

These models will be adequate, provided that the sum of the weights or coefficients of the model approximate the value of a, that is $b + c \approx a$. The actual values of b and c will differ from one model to another, owing to differences in initial conditions during training. This can result in a large spread in the variance, i.e. whereas the actual variance of z will be $var(z) = a^2var(x)$, the variance of the estimated z will be $var(z_{est}) = var\{bx+cy\} = b^2var(x) + c^2var(x) = (b^2+c^2)var(x)$. This is not the case with partial least squares models, for example, where the values of b and c tend to be equal.

4.4. OUTLIERS AND INFLUENTIAL OBSERVATIONS

Several assumptions about the residuals are made when regression analysis is performed. In particular, it is assumed that the residuals are independent, have a zero mean and constant variance, and for statistical testing, have a normal distribution.

The presence of influential observations violates these assumptions, and it is therefore important to identify these observations. This is usually done by preliminary data screening by use of scatter diagrams or examination of the univariate statistics of the variables, followed by examination of the residuals of the model, and the explanatory variables.

4.4.1. Identification of influential observations

It is important to examine the observations in order to identify extreme cases, which have an undue influence on the regression model. These influential observations should either be deleted, rescored, or the variable should be transformed.

In regression analysis, the univariate extremity of observations is analyzed with respect to the response and predictor variables. These observations show up as cases far from the mean of the variable, and unconnected to the other observations. Influential observations include outliers (observations with large residual values with respect to the regression model), leverage points (distinct from other observations owing to their different values on the independent variables), as well as other points that have a disproportionate effect on the model.

Table 4.5. Diagnostics for identification of influential observations (p = number of predictor variables, n = number of observations).

Diagnostic (D)	Criterion	Comment
Standardized residual	(1)$D > \|3.3\|$	$(y_{pred} - y_{obs})/\sigma$
Deleted studentized residual	(1) $D > \|2\|$	$(y_{pred,i} - y_{obs,i})/\sigma$, when i'th observation not used in parameter estimates (PRESS)
Cook's distance	$D > 1$	$(y_{pred,i} - y_{obs,i})/d_i$, (a measure of overall fit)
DFBETA	(2)$D > \|1\|$ or (3) $D > \|2/n^{½}\|$	$\Delta\beta_{ij}$ (change in j'th coefficient of model, when i'th observation is deleted)
DFFIT	$D > \|2/(p/n)^{½}\|$	The change in the predicted values (DFFit) that results from exclusion of the i'th observation.
Covariance ratio – 1	$D > \|3p/n\|$	Ratio of the determinant of the covariance matrix with a particular case excluded, to the determinant of the covariance matrix with all cases included.
Mahalanobis distance (Penny, 1996)	$D > p(n-1)^2F_{p,n-p-1,\alpha/2}/ n(n-p-1+p\ F_{p,n-p-1,\alpha/2})$	$d_i^2 = (x_i-x_{avg})^TS^{-1}(x_i-x_{avg})$, search for single outlier
Jack-knifed Mahalanobis distance	$D > np(n-2)F_{p,n-p-1,\alpha/2}/ (n-1)(n-p-1)$	$d_{(i)}^2 = (x_i-x_{avg,(i)})^TS_{(i)}^{-1}(x_i-x_{avg,(i)})$, search for single outlier

(1)Based on a significance level of 0.001 and n < 1000, (2)For relatively large samples (50 or more), (3)Small data sets, (4)Large data sets

The most important measures of the impact of observations on the model are *leverage*, *discrepancy* and *influence*. As the name suggests, leverage points may lever the model in their direction, owing to their differences from the other observations on which the model is based. Leverage assesses outliers in the set of predictor variables. Observations with high leverage are far from the others. However, they can be far from the other observations, but more or less in line, or off-line, as indicated in Figure 4.3. Discrepancy is a measure of the extent to which an observation is in line with the others, while influence is defined as the product of leverage and discrepancy. Influence assesses the change in the coefficients of the regression model when one of the observations is deleted. Measures of influence are summarized in Table 4.5.

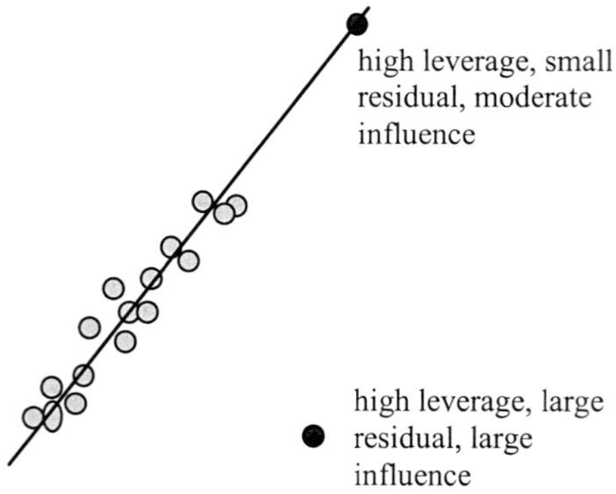

Figure 4.3. Relationships among leverage, residual values and influence.

Screening for outliers can be performed before or after regression (analysis of the residuals). The problem with screening after the regression model has been fitted, is that the practitioner may be biased towards screening observations that may favour the desired outcome of the analysis. It may therefore be safer to deal as far as possible with outliers prior to fitting the model.

These diagnostics all measure some aspect of influence, but unfortunately it is not always clear which aspects are to be preferred, especially since the different diagnostics may yield vastly different results. In view of this, it is often prudent to consider only those observations that score high on several diagnostics simultaneously.

4.4.2. Illustrative case study: Consumption of an additive in a leach plant

As an illustration of the identification of influential observations, consider the following set of data, where the consumption of an additive in a leaching plant is modelled as a function of plant variables.

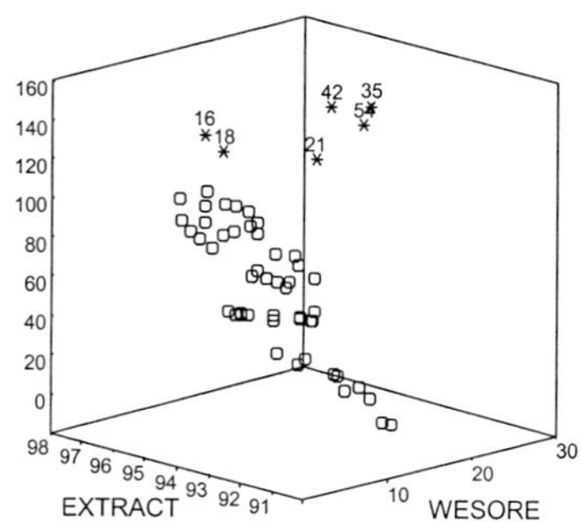

Figure 4.4. Sakresote as a function of EXTRACT and WESORE.

Owing to geological unconformity, the Kimberley Reef at Leslie gold mine in South Africa lies on a series of four different shales (De Ruijter, 1989).

Table 4.6. Regression model of reagent consumption based on all 54 observations ($R^2_{adj} = 0.584$).

Model		Unstandardized Coefficients B	Unstandardized Coefficients Std. Error	Standardized Coefficients Beta	t	Sig.
1	(Constant)	-2011.226	240.689		-8.356	.000
	EXTRACT	21.335	2.503	.787	8.524	.000
	WESORE	4.354	1.032	.390	4.221	.000

a. Dependent Variable: SAKRESOT

Table 4.7. Regression model of reagent consumption after omission of observations 16, 18, 21, 35, 42 and 54 ($R^2_{adj} = 0.664$).

Model		Unstandardized Coefficients B	Unstandardized Coefficients Std. Error	Standardized Coefficients Beta	t	Sig.
1	(Constant)	-1421.732	192.540		-7.384	.000
	EXTRACT	15.858	1.956	.763	8.110	.000
	WESORE	-1.714	1.317	-.122	-1.301	.200

a. Dependent Variable: SAKRESOT

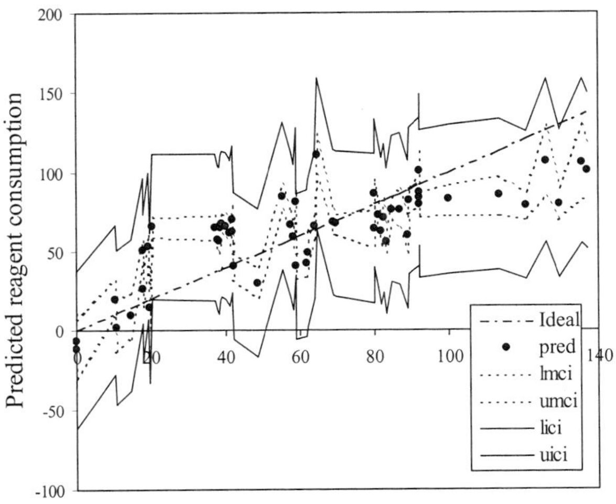

Figure 4.5. Model fit to data, with individual (LICI and UICIC, broken lines) and mean (LMCI and UMCI, solid lines) lower and upper 95% confidence limits.

Figure 4.4 shows the reagent consumption as a function of the variables EXTRACT and WESORE. As can be seen from this figure, most of the 54 observations (indicated with empty circles) form an ellipsoidal cluster in the 3D-variable space. However, 6 of these observations, viz. 16, 18, 21, 35, 42 and 54 (indicated with asterisks) appear to be outliers or leverage points. Although not clear from Figure 4.4, these 6 observations deviate from the principal axis of the ellipsoid mainly along the WESORE axis.

When the reagent consumption (Y) is regressed onto WESORE and EXTRACT based on all the observations, the model can explain approximately 58.4% of the variance in the data as indicated in Table 4.6. The various confidence limits of the model are indicated in Figure 4.5. Regression without observations 16, 18, 21, 35, 42 and 54 yield a more accurate model (R^2_{adj} = 66.4%), as indicated in Table 4.7. Also note that with the omission of these observations, the significance of the WESORE variable decreases sharply.

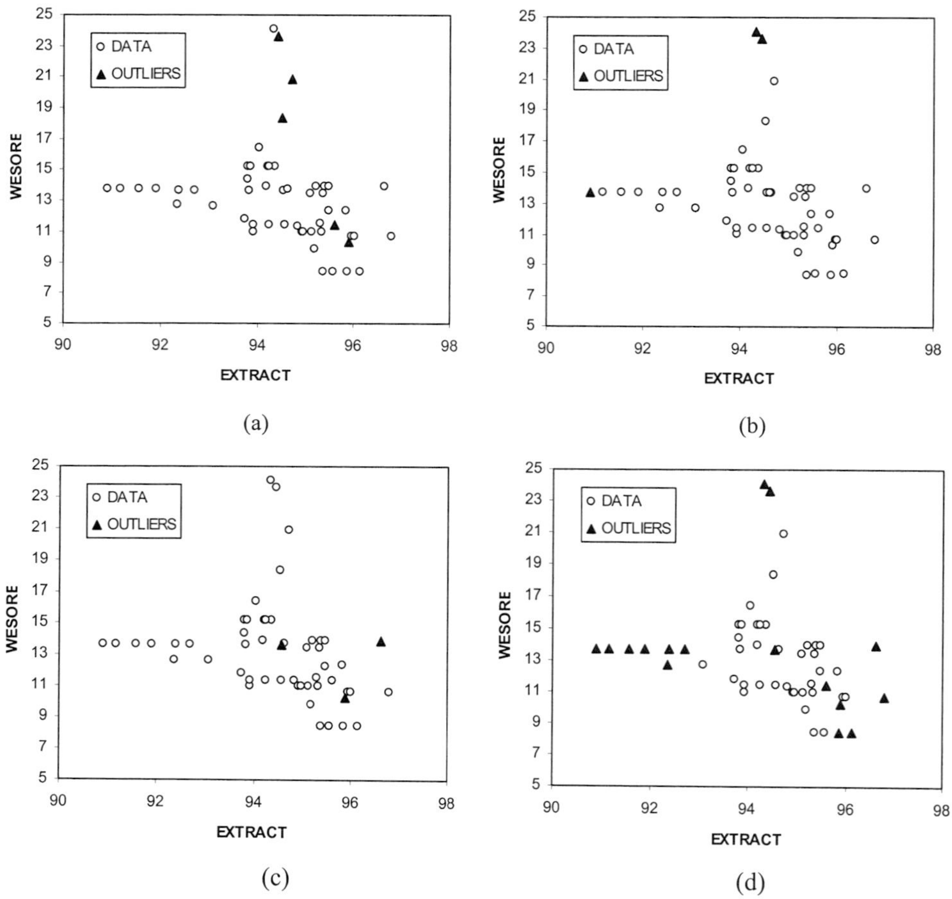

Figure 4.6(a) Outliers as indicated in Figure 4.4, (b) Mahalanobis distance, (c) deleted Studentized residuals and (d) covariance ratio criterion.

The model diagnostics calculated for all the observations are summarized in Table 4.8. Shaded entries in the table indicate observations that can be considered influential, based on the criteria outlined in Table 4.5. Although no influential observations are indicated by Cook's criterion (Cook and Weisberg, 1982), it is worth noting that a number of observations have relatively large Cooke values compared to the other observations. These include observations 35, 42, 12, 54 and 16 in that order. Other outliers indicated by different diagnostics are shown in Figure 4.6. Based on the Mahalanobis distance, only one of the observations indicated with asterisks in Figure 4.4 is an outlier, while three of these observations are outliers according to the covariance ratio criterion.

Table 4.8. Residual diagnostics, showing the observation number (NO), the predictor variables (EXTR, WESORE), the predicted observations (SAKRES), unstandardized residuals (RES), deleted unstandardized residuals (DRES), studentized residuals (STRES), deleted studentized residuals (DSTRES), Mahalanobis distance (MAH), Cook's distance (COOK), adjusted covariance ratio (COVR-1) and the change in the model fit (DFIT). Outliers are indicated by shaded entries.

NO	EXTR	WESORE	SAKRES	RES	DRES	STRES	DSTRES	MAH	COOK	COVR-1	DFIT
1	94.56	13.70	20.40	-45.44	-46.33	-1.99	-2.08	0.03	0.03	0.84	-0.89
2	94.22	15.24	37.00	-28.29	-29.05	-1.24	-1.27	0.40	0.01	0.99	-0.76
3	93.94	11.01	42.20	1.30	1.35	0.06	0.06	0.91	0.00	1.10	0.05
4	95.36	13.45	89.20	7.38	7.59	0.32	0.33	0.51	0.00	1.09	0.21
5	91.57	13.71	10.60	8.50	9.64	0.37	0.39	5.28	0.01	1.19	1.14
6	94.21	15.24	38.50	-26.58	-27.30	-1.17	-1.19	0.40	0.01	1.00	-0.71
7	95.39	13.96	55.50	-29.18	-30.12	-1.28	-1.31	0.67	0.02	0.99	-0.94
8	91.17	13.71	.00	6.44	7.55	0.28	0.30	6.84	0.01	1.24	1.12
9	95.20	9.88	81.90	19.04	19.85	0.84	0.85	1.19	0.01	1.06	0.81
10	93.82	14.44	19.20	-34.08	-34.94	-1.50	-1.53	0.33	0.02	0.95	-0.86
11	94.64	13.74	39.00	-28.72	-29.30	-1.26	-1.28	0.05	0.01	0.98	-0.57
12	96.63	13.96	64.60	-46.54	-50.53	-2.04	-2.21	3.21	0.13	0.87	-4.00
13	94.93	10.99	40.90	-21.04	-21.65	-0.92	-0.94	0.52	0.01	1.04	-0.62
14	93.85	13.70	17.80	-32.90	-33.67	-1.44	-1.48	0.24	0.02	0.96	-0.78
15	95.38	8.44	89.10	28.67	30.60	1.26	1.31	2.37	0.04	1.02	1.94
16	95.90	10.30	129.60	49.97	52.52	2.19	2.35	1.59	0.09	0.82	2.55
17	95.33	11.00	41.70	-28.81	-29.77	-1.26	-1.29	0.72	0.02	0.99	-0.95
18	95.62	11.41	120.80	42.31	43.85	1.86	1.94	0.87	0.04	0.88	1.53
19	92.40	13.70	10.20	-9.56	-10.27	-0.42	-0.43	2.68	0.00	1.13	-0.71
20	95.84	12.36	92.00	4.68	4.87	0.21	0.21	1.07	0.00	1.10	0.19
21	94.53	18.34	113.70	28.29	30.51	1.24	1.30	2.86	0.04	1.04	2.21
22	95.23	13.97	58.90	-22.41	-23.05	-0.98	-1.00	0.49	0.01	1.03	-0.64
23	94.25	11.42	62.30	13.00	13.36	0.57	0.57	0.46	0.00	1.07	0.36
24	94.05	16.43	57.60	-9.25	-9.61	-0.41	-0.41	1.03	0.00	1.09	-0.36
25	94.97	10.99	41.70	-21.09	-21.71	-0.93	-0.94	0.53	0.01	1.04	-0.62
26	94.63	13.76	69.70	2.10	2.14	0.09	0.09	0.05	0.00	1.08	0.04
27	95.11	13.45	84.70	8.21	8.41	0.36	0.36	0.27	0.00	1.08	0.20
28	91.91	13.71	14.60	5.25	5.81	0.23	0.24	4.10	0.00	1.17	0.56
29	94.19	13.97	58.10	-1.02	-1.05	-0.04	-0.04	0.08	0.00	1.08	-0.02
30	94.83	11.37	41.00	-20.46	-20.99	-0.90	-0.91	0.36	0.01	1.04	-0.53
31	93.82	15.25	38.20	-18.60	-19.15	-0.82	-0.83	0.53	0.01	1.05	-0.55
32	95.48	13.97	80.10	-6.55	-6.77	-0.29	-0.29	0.79	0.00	1.09	-0.23
33	96.14	8.45	86.80	10.11	10.94	0.44	0.46	3.06	0.01	1.13	0.83
34	95.96	10.70	99.90	17.25	18.11	0.76	0.77	1.53	0.01	1.07	0.86
35	94.33	24.05	135.60	29.60	39.66	1.30	1.52	12.46	0.26	1.24	10.06
36	95.48	12.34	92.30	12.75	13.14	0.56	0.56	0.58	0.00	1.07	0.39
37	96.80	10.70	92.00	-8.57	-9.31	-0.38	-0.39	3.24	0.00	1.14	-0.74
38	95.31	11.55	81.00	8.52	8.77	0.37	0.38	0.53	0.00	1.08	0.25
39	93.74	11.86	59.00	18.66	19.28	0.82	0.83	0.72	0.01	1.05	0.62

40	90.91	13.71	.00	11.99	14.42	0.53	0.57	7.97	0.02	1.25	2.44
41	93.88	15.24	37.80	-20.24	-20.82	-0.89	-0.90	0.50	0.01	1.04	-0.58
42	94.72	20.88	137.00	36.48	42.61	1.60	1.77	6.64	0.17	1.03	6.13
43	92.71	13.70	18.00	-8.37	-8.86	-0.37	-0.37	1.93	0.00	1.11	-0.49
44	94.57	11.41	83.20	27.11	27.81	1.19	1.21	0.34	0.01	1.00	0.70
45	93.09	12.69	48.70	18.62	19.49	0.82	0.83	1.40	0.01	1.07	0.88
46	94.37	15.23	69.00	0.55	0.56	0.02	0.02	0.41	0.00	1.09	0.01
47	96.01	10.69	92.00	8.33	8.75	0.37	0.37	1.61	0.00	1.11	0.43
48	94.26	15.23	64.00	-2.10	-2.16	-0.09	-0.09	0.40	0.00	1.09	-0.06
49	95.87	8.44	82.40	11.51	12.38	0.51	0.52	2.74	0.01	1.12	0.87
50	95.13	11.00	40.00	-26.25	-27.05	-1.15	-1.17	0.59	0.01	1.01	-0.80
51	92.36	12.71	19.60	5.00	5.42	0.22	0.23	3.10	0.00	1.15	0.42
52	95.57	8.45	79.90	15.37	16.44	0.67	0.69	2.47	0.01	1.10	1.07
53	93.93	11.42	61.90	19.43	20.07	0.85	0.86	0.71	0.01	1.05	0.64
54	94.44	23.61	126.10	19.67	25.80	0.86	0.99	11.62	0.10	1.31	6.13

The use of regression diagnostics, such as highlighted in the above case study, can be complemented by the use of a form of dummy variable regression (as discussed below). With this approach, a dummy variable is added for each observation and the model reestimated. Dummy variables with significant coefficients indicate influential observations. In order to avoid problems with too few degrees of freedom, the approach can be implemented in stages, by adding dummy variables for selected observations at each stage (Hair et al., 1995).

4.5. ROBUST REGRESSION MODELS

Robust regression methods constitute an important complement to simple linear regression techniques, in that they produce results similar to least squares regression when the data are linear with normally distributed errors. However, they yield significantly different results from least squares fits when the errors do not satisfy the normality conditions. An example of these types of techniques is *least trimmed squares regression*.

The least trimmed squares method thus differs from the ordinary least squares approach, where all the residuals are taken into account in the estimate of the regression coefficient β_{LS}, that is $\Sigma_{i=1}^{n} r^2_{(i)}\beta$. Since a single observation can have a very large effect on the least squares estimator β_{LS}, it is said to lack robustness. The least trimmed squares estimator has the highly appealing property that its breakdown point[32] is approximately ½ (if q is the right fraction of n). The breakdown point of a regression estimator is the largest fraction of data which can be replaced by arbitrary large values without making the Euclidean norm $||\beta|| = \Sigma_{i=1}^{p}\beta_i^2$ of the resulting estimate tend to ∞. For example, consider the use of the sample mean $z_{avg} = \Sigma_{j=1}^{n}z_j$ to estimate location. The breakdown point of the mean estimator is 0, since if any single $z_j \rightarrow \infty$, then $z_{avg} \rightarrow \infty$. The sample median on the other hand, does not suffer from this drawback.

[32] These types of estimators are known as high breakdown point estimators.

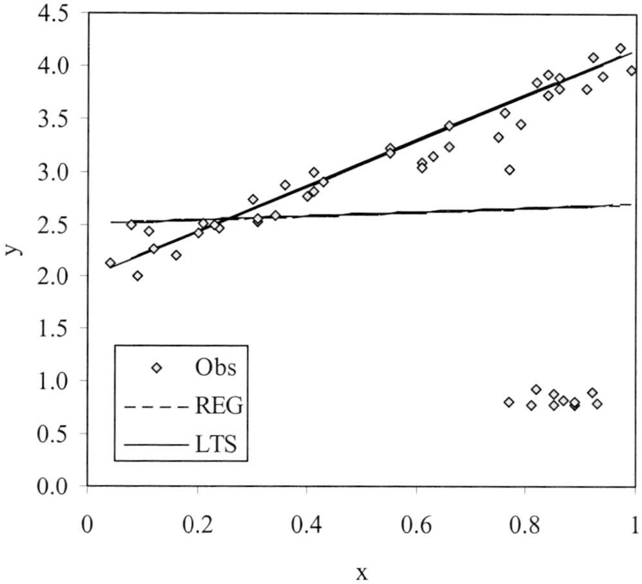

Figure 4.7. A least squares (REG), y = 2.51+0.191x, and least trimmed squares (LTS), y = 2.00+2.15x regression model fitted to a set of clustered data. The LTS model is based on 80% of the data.

The least trimmed squares estimate β''_{LTS} minimizes the sum of the q smallest squared residuals $\Sigma_{i=1}^{q} r^2_{(i)}\beta$, where $r_{(i)}\beta$ is the i'th ordered residual. The value of q is often set to be slightly larger than n/2 (where n is the number of observations). Figure 4.7 shows an example of a least squares (REG) least trimmed squares fit (LST). Figures 4.8-4.9 show the residuals, which indicate a close to optimal fit. The high breakdown point of the estimator means that the least trimmed squares regression model fits the bulk of the data well, even when these data consist of little more than 50% of the total number of observations.

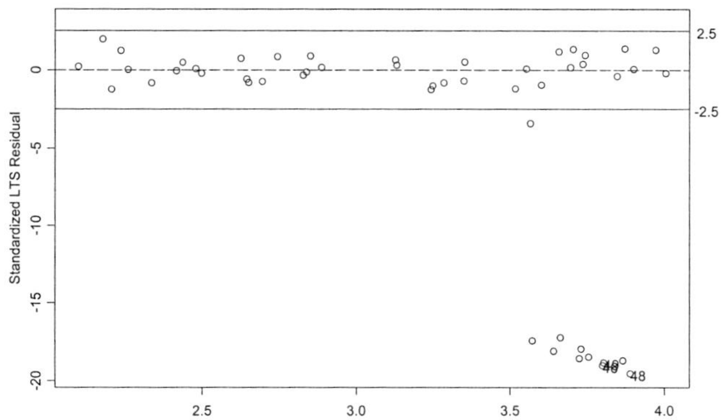

Figure 4.8. Standardized LTS residuals versus the predictor variable.

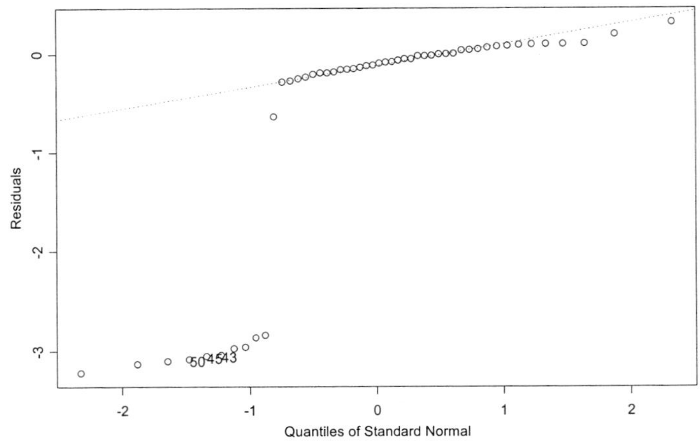

Figure 4.9. Residuals versus quantiles of the standard normal distribution.

4.6. DUMMY VARIABLE REGRESSION

The exploratory variables that enter a regression model are usually assumed to be precise quantities measured on scales having well-defined properties. For example, income measured in dollars, output measured in tons, or temperature on the Celsius scale. It frequently happens, however, that the variables under study are qualitative and serve as label descriptions or indicators of the absence or presence of a quality or attribute. For example, a process may either be *in control* or *out of control*, a temperature may be *high*, *intermediate* or *low*, a valve may be *open* or *close*, etc. In such instances, it is not possible to set up a natural scale to differentiate the various categories or levels of each qualitative variable. However, some strategy for including such variables needs to be developed, for the obvious reason that qualitative variables can influence a dependent measure like any other explanatory variable in the regression model.

The researcher can deal with these situations by making use of dummy variables. In general, if a particular variable has r levels, we need to construct r – 1 dummy variables. Suppose we are investigating which factors, if any, influence the conversion in a reactor. As a first step, we hypothesise that average temperature (200-300°C), type of catalyst (Fe, Co), form of catalyst (powder, pellet), and impeller-stator geometry (*large with small slots, large without slots, small with large slots, small without slots*) all should have an influence. Consider the following coding scheme:

1. Let x denote the average temperature (continuous variable).

2. Define the two categories of catalyst by

$$D_1 = \begin{cases} 1 & \text{if the catalyst is Fe} \\ 0 & \text{if the catalyst is Co} \end{cases}$$

3. Define the form of catalyst by

$$D_2 = \begin{cases} 1 & \text{if the form of the catalyst is powder} \\ 0 & \text{if the form of the catalyst is pellet} \end{cases}$$

4. Define the four categories of impeller-stator assembly by the three dummy variables D_3, D_4, and D_5, where the binary vectors (D_3, D_4, D_5,) denote the following

$D_3 = (1, 0, 0)$ large stator with small slots

$D_4 = (0, 1, 0)$ large stator without slots

$$D_5 = \begin{cases} (0, 0, 1) & \text{small stator with large slots} \\ (0, 0, 0) & \text{small without slots} \end{cases}$$

Letting y denote the conversion in the reactor, the regression model is of the form

$$y_i = \alpha_0 + \alpha_1 D_{1i} + \alpha_2 D_{2i} + \alpha_3 D_{3i} + \alpha_4 D_{4i} + \alpha_5 D_{5i} + \beta_1 x_i + e_i \qquad (4.44)$$

The α-coefficient assigned to each dummy variable is often called a *differential intercept coefficient*, since they reveal by how much the value of the intercept term of the category denoted by the value 1 differs from the intercept coefficient of the category denoted by the value zero (the *base* or *comparison category*).

For example, assuming as usual that $E(e_i) = 0$, we see that

1. The mean conversion in the reactor using pelletized cobalt catalyst and a small stator assembly without slots is

$$E(y_i|x_i, D_1 = D_2 = D_3 = D_4 = D_5 = 0) = \alpha_0 + \beta_1 x_i \qquad (4.45)$$

2. The mean conversion in the reactor using pelletized iron catalyst and a small stator assembly without slots is

$$E(y_i|x_i, D_1 = 1, D_2 = 1, D_3 = D_4 = D_5 = 0) = (\alpha_0 + \alpha_1 + \alpha_2) + \beta_1 x_i \qquad (4.46)$$

3. The mean conversion in the reactor using powdered iron catalyst and a small stator assembly with large slots is

$$E(Y_i|X_i, D_1 = 1, D_2 = D_3 = D_4 = D_5 = 1) = (\alpha_0 + \alpha_1 + \alpha_2 + \alpha_3 + \alpha_5) + \beta_1 X_i \qquad (4.47)$$

In order to assess the significance of the type of catalyst, the significance of α_1 has to be considered, while the effect of the form of the catalyst would depend on the significance of α_2. Evaluation of the effect of the stator-assembly would be based on the individual significance levels of α_3, α_4, and α_5, or a simultaneous test of $\alpha_3 = \alpha_4 = \alpha_5 = 0$ by use of standard model comparisons.

While the construction of dummy variables as defined above is common practice, occasionally researchers prefer the columns in the data matrix **X** corresponding to the qualitative predictors to be orthogonal to the column of ones associated with the usual intercept term. This can be achieved as follows. Suppose that of n observations, n_1 come from the first category of a qualitative variable denoted quite generally as C_1, and n_2 come from the second category, C_2. To ensure orthogonality, the variable D is defined by

$$D = \begin{cases} \dfrac{-n_2}{n_1 n_2 (n_1 + n_2)} & for\ C_1 \\ \\ \dfrac{n_1}{n_1 n_2 (n_1 + n_2)} & for\ C_2 \end{cases} \tag{4.48}$$

This idea is readily extended to more than one qualitative variable.

4.7. RIDGE REGRESSION

Ridge regression is designed to overcome the problems associated with multicollinearity. As discussed previously, unstable model parameter estimates arises when the matrix $\mathbf{X}^T\mathbf{X}$ becomes close to singular. With ridge regression, the parameter estimates are stabilized by trading off unbiasedness in estimation for a reduction in the variance of the estimates. The procedure starts with a linear regression model of the form of equation (4.51).

$$\mathbf{y} = \mathbf{X}\beta + e \tag{4.49}$$

where

$$\mathbf{b} = (\mathbf{X}^T\mathbf{X})^{-1}\mathbf{X}^T\mathbf{y} \tag{4.50}$$

and

$$\text{cov}(\mathbf{b}) = \sigma^2\mathbf{I}(\mathbf{X}^T\mathbf{X})^{-1} \tag{4.51}$$

as discussed previously.

The distance between the actual values (β) of the model coefficients and the estimated values (**b**) can be defined as

$$L^2 = (\mathbf{b}-\beta)^T(\mathbf{b}-\beta) \tag{4.52}$$

It can be shown that the mean square error of the regression estimators is

$$\text{MSE} = E(L^2) = \text{tr}\{\text{cov}(\mathbf{b})\} = \sigma^2\text{tr}(\mathbf{X}^T\mathbf{X})^{-1} \tag{4.53}$$

$$= \sigma^2\Sigma_{j=1}{}^{P}(\lambda)^{-1}$$

with λ_j the j'th eigenvalue of the matrix $\mathbf{X}^T\mathbf{X}$. Nonorthogonality of the predictor variables will cause some of the λ_j values to be small, and the distance between **b** and β to be correspondingly large.

Ridge regression estimates can be defined via the introduction of a scalar value k[33], so that as a function of k, the ridge regression estimates become

$$\mathbf{b}_{(k)} = (\mathbf{X}^T\mathbf{X} + k\mathbf{I})^{-1}\mathbf{X}^T\mathbf{y} \tag{4.54}$$

From this follows that

$$E\{\mathbf{b}_{(k)}\} = (\mathbf{X}^T\mathbf{X} + k\mathbf{I})^{-1}(\mathbf{X}^T\mathbf{X})\beta \tag{4.55}$$

So that when $k = 0$, $(\mathbf{X}^T\mathbf{X} + k\mathbf{I})^{-1}(\mathbf{X}^T\mathbf{X}) = \mathbf{I}$ and $E\{\mathbf{b}_{(k)}\} = \beta$. When $k \neq 0$, $\mathbf{b}_{(k)}$ becomes a biased estimate of β. The covariance matrix of the ridge regression estimators is

$$\mathrm{cov}(\mathbf{b}_{(k)}) = \sigma^2(\mathbf{X}^T\mathbf{X} + k\mathbf{I})^{-1}(\mathbf{X}^T\mathbf{X})(\mathbf{X}^T\mathbf{X} + k\mathbf{I})^{-1} \tag{4.56}$$

Compared with the ordinary least squares estimator of b,

$$L_{(k)}^2 < L^2 \text{ or } (\mathbf{b}_{(k)}-\beta)^T(\mathbf{b}_{(k)}-\beta) < (\mathbf{b}-\beta)^T(\mathbf{b}-\beta) \tag{4.57}$$

$$MSE_{(k)} = E(L_{(k)}^2) = \mathrm{tr}\{\mathrm{cov}(\mathbf{b}_{(k)})\} = \sigma^2\{\lambda j/(\lambda j + k)\}^2 + k^2\beta^T(\mathbf{X}^T\mathbf{X} + k\mathbf{I})^{-2}\beta \tag{4.58}$$

The first term on the right-hand side of equation (4.58) is the sum of the variances of the elements of the ridge estimator $\mathbf{b}_{(k)}$, which decreases monotonically with an increase in k. In contrast, the second term on the right-hand side of equation (4.60) represents the squared bias, which is zero when $k = 0$, and which increases monotonically with an increase in k. For some value of $k = k" > 0$, the mean square error of the biased estimates will be smaller than the mean square error of the ordinary least squares estimates, i.e. MSE(k") > MSE(0).

Although a k"-value always exists, it is dependent on β and σ^2, which are usually not known in practice and therefore in practice it is also usually impossible to know whether for a given k"-value, MSE(k") < MSE(0). One of the most frequently used methods for the selection of k is the ridge trace method. With this method, ridge regression estimates are plotted as a function of k, and the minimum value of k that leads to stable regression estimates is then selected. The method is neither foolproof, nor the only method of estimating k (Hoerl, 1962, Vinod, 1978; Hoerl and Kennard, 1968, 2000). The application of ridge regression is illustrated by the following case study on the consumption of an additive on a gold leach plant, i.e. the case study considered in section 4.4.2 is revisited.

As was shown previously with the quadratic modeling of the consumption of an additive to a leach plant, the variables were highly correlated, which led to collinearity problems and unreliable estimates of the model coefficients. Figure 4.10, shows the ridge trace of the problem, relating the four estimated model coefficients with the scalar value of k.

Table 4.9 shows the values of the ordinary least squares (OLS) and ridge regression coefficients. As can be seen from this trace, the ordinary least squares coefficients of the model are likely to be overestimated, or at least they are not collectively stable. The variables have all been scaled to zero mean and unit variance prior to fitting the models.

---

[33] Interesting values of k usually lie in the range (0,1). This is always the case for standardized values.

Table 4.9. Ordinary least squares (OLS) and ridge regression model coefficients. The ridge trace coefficients are estimated from Figure 4.10 at k = 0.06 and the last column indicates estimates based on the method of Hoerl et al. (1975).

Variable	OLS (k=0)	Ridge (k=0.06)	Ridge (k=0.000576)
EXTRACT	-3.815	0.370	0.056
EXTRACT2	4.469	0.372	0.598
WESORE	1.609	-0.19	-1.564
WESORE2	2.005	0.57	1.949

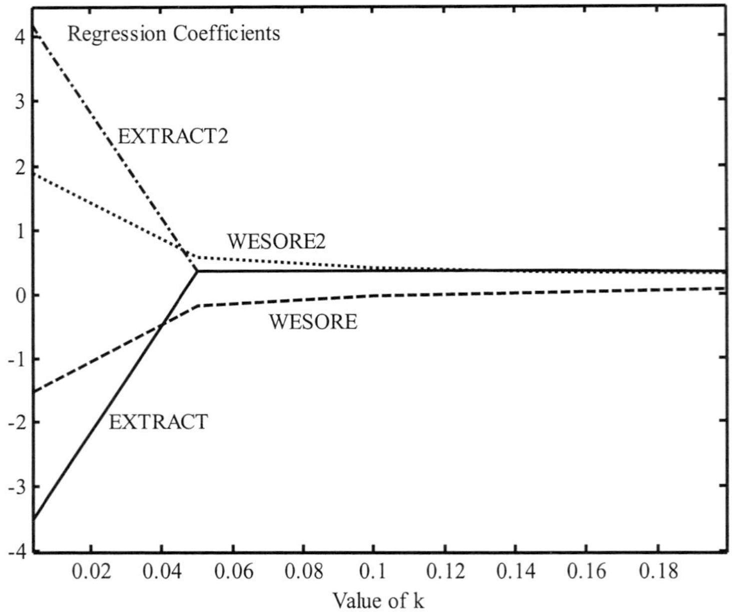

Figure 4.10. Variation of regression coefficient as a function of k.

Moving a short distance from the least squares point k = 0, causes a rapid decline in the absolute values of all the variables. Moreover, the coefficients stabilize from k > 0.05. As can be seen from Table 4.9, two of the coefficients (EXTRACT and WESORE) have changed sign after the introduction of the small bias of 0.000576 and it is doubtful whether they are significant. Figure 4.11 shows the fit of the ordinary least squares (indicated by crosses) and ridge regression models (indicated by empty circles) to the data. As can be seen from this figure, the performance of the two models is virtually identical, despite the large differences in the model coefficients summarized in Table 4.8.

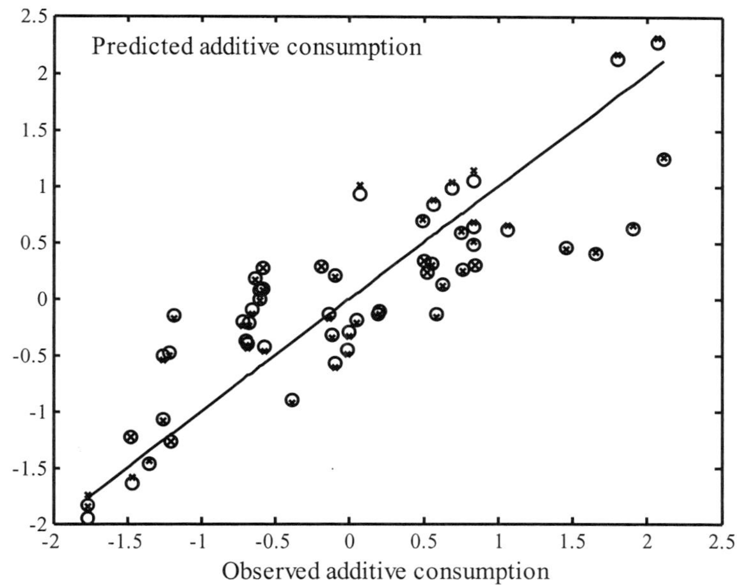

Figure 4.11. Fitting of OLS (x) and RR models (o).

4.8. CONTINUUM REGRESSION

Multiple linear regression, principal component regression and partial least square modelling can all be combined in a single unified approach known as continuum regression (Lorber et al., 1987; Stone and Brooks, 1990; Wise and Ricker, 1993). Continuum regression attempts to satisfy two objectives, namely to provide a parsimonious description of the input data and to correlate the input and output data. Multiple linear regression is therefore an extreme case of continuum regression, seeking to correlate the **X**- and **Y**-blocks of data, without taking the structure of the **X**-block into account. Principal component regression starts with a model of the structure of the **X**-block and correlates this step-by-step with the **Y**-block.

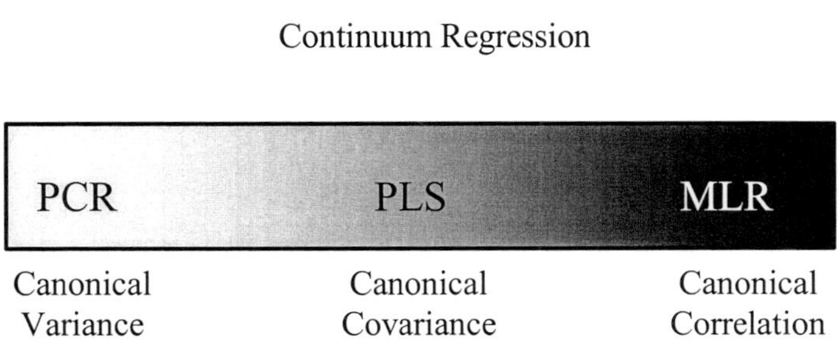

Figure 4.12. Relationship between PCR, PLS, MLR and CR.

In contrast, the conventional method of partial least squares is a compromise between multiple linear regression and principal components regression in that it accounts for the structure of the **X**-block, as well as the covariance between the **X**- and **Y**-blocks of data, as indicated in Figure 4.12. The idea behind continuum regression is to change this balance between the description of variance and the captured correlation continuously.

Although an order (continuum parameter) of 0, 1 or ∞ indicates a multiple linear regression (MLR), partial least squares (PLS) or principal component regression (PCR) model respectively, the present continuum regression algorithms are unstable at extreme values of the continuum parameter and for practical purposes, these values are usually bounded (e.g. $0.001 <$ continuum parameter < 10). On one end of the continuum, correlation between the variables in either the **X**- or the **Y**-block is negligible (i.e. model is equivalent to that of multiple linear regression). On the other end of the continuum, the correlation between the input variables (**X**-block) dominates, yielding a principal component regression model.

4.9. CASE STUDY: CALIBRATION OF AN ON-LINE DIAGNOSTIC MONITORING SYSTEM FOR COMMINUTION IN A LABORATORY-SCALE BALL MILL

The high-intensity vibrations emitted by comminution equipment are a useful indication of the state of the process and plant operators responsible for controlling comminution circuits often exploit this knowledge. The feasibility of diagnostic monitoring and control systems based on interpretation of the acoustic signals emitted from milling equipment was already established in the early 1960s (Anon, 1963; Rowland, 1963). The first controllers merely caused an on-off action with the feeders, but later incremental control was used. The largest drawback of these controllers was their vulnerability to extraneous noise that was difficult to eliminate at the time. Furthermore, the equipment was not particularly sensitive and could only detect large changes reliably. In these early stages it was noted that acoustic control of pebble mills was easier than that of mills with steel grinding media that had higher and more varied noise levels.

Since then, acoustic signal analysis in comminution equipment has been used to monitor the performance of industrial SAG mills (Lynch, 1977), the detection of particle size, breakage rates, slurry properties, mill load, liner wear, etc. (Watson, 1985; Watson and Morrison, 1986; Zeng and Forssberg, 1992, 1993a, 1993b, 1993c, 1996; Zeng et al., 1993). In these and earlier studies, the digital acoustic signals were usually transformed to power spectral densities, typically resulting in several hundred variables that had to be related to the parameters of interest in the comminution process. Although previous investigations have demonstrated the potential advantages of using acoustic emissions to monitor a wide range of equipment and comminution conditions, much work still needs to be done with regard to the interpretation of these acoustic spectra. The accuracy of process and measurement models is an important issue. For example, model maintenance becomes a problem as machinery wears and operating conditions change. Also, more accurate models allow more sensitive control systems, capable of detecting changes in grinding parameters, ore types, etc.

Continuum regression is well-suited to these analyses, as is illustrated by considering the different comminution behaviours of two complex sulphide ores from the Bushveld Igneous complex in South Africa (Theron and Aldrich, 2000).

4.9.1. Experimental setup

A diagrammatic representation of the experimental setup is shown in Figure 4.13. The laboratory mill used in all experiments was lined with rubber and fitted with a flanged lid to facilitate loading and to provide watertight closure during wet grinding. The mill had an inner diameter of 200 mm and an inner length of 280 mm. It was fitted with a 1.1 kW motor that could be controlled by means of a variable speed drive, as well as a set of pulleys that permitted mill speeds of up to 98 rpm as measured by a tachometer.

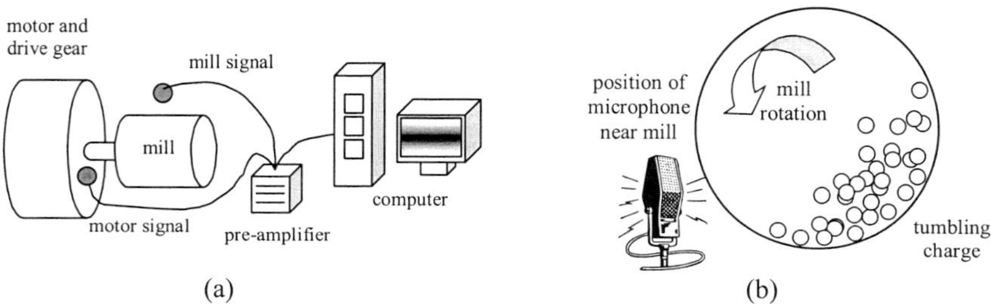

Figure 4.13. Schematic diagram of (a) experimental setup and (b) the position of the microphone with regard to the mill charge.

An Intel Pentium II, 200 MHz with 32 megabyte RAM and a 2 GB hard disk was used for signal and data capturing purposes. The amplified stereosignal from the pre-amplifier was routed to a sound card in the PC for processing.

4.9.2. Experimental procedure

Pre-crushed Merensky and UG2 ores were milled in order to study the effect of changing particle size on the sound emission. In all experiments the mill was loaded with a charge of approximately 3 L, that is approximately 38% of the total volume of the mill. The charge consisted of 290 steel balls with a diameter of 25 mm, 1 kg of ore and 667 mL of water. This resulted in a slurry concentration of approximately 60% by mass.

Since sampling of the particle sizes during runs led to significant disturbance of the mill load through loss of fines, mill loads were only sampled at the end of each run. The runs differed in duration, i.e. runs could last 5, 10, 15, 20, 25, 30, 35, 40, 45 or 50 minutes. In all, 63 runs were recorded for the Merensky ore, while 39 runs were recorded for the UG2 ore, i.e. runs were replicated approximately 6 and 4 times on average for the Merensky and UG2 ores respectively. The large particle size fractions, e.g. 425 μm, 600 μm, 1400 μm, 2800 μm and 4000 μm, were sieved by hand to reduce fines build-up on the wire mesh, while the smaller fractions, e.g. < 75 μm, 75 μm, 106 μm, 150 μm, 212 μm and 300 μm were sieved using a mechanical shaker.

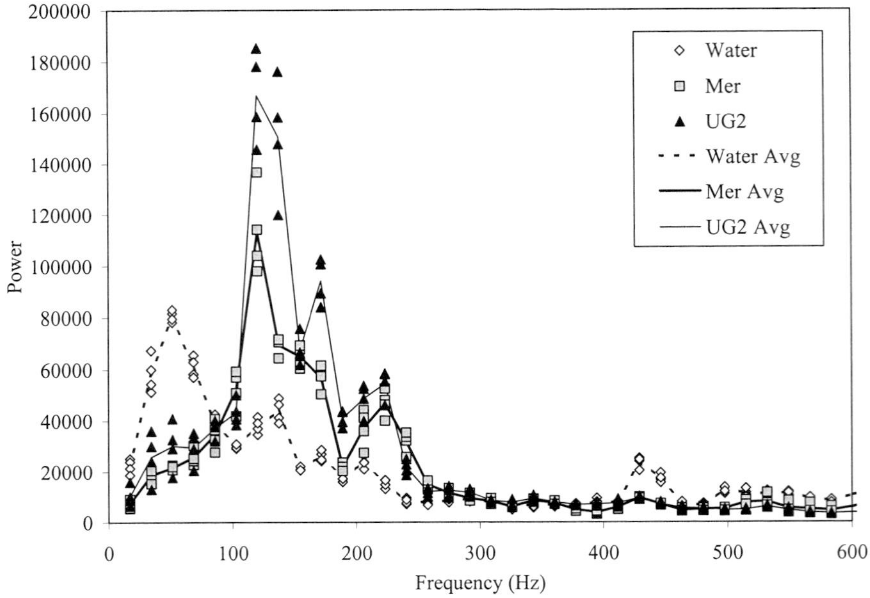

Figure 4.14. Average power spectra of Merensky ore (bold solid line), UG2 ore (thin solid line) and water (broken line) obtained after 5 minutes of milling. The diamonds, squares and triangles indicate the individual observations on which the respective averages were based. Average standard deviations over the frequency spectrum for Merensky ore, UG2 ore and water are 9.47%, 12.24% and 12.25% respectively.

Acoustic signals were collected for a duration of one minute just before the end of a particular run. For example, in runs that were terminated after 10 minutes, recording would have taken place from the 9th to the 10th minute of the run, for 20 minute runs, recording would have taken place from the 19th to the 20th minutes, etc. The amplified signal of the microphone fixed close to the mill shell was recorded onto the hard disk of a personal computer via a sound card. By making use of Nyquist's sampling theorem, requiring that the sampling rate should be at least 2½ times higher than the highest desired frequency, a sampling rate of 11025 Hz was selected. This gave a highest recorded frequency of 4410 Hz. Analysis of the signal showed that the lower frequencies contained most of the information from the grinding process. Samples were thus recorded at 11025 Hz with 16 bit linear quantization. All sound files were saved in the standard Windows PCM *.*wav* format.

The recorded signals were converted by means of Fast Fourier transforms to a series of power spectral densities, where each density was associated with a different frequency (equations 4.61-4.64). The power spectral densities of typical samples obtained after 5 minutes of milling are shown in Figure 4.14. These spectra tended to flatten out after prolonged milling (i.e. smaller particles), as indicated in Figure 4.15. Figure 4.15 shows the trend for the power densities associated with the 125 Hz frequency, as well as the associated trend in the particle sizes. As indicated by the figure, the power decreased approximately 67% over a 45 minute period, while the percentage of particles smaller than 106 μm showed a concomitant increase. These samples were saved in ASCII format from where they could be exported to a spreadsheet (Microsoft Excel®). Owing to limitations in the spreadsheet software, only 239 samples of the power spectrum vectors were used in subsequent analyses.

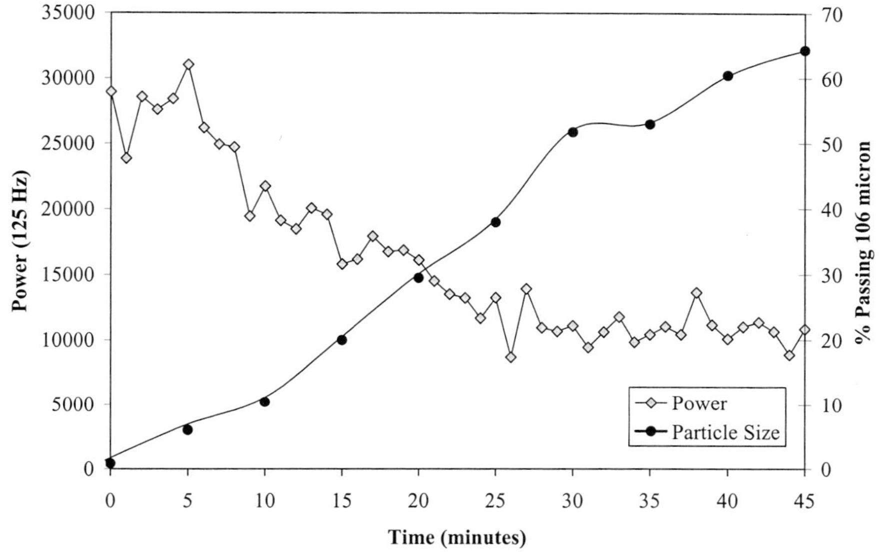

Figure 4.15. Decrease in the power associated with the 125 Hz frequency over time (shaded diamonds, left axis) versus the increase in the percentage of particles smaller than 106 μm (solid circles, right axis). Note that particle sizes were sampled at five-minute intervals only, while the power spectra were sampled at one-minute intervals. Lines simply connect observations to emphasise trends.

4.9.3. Processing of acoustic signals

The Fast Fourier transform (FFT) was used to decompose the signal into its frequency components (Proakis et al., 1996), that is

$$F(w) = (1/T_p)\int_0^{T_p} f(t)\exp(-j2\pi 2\pi)dt \tag{4.59}$$

with T_p the period and F(0) the fundamental frequency. Parseval's theorem gives the relation between a time signal f(t) and its Fourier transform F(w) as

$$\int_{-\infty}^{\infty}|f(t)^2|dt = (1/2\pi)\int_{-\infty}^{\infty}|F(w)^2|dw \tag{4.60}$$

This relation makes it possible to obtain the frequency components of the signal from the time dependent form.

Stremler (1990) summarized the power spectral density of a signal as the energy per unit of frequency, which displays the relative energy contributions of the various frequency components. The area under the power spectral density function gives the energy within a given band of frequencies.

Welch's method (Proakis et al., 1996) for averaging periodograms was used to estimate the power spectra of the signals, since the signals recorded were not all of the same duration. The method allows data sections to overlap and windows the data segments before computing the periodogram, viz.

$$P(f) = (2/T_p)|\int_0^{T_p} w(t)f(t)\exp(-j2\pi 2\pi dt)| \quad (4.61)$$

with w(t) the windowing function. The final periodogram is averaged, unbiased and consistent. The windowing function used in this investigation was the triangular Bartlett function (Proakis et al., 1996)

$$w(n) = 1 - |2n\text{-}M\text{-}1|/(M\text{-}1), \text{ for } 0 \leq n \leq M\text{-}1 \quad (4.62)$$

where n is the bin number and M-1 the length of one section for the calculation of the FFT.

The signal was divided into 512 overlapping sections, each of which was detrended in MATLAB, then windowed by using the Bartlett window. The signal was zero-padded to a length of 512 points. The power spectrum was formed by squaring the length of the overlapping sections and averaging their discrete Fourier transforms. The sections were overlapped with 128 points per neighbouring section. The result was a power spectrum vector with 257 samples. These samples explain the intensity or volume of each respective frequency in the mill signal.

4.9.4. Results and discussion

Figure 4.16 shows the regression coefficients of the Merensky and UG2 ore models, reconstructed from the latent variables. It is interesting to note the five or six large regression coefficients associated with the spectral frequencies of less than 1400 Hz in the Merensky ore. Figure 4.17 shows the cumulative predicted error sum of squares (PRESS) as functions of the number of latent variables and continuum parameters for the Merensky (top) and UG2 (bottom) ore models respectively. Both these surfaces are typical for continuum regression models. The level regions to the right of these surfaces are sometimes referred to as multivariate regression planes and represent models identified with so many latent variables, that they have converged to the multiple linear regression (MLR) solution. On the left of these figures are models with large errors, sometimes referred to as principal component regression (PCR) mountains, i.e. models identified with too few latent variables. These regions include a so-called valley of best models.

In the case of the UG2 ore, the best model was able to explain 83.0% of the overall variance in the particle size distributions. The smaller particles could be modelled more accurately (explained variance of 83.4%) than the larger particles (explained variance of 76.8%). The lower accuracy of the UG2 models could possibly be attributed to the smaller data set that was used in the development of the model and future work may include a more rigorous comparison of the models done on larger data sets.

In addition, the mechanisms controlling the characteristics of the noise signals also need to be addressed by further experimental work. For example, the overall decrease in the noise emitted from the mill (similar to the trend indicated in Figure 4.15) can possibly be explained by the increase in the pulp viscosity associated with the build-up of fines in the slurry. As the viscosity of the pulp increases, the ball-ball collisions inside the mill are damped and the noise inside the mill is decreased. However, this needs to be investigated more extensively, as the temperature would also rise with prolonged grinding, which would tend to decrease the viscosity of the pulp.

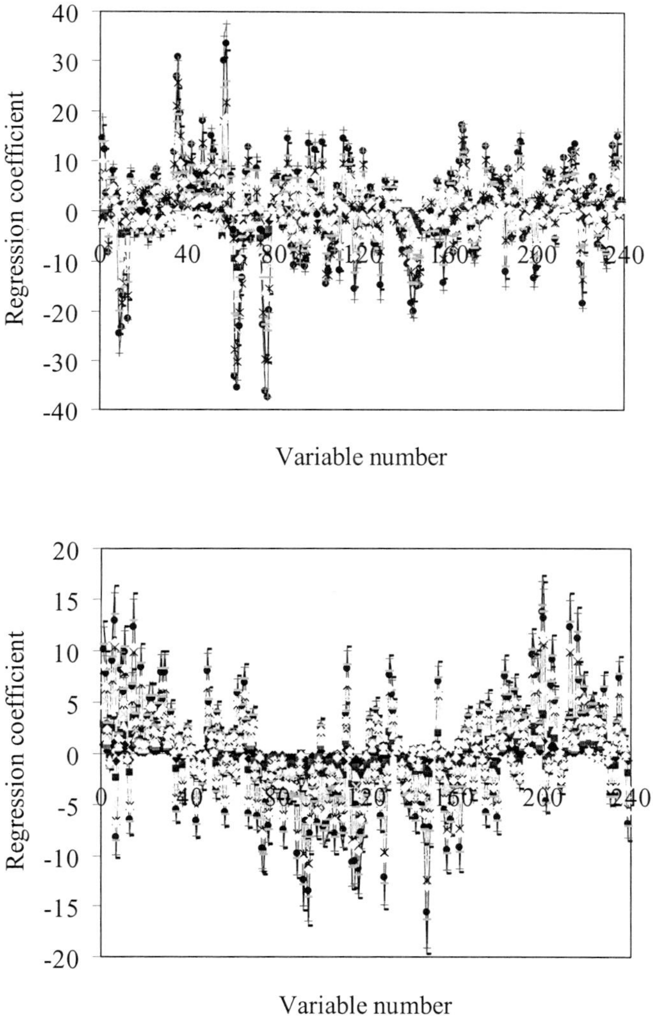

Figure 4.16. Regression coefficients for the best PRESS model for the Merensky (top) and UG2 ores (bottom). In the case of the Merensky ores, the best model was one based on 7 latent variables and a power of 1, i.e. a partial least squares (PLS) model. The best UG2 model was obtained with 8 latent variables and a power of 5, i.e. closely resembling an PCR model in character.

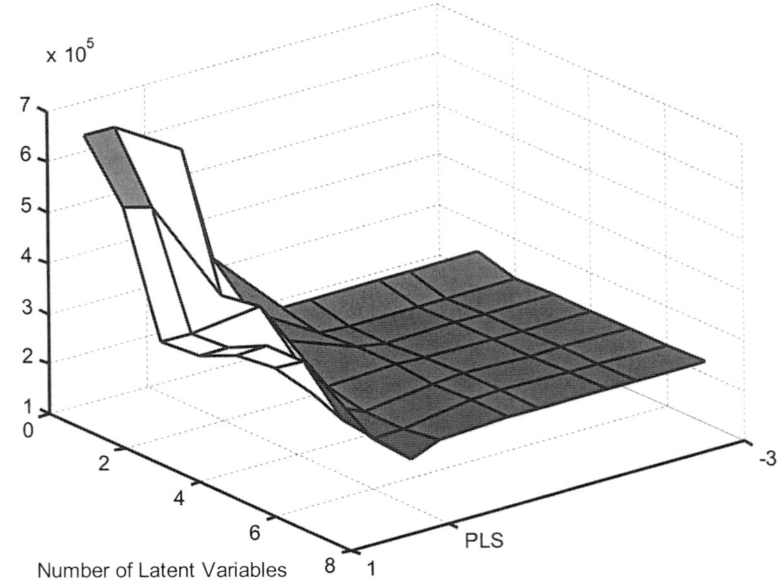

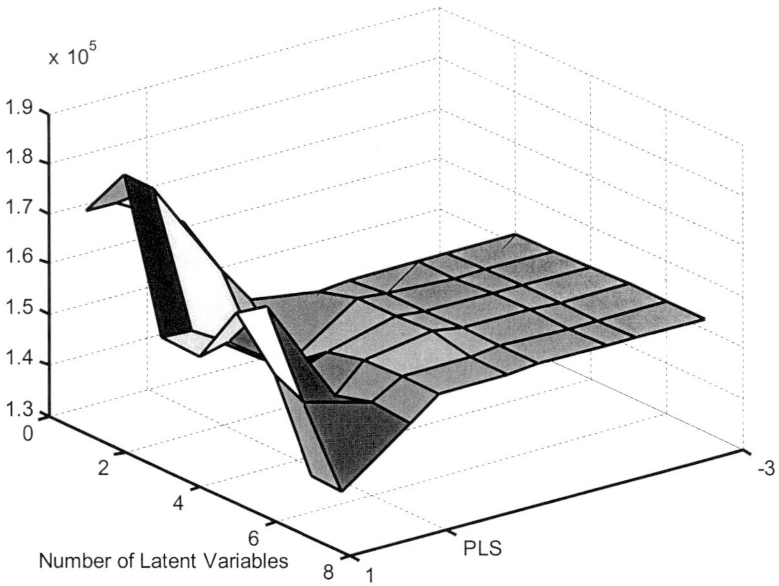

Figure 4.17 shows the cumulative predicted error sum of squares (PRESS) as functions of the number of latent variables and continuum parameters for the Merensky (top) and UG2 (bottom) ore models respectively.

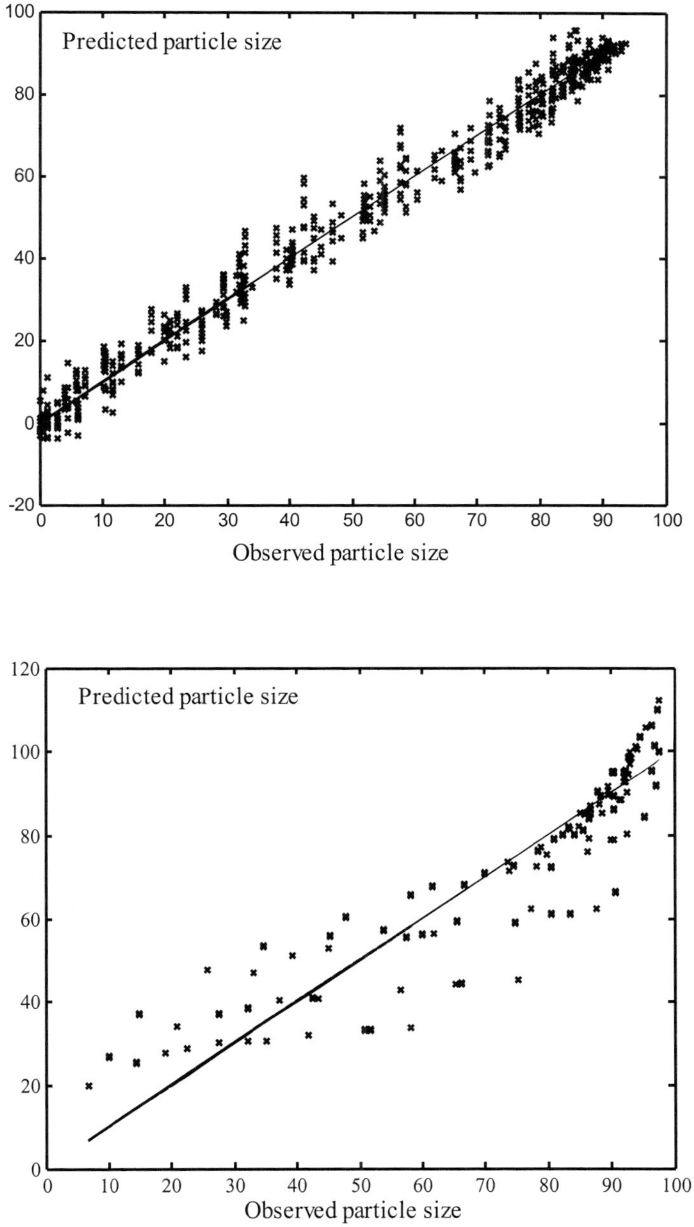

Figure 4.18. Prediction of particle size for the Merensky ore (top) and the UG2 ore (bottom). The Merensky and UG2 models could explain 97.7% and 92.7% respectively of the variance in the particle size data.

Figure 4.18 shows the prediction of particle size for the Merensky ore (top) and the UG2 ore (bottom). The Merensky and UG2 models could explain 97.7% and 92.7% respectively of the variance in the particle size data. Although the models developed in this investigation could be used to monitor the performance of a tumbling mill, it seems reasonable that they could be extended to other systems where on-line particle size analysis is important. For example, by disturbing the particles on a conveyor belt or in other particle feeder systems, it should be possible to measure particle size distributions on-line via the acoustic signals emitted by these particles.

In summary

- The covariance structures of the power spectra and particle size data of the two ores (Merensky and UG2) differed appreciably. In the Merensky ore, a better model could be derived by use of partial least squares, rather than a principal component regression model. In fact, a principal component regression model for the Merensky ore could only explain 82.4% of the total variance in the size distributions, compared with 93.4% for the partial least squares model for the same data. In contrast, in the UG2 ore the continuum regression model was similar to a principal component regression model (both of which could explain approximately 82%-83% of the variation in the data).

- In the Merensky model, a number of frequency bands ranging up to 1400 Hz appeared to dominate in their contribution to the size distribution model. This was not the case with the UG2 ore, where the entire range of frequencies up to 4000 Hz appeared to contribute towards the particle size models.

- In addition, the UG2 ore differed from the Merensky ore in that the smaller particle fractions (< 400 μm) could be modelled significantly more accurately than the coarser particle size fractions (400 μm - 4000 μm).

4.10. NONLINEAR REGRESSION MODELS

As discussed previously, linear regression models are equivalent to single (output) layer feedforward neural networks (perceptrons) with linear activation functions. Nonlinear models can be obtained by use of nonlinear activation functions in the output layer, as well as the addition of hidden layers with nonlinear activation functions. However, it makes little sense to use a neural network to identify a linear relationship, since least squares models are optimal, as was shown previously, and use of a (more complicated) neural network model will not afford any advantages over the simple linear model, and may even perform worse, since an optimal model may not be guaranteed in a least squares sense. In the next sections, a few alternative approaches to the building of nonlinear models are considered and compared with neural network models.

4.10.1. Regression splines

Spline functions are constructed by joining pieces of local polynomials at certain values or *knots*. Functions of any shape can be fitted by selecting appropriate knot positions and polynomials of a suitable order. Once splines are used to regularize a set of data, they are referred to as *smoothing* splines. In this case the objective is not so much to model the underlying trends in the data themselves, but to filter the noise from the data and to summarize the main trends in the data. With regression splines, the objective is the same as that of other regression models, viz. to develop a flexible, yet parsimonious model representing the behaviour of the *population* from which the observations are derived, rather than the *sample* of observations.

In spline regression, the ranges of the predictor variables are subdivided into a number of intervals and each interval is represented by a low-order polynomial, usually with the constraint that the functions are continuous at the junctions. It is clear that even complex curves can be approximated by a sufficiently large number of straight lines. Smooth curves can be obtained by replacing the straight lines with second or third order polynomials. Thus not only the fitted function will be continuous, but also its first or second derivatives. The idea is illustrated in Figure 4.19.

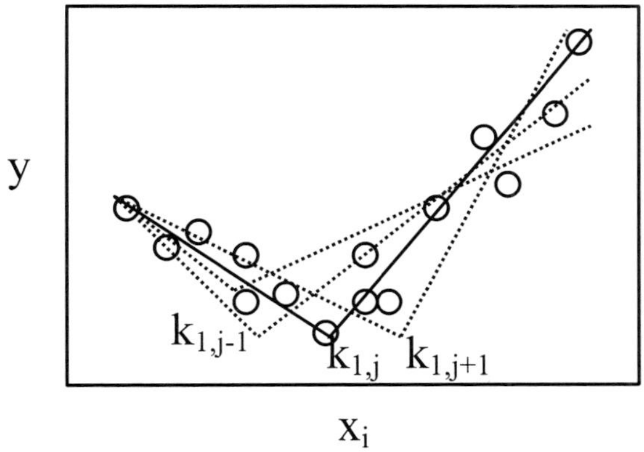

Figure 4.19. Search for the optimal knot position for explaining the response y from the predictor variable x_i. The solid line represents the best fit.

The first step in the above procedure is to determine the number and width of the relevant intervals, i.e. the number of knots and their positions. The best knot positions can be determined via an examination of a scatter plot of the data, otherwise the knots can simply be spread evenly over the range of variables. As an example, consider the following data in Table 4.10.

The spline curve can be expressed as

$$x_2 = b_0 + b_1x_1 + c_1p_1 + c_2p_2 + c_3p_3 = 0 \qquad (4.63)$$

where the coefficients c_k represent the change in slope of the line at the kth knot. The values of the additional variables p_j are simply $x_{1,i}$ - k_j for $x_{1,i} > k_j$ and 0 otherwise. For the data in Table 4.10, shown in Figure 4.12, the first four points can be represented by $x_2 = 3x_1$. From the fifth observation onwards, the variable p_1 takes the value (x_1-4), while p_2 and p_3 remain zero. The value of c_1 is the change in slope at that point, that is -6-3 = -9.

Likewise, from the sixth observation onwards the variable p_2 takes the value $(x_1\text{-}5)$, while p_3 remains zero. The value of c_2 is the change in slope at that point, that is -2-(-6) = 4. Finally, from the 8th observation onwards the variable p_3 takes the value $(x_1\text{-}7)$. The value of c_3 is again the change in slope at that point, that is 1-(-2) = 3. This gives

$$x_2 = 3x_1 - 9(x_1-4) + 4(x_1-5) + 3(x_1-7) = 0$$

The fitted spline curve is shown in Figure 4.20.

Table 4.10. Data for construction of spline model in Figure 4.20.

i	x_1	x_2	$p_1 = (x_1-4)_+$	$p_2 = (x_1-5)_+$	$p_3 = (x_1-7)_+$
1	1	3	0	0	0
2	2	6	0	0	0
3	3	9	0	0	0
4	4	12	0	0	0
5	5 (p_1)	6	1	0	0
6	6 (p_2)	4	2	1	0
7	7	2	3	2	0
8	8 (p_3)	3	4	3	1
9	9	4	5	4	2
10	10	5	6	5	3
11	11	6	7	6	4
12	12	7	8	7	5

In practice, the line fits will not be exact, but can be determined by least squares fits, provided that the knot positions are known. Consequently, the coefficients (c_j) of the variables (p_j) can be tested for significance and if not significant, the variable can be omitted. If the knot positions are not known in advance, they also have to be estimated, which means model estimation becomes a nonlinear regression problem.

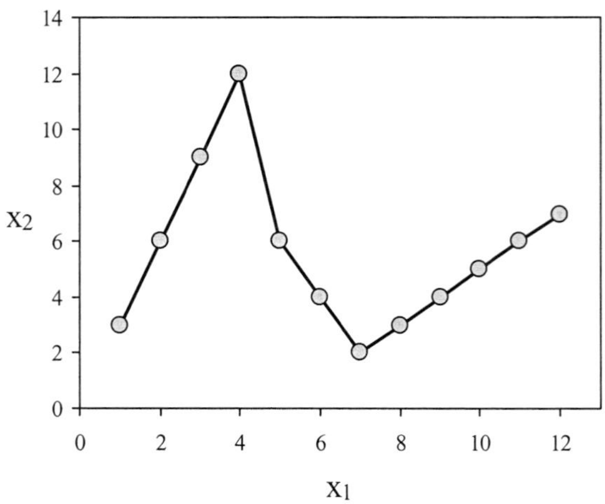

Figure 4.20. Spline curve fitted to $(x_1;x_2)$-data in Table 4.10.

Moreover, in practice higher order polynomials are often fitted to obtain smoother curves. When quadratic polynomials are fitted, it means that two parameters have to be estimated for each line segment, i.e. one for each term of the polynomial, namely $p_{j,1} = (x-k_j)_+$ and $p_{j,1} = (x-k_j)_+^2$. The resulting curve will be smooth, as will be its first derivative, which will be continuous at the knot position.

4.10.2. Alternating Conditional Expectation (ACE)

Nonparametric regression by means of alternating conditional expectations (ACE) was introduced by Breiman and Friedman (1985). The idea is to find nonlinear transformations of the response $\theta(y)$ and predictor variables $\phi(x)$, so that the additive model

$$\theta(y) = \Sigma_{i=1}^{p}\phi_i(x_i) + \varepsilon \quad (4.64)$$

approximates the data satisfactorily (Frank and Lanteri, 1988). If y and x are random variables with a joint distribution F, then the goodness of fit can be expressed as

$$e_{est}^2 = \Sigma_{i=1}^{n}[\theta_{est}(y_i) - \Sigma_{i=1}^{p}\phi_{est,k}(x_{ik})]^2/\Sigma_{i=1}^{n}\theta_{est}^2(y_i) \quad (4.65)$$

where θ_{est} and $\phi_{est,k}$ are the estimated values of θ and ϕ_k and are standardized to have zero means.

The ACE algorithm typically starts with the simple linear transformations $z_i = b_i x_i$ as a first approximation to the optimal nonlinear transformation. This yields the model $y = \Sigma z_i^{(1)} + e$, with the superscript indicating the iteration number. Better fits are developed from this point by systematically treating each predictor variable in turn as a response variable, i.e.

$$z_i^{(1)} = y - z_1^{(1)} - z_2^{(1)} \ldots - z_{i-1}^{(1)} - z_{i+1}^{(1)} \cdots - z_m^{(1)} - e \quad (4.66)$$

The predictor variables z_i are updated through a process of backfitting by fitting a smooth curve through the scatter plot of z_i vs $(y - z_1^{(1)} - z_2^{(1)} \ldots - z_{i-1}^{(1)} - z_{i+1}^{(1)} \cdots - z_m^{(1)} - e)$. This fit does not have to be parametric and can be a simple moving average along the z_i-axis, spanning 10%-20% of the data points, for example. The expectation of z_i is thus calculated for each observation as the local average of $(y - z_1^{(1)} - z_2^{(1)} \ldots - z_{i-1}^{(1)} - z_{i+1}^{(1)} \cdots - z_m^{(1)} - e)$. The updated information is then used to fit z_{i+1} to the other variables. This is by no means the only way in which the expectation functions $f_i(x_i)$ can be estimated. For example, spline regression is used in the multiple optimal regression by alternating least squares (MORALS) algorithm (Young et al., 1976) that is related to the ACE algorithm. Good transformations of ACE typically depend on a large ratio of observations to predictor variables. Although these transformations can be very flexible, the model remains additive and consequently interactions between variables cannot be accommodated.

4.10.3. Additive models based on variance stabilizing transformation (AVAS)

Like ACE, the additive and variance stabilizing (AVAS) approach (Tibshirani, 1988) tries to find transformations

$$\theta(y) = \Sigma_{i=1}^{p}\phi_i(x_i) + \varepsilon \quad (4.67)$$

but differs from ACE in that it chooses $\theta(y)$ to achieve a special variance stabilizing feature. Specifically, θ, ϕ should have properties so that

$$E[\theta(y)|x] = \Sigma_{i=1}^{P}\phi_i(x_i) \tag{4.68}$$

and

$$\mathrm{var}[\theta(y)|\ \Sigma_{i=1}^{P}\phi_i(x_i)] = \mathrm{constant} \tag{4.69}$$

The transformations of the predictor variables are determined by means of backfitting, similar to that in the ACE algorithm, i.e. $\phi_k(x_k) = E[\theta(y) - \Sigma_{i\neq k}^{P}\phi_i(x_i)|x_k]$. AVAS do not suffer from the anomalies of ACE, e.g. not finding good estimates of the parameters of a true additive model, when the normality of ε and the joint normality of $\phi(x)$ fail to hold.

4.10.4. Projection pursuit regression (PPR)

With centred y the response variable and $\mathbf{x} = \{x_1, x_2, \dots x_p\}^T$ the centred predictor variables as before, and $\mathbf{a}_1$, $\mathbf{a}_2$, ... denoting p-dimensional unit or directional vectors, then projection pursuit regression (Friedman and Stuetzle, 1981; Huber, 1985) searches for M direction vectors, so that

$$y = y_{avg} + \Sigma_{m=1}^{M}\beta_m\phi_m(\mathbf{a}_m{}^T\mathbf{x}) \tag{4.70}$$

The $\mathbf{x}$ vectors are projected onto the direction vectors $\mathbf{a}_1$, $\mathbf{a}_2$, ... $\mathbf{a}_M$ to get lengths $\mathbf{a}^T\mathbf{x}$, while good direction vectors $\mathbf{a}$ are pursued. The true model parameters minimize the mean square error

$$S = E[y - \mu y - \Sigma_{m=1}^{M}\beta_m\phi_m(a_m{}^Tx)]^2 \tag{4.71}$$

over all possible β_m, ϕ_m and a_m. Projection pursuit regression subsumes the ACE model, under the restriction that $\theta(y) = y$, which occurs when $M = p$, $\mathbf{a}_1 = (1, 0, \dots 0)^T$, $\mathbf{a}_2 = (0, 1, \dots 0)^T$, ... $\mathbf{a}_p = (0, 0, \dots 1)^T$, and β_m is incorporated into ϕ_m. Likewise, the ordinary linear model is obtained when $M = 1$ and $\beta_1 = (b_1^2 + b_2^2 + \dots + b_p^2)^{½}$, $\phi_1(x) = x$ and $\mathbf{a}^T = (b_1, \dots b_p)/(b_1^2 + b_2^2 + \dots + b_p^2)^{½}$. Unlike the ACE and AVAS models, projection pursuit regression can accommodate interaction between variables.

Projection pursuit regression is closely related to multilayer perceptron neural networks. On this basis, projection pursuit learning for neural networks has been proposed by various authors (Maechler et al., 1990; Zhao and Atkeson, 1991; Hwang et al., 1991, 1992, 1994). The parameters $\mathbf{a}_m$ define the projection of the input vectors onto a set of planes ($m = 1, 2, \dots M$). As is the case with multilayer perceptrons, these projections are transformed by nonlinear activation functions (ϕ_m). In turn, the activation functions are linearly combined to give the output y. However, unlike a multilayer perceptron, the nonlinear transformation or activation functions are neither all the same, nor prescribed a priori, but are fitted to the data. In addition, the parameters in the projection pursuit regression model are optimized cyclically in groups, unlike the simultaneous optimization of the weights of a multilayer perceptron. In short, projection pursuit regression can be seen as a generalization of multilayer perceptrons, i.e. with more flexible activation functions. As such, it is also a universal approximator (Diaconis and Shahshahani, 1984).

4.10.5. Multivariate Adaptive Regression Splines (MARS)

Multivariate adaptive regression splines (MARS) represent an extension to the class of additive models, for which the mapping can be expressed as

$$y = \Sigma_{j=1}^{M} w_j \Pi_{k=1}^{K(j)} \phi_{jk} x_{i(k,j)} + w_j \quad (4.72)$$

i.e. the j'th basis function is represented by a product of one-dimensional spline basis functions ϕ_{jk} (Press et al., 1992), each of which depends on one of the input variables x_i. The basis functions are adaptive, i.e. the number of functions K(j) and the knots for the one-dimensional spline function are all determined from the data.

The method of multivariate adaptive regression splines (Friedman, 1991) is based on the idea of spline-fitting in a multivariate context. In general the fitting of multivariate splines is considerably more difficult than its single-variable counterpart, since the placing of the knots is severely affected by the dimensionality of the data. For example, a problem with 6 variables and 2 knots (or 3 regions) per variable would mean that $3^6 = 729$ subregions would have to be considered in the four-dimensional predictor space. This would require at least as many observations, if each subregion is represented by a single observation only. In this case each region would be fitted by a constant value, i.e. the model would be equivalent to an ANOVA model containing all the higher-order interactions and a model that is discontinuous at the boundaries. If each subregion is fitted with a first-order model as a first approach to a continuous response surface, 6 parameters would have to be estimated per subregion. A conservative 10-20 observations would be required to estimate such a local model, amounting to between 7000-14 000 observations for the overall model. Often such a large number of observations may not be available.

The technique of multivariate adaptive regression splines gets around this problem by fitting a multivariate spline model that is more parsimonious in terms of the number of subregions considered. Even so, MARS can still only be applied when a reasonably large number of observations are available. Instead of considering each subregion explicitly, the model is expressed as a summation of a few basis functions. Each basis function is a piece-wise polynomial representing a certain variable over a certain range. Interaction between variables can also be accommodated over localized regions of the predictor space. With MARS, the important variables are selected and the subregions are represented by optimal location of each knot.

For example, consider a simple two-dimensional response surface that can be decomposed into two separate univariate contributions, as well as a localized bivariate interaction term, as follows

$$y = f_1(x_1-k_1)_+ + f_2(x_1-k_1)_- + f_3(x_2-k_2)_+ + f_4(x_2-k_2)_- + f_5[(x_1-k_1)_+(x_2-k_3)_+] \quad (4.73)$$

The subscript "+" indicates that the argument is non-negative, that is negative arguments are set to zero. Likewise, the subscript "-" indicates that the argument is non-positive, that is positive arguments are set to zero. Knot positions are indicated by k_1, k_2 and k_3. The knot position k_1 belongs to the variable x_1, while the knot positions k_2 and k_3 belong to the variable x_2.

Variables and knot positions are selected systematically as follows. Each of the n values of each of the m predictor variables x_{ij} ($i = 1, 2, \ldots m$; $j = 1, 2, \ldots n$) is considered for the first pair of basis functions. The first knot k_1 splits the first variable x_1 into two segments, viz. observations with values higher than k_1 and observations with values lower than k_1. A simple regression of y on x_1 is done for both segments of the data space and the knot position corresponding with the best fit (least sum of squares) is selected. This is repeated for all the other variables and the variable x_i with knot position $k = x_{ij}$ is used to establish the first two basis functions.

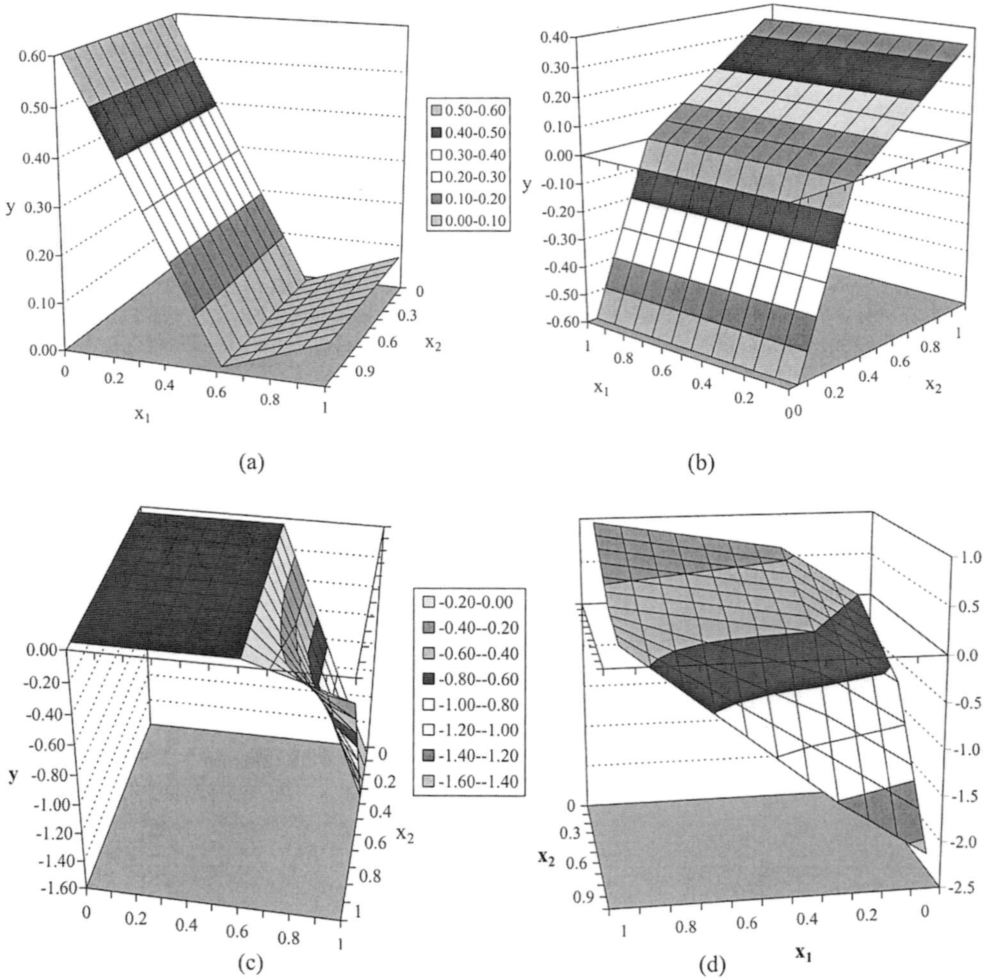

Figure 4.21. Univariate spline models, (a) $f_1 + f_2$ and (b) $f_3 + f_4$, as well as (c) bivariate spline model, f_5 and (d) MARS model as a superposition of the univariate and bivariate splines in (a), (b) and (c).

The process is repeated with the residual values of the response of the first two basis functions to select the next pair of basis functions. Once all subregions have been formed, subregions may be split further and the model may then be overfitted deliberately prior to making use of a backward or pruning procedure to find the optimal model. A typical criterion for optimal model complexity is $(SS/n)(1-kM/n)^2$, where SS is the sum of squares of the model residuals, M is the final number of basis functions and n is the number of observations. k is an empirical factor ($k \approx 4$) that penalises additional basis functions. MARS allows the construction of models in terms of a number of additive terms. An example of a MARS model and its univariate and bivariate components is shown in Figure 4.21.

4.10.6. Classification and regression trees

a) Binary decision trees

As an alternative to fitting nonlinear multivariate mappings, the input space can also be partitioned into regions to which different mappings can be fitted. Many partitioning algorithms are based on hyperplanes parallel to the axes of the input variables. In the simplest case, the output variable is taken to be constant within each region.

The inductive classification model is a form of empirical learning through which general conclusions are inferred from specific examples. In particular, the induction algorithm forms generalized rules linking the features, or independent variables, of a set of exemplars to the predefined classes or dependent variables of each of the exemplars. This decision tree is equivalent to a set of IF-THEN rules, which can be interpreted by the data analyst, or can be incorporated into an expert system shell to serve as a knowledge-based decision support system.

For example, an expert system called GASOIL was employed by British Petroleum in 1986, to design gas-oil separation systems for offshore oil platforms. Gas-oil separation is done at the wellhead by a large complex and expensive separation system, the design of which depends on among other the relative proportions of gas, oil and water, flow rates, pressures, temperatures and fluid rheology. At the time, GASOIL was the largest commercial expert system in the world, containing approximately 2500 rules. Manual design of such a system would have taken approximately 10 man-years to complete. However, by using inductive techniques, applied to a database of existing designs, the system was reportedly developed in a 100 man-days. It is reputed to perform better than human experts, and has apparently saved BP many millions of dollars.

To gain a better understanding of the way in which decision tree algorithms extract rules from data, consider the following system, where 2k + 1 discrete events are possible. Assume that the probability of event i occurring, is $p_i \in [0,1]$. According to basic information theory, the information gained when event i is observed, is

$$I = -\log(p_i) \tag{4.74}$$

Although the base of the logarithm is arbitrary, 2 is often used for convenience. Thus if the probability of an event occurring is very small, and the event actually occurs, this is deemed highly significant, and translates to a large gain in information. Conversely, if the event is highly likely to occur in the first place, then its realization does not translate to a large gain in information.

The total entropy of a system is consequently defined as the total information of the system, i.e.

$$E = -\Sigma_i p_i \log(p_i) \quad (4.75)$$

For example, consider the situation where only two types of events are possible (e.g. two classes, such as is the case with plant operation, which can either be *normal* or *abnormal*).

This is portrayed in Figure 4.22(a), where a random sample within the boundaries of the system (the circle), can yield either class 1 (the solid markers), or class 2 (the empty markers). Since there are as many solid as empty markers (observations), the probability of each occurring can be estimated, i.e. $p_1 = p_2 = 0.5$. The total entropy of the system is therefore $E = -\frac{1}{2}\log(\frac{1}{2})-\frac{1}{2}\log(\frac{1}{2}) = 1$. If this system is split into two new sets, along the boundaries of the two events, as indicated in Figure 4.22(b), the resultant entropy of the split system changes. This entropy is the sum of the entropies of both subsystems, i.e. $E = -1.\log(1)-1.\log(1) = 0$. Entropy can thus be seen as a measure of uncertainty. In the latter case all uncertainty has been removed, since the behaviour (classes) of each subsystem is known perfectly in advance.

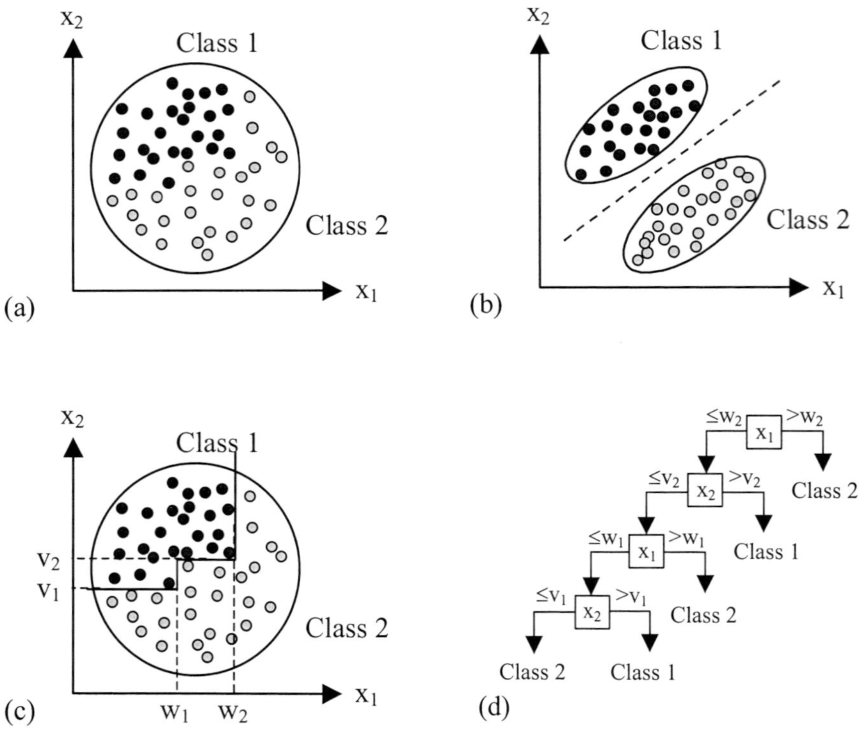

Figure 4.22. Basic principles of entropy-driven rule induction from data, (a) System as a whole, with unit entropy, (b) system with zero entropy, (c) binary splitting of feature space and (d) decision tree resulting from binary splitting of feature space.

By recursively splitting the feature space, so that entropy is reduced maximally at each split (defined by a split variable and split point), as indicated in Figure 4.22(c), binary decision trees can be generated, such as shown in Figure 4.22(d). These trees are equivalent to IF-THEN rules. The tree in Figure 4.22(d) being equivalent to

IF $x_1 > w_2$ THEN Class = 2

IF $x_1 \leq w_2$ AND $x_2 > v_2$ THEN Class = 1

IF $x_1 \leq w_2$ AND $x_2 \leq v_2$ AND $x_1 > w_1$ THEN Class = 2

IF $x_1 \leq w_2$ AND $x_2 \leq v_2$ AND $x_1 \leq w_1$ AND $x_2 > v_1$ THEN Class = 1

IF $x_1 \leq w_2$ AND $x_2 \leq v_2$ AND $x_1 \leq w_1$ AND $x_2 \leq v_1$ THEN Class = 2

Rule induction systems therefore attempt to partition the data space in such a way that the entropy of all the subspaces is progressively reduced (also resulting in reduced entropy for the system as a whole).

As an example, consider the following set of data (S) in Table 4.11 below, from which a decision tree has to be constructed, and the equivalent rules derived. The set S contains two messages (classes), namely "+" and "-". The probability of each message being sampled over S is $p_+ = p_- = 0.50$. The information entropy (average or expected information content) of the training set S can be calculated as

$$E(S) = -\Sigma_{i=+,-} p_i \log_2(p_i)$$

with p_i the probability of class i in the training set S.

Thus, for the data in Table 4.11

$$E(S) = -(3/6)\log_2(3/6) - (3/6)\log_2(3/6) = 1$$

Table 4.11. Example of a set of data (S)

Attributes				Class
A	B	C	D	
0	0	1	0	+
1	0	0	0	+
0	1	1	1	+
1	1	0	1	-
1	1	1	1	-
1	1	1	0	-

The next step is to search for the variable (attribute) and split point that will yield the largest reduction in the entropy of the system. Starting with attribute A, it is clear that only one split point (between 0 and 1) is possible (as is the case with attributes B, C and D). The reduction in entropy upon splitting the data set (S) on attribute A, between 0 and 1, is as follows

$$E(S_{A,j}) = \Sigma_{A,j=0,1} p_{A,j} E(S_{A,j})$$

where pA,i is the weight of the subtree where all the instances have attribute $A = A_j$. $E(S_{A,j})$ is the entropy of the subtree under $A = A_j$. Therefore, for A

$E(S_A) = p_{A=0}E(S_{A=0}) + p_{A=1}E(S_{A=1})$

$E(S_A) = (2/6)[-(2/2)\log_2(2/2)] + (4/6)[-(1/4)\log_2(1/4)-(3/4)\log_2(3/4)]$

$E(S_A) = 0.541$

The entropy gain when splitting is done on attribute A is therefore

$G(A) = E(S) - E(S_A)$

$G(A) = 1 - 0.541$

$G(A) = 0.459$

Similar calculations can be made for the other attributes, so that

$E(S_B) = 0.541$ and $G(B) = 0.459$

$E(S_C) = 1$ and $G(B) = 0$

$E(S_D) = 0.918$ and $G(B) = 0.082$

Initial splitting can therefore take place on either attribute A or B, since both provide the same entropy gain, or can be seen as delivering the largest amount of information in the root node. The split generates two subtrees, each of which is treated as an individual tree and which is recursively subdivided until a terminal node is reached. At the terminal nodes of the tree, the instances all belong to a single class, and the entropy of the terminal node is zero. The fully constructed decision tree shown below in Figure 4.23, can subsequently be converted to the following production rules:

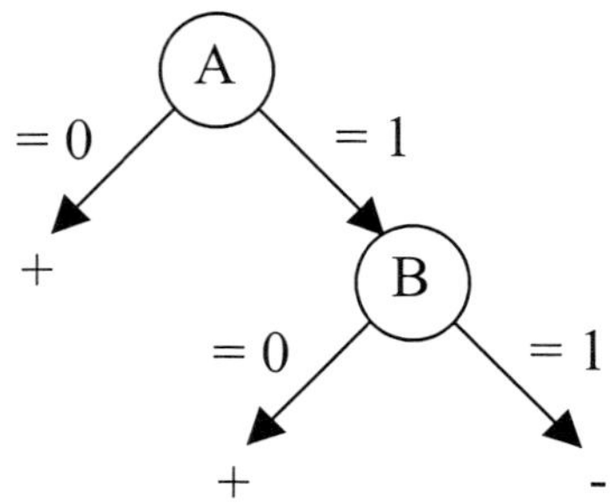

Figure 4.23. Decision tree derived from the data in Table 4.11.

If A = 0, then Class = "+"

If A = 1 and B = 0, then Class = "+"

If A = 1 and B = 1, then Class = "-"

Ideally, the algorithm used in this example deals with discretized attributes. If the attributes have continuous or numerical values, instead of discrete or categorical values, the algorithm will still treat the data as discrete, and search for split points between each numerical value (which it will see as discrete). If all n the instances have different numerical values, then n - 1 split points will have to be considered.

b) Regression trees

Regression trees are constructed in much the same way as binary decision trees, except that the terminal nodes do not represent classes, but continuous values. This is referred to as a zero'th order regression tree. Higher order regression trees contain models (linear or higher order) in their terminal nodes that model the continuous response function in the domain represented by the specific branch of the tree. Trees are referred to as neural regression trees when the response function is represented by a neural network.

Tree-based models such as these provide an alternative to linear and additive models for regression problems and to linear and additive logistic models for classification problems. As described above, tree models are fit by successively splitting the data to form homogeneous subsets of the data. The result is a hierarchical tree of decision rules, which can be used for prediction or classification.

With noisy data, or by using irrelevant attributes, it is possible, or even likely, that the algorithm will build a decision tree that is consistent with all the data in the training data file. This problem of overfitting is very common and can be surmounted by making use of decision tree pruning (Mingers, 1989), which prevents splitting of nodes that are not clearly relevant. For example, when a set of examples is split on an attribute, and the information gain is close to zero, it means that the resulting subsets have roughly the same proportions of each class as the original data set. The information gain is therefore a good indication of irrelevance.

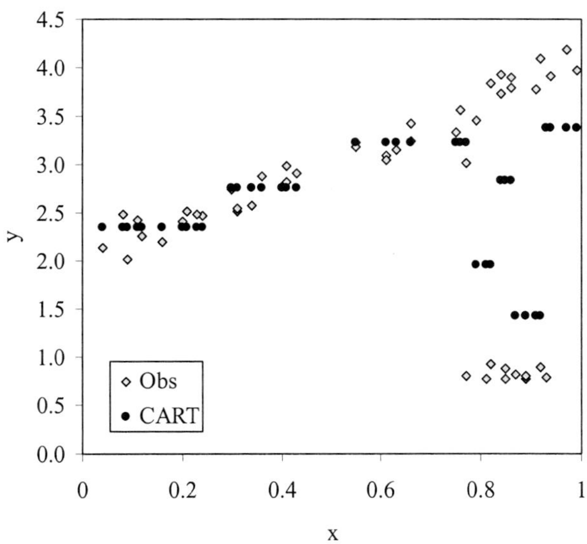

Figure 4.24. Regression tree model (CART) for clustered data (Obs).

The actual size of the gain required to split on an attribute, is determined by classical tests of statistical significance. A significance test begins by assuming that there is no underlying patterns in the data (the null hypothesis) and then analyzes the data to calculate the extent to which the data deviates from a perfect absence of pattern. If the actual deviation is statistically unlikely (usually a probability of 5% or less), it is taken as substantial evidence of the presence of a significant pattern in the data.

More specifically, the null hypothesis states that the particular attribute being considered is irrelevant, and that the information gain of an infinitely large sample would be zero. It is therefore necessary to calculate the probability that, under the null hypothesis, a sample of size ν would exhibit the observed deviation from the expected distribution of positive (belonging to class) and negative (not belonging to class) examples. The deviation can be measured by comparing the actual numbers of positive (p_i) and negative (n_i) examples in each subset, to the expected numbers p_i" and n_i", assuming irrelevance. That is

$$p_i" = p(p_i + n_i)/(p + n), \text{ and } n_i" = n(p_i + n_i)/(p + n) \qquad (4.76)$$

A convenient measure of the total deviation is provided by

$$D = \Sigma_{i=1}^{\nu}\{(p_i-p_i")^2/p_i"+ (n_i-n_i")^2/n_i"\} \qquad (4.77)$$

Under the null hypothesis, D is distributed according to the χ^2 (chi-squared) distribution with ν-1 degrees of freedom. Therefore the probability that the attribute is irrelevant can be calculated with the help of χ^2-tables, or with statistical software. This form of pruning is known as χ^2-pruning.

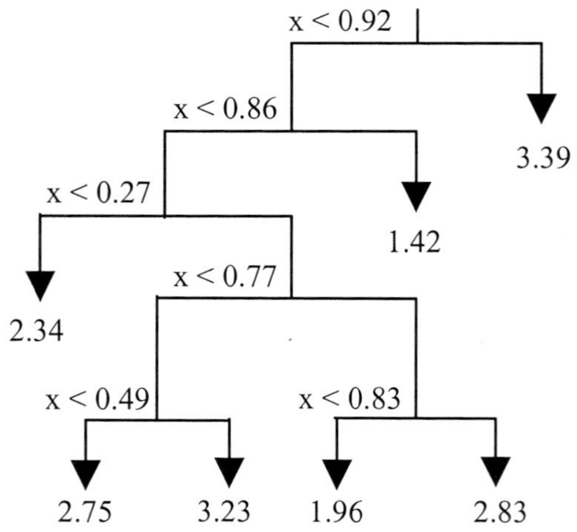

Figure 4.25. Regression tree structure used to predict data in Figure 4.24. Branching takes place towards the left when the branching condition is satisfied.

Another way of countering overfitting is by means of cross-validation, where the performance of the decision tree or equivalent set of rules, is tested against an independent set of data with a distribution similar to the data used to construct the decision tree. This method is often used in conjunction with pruning methods, such as the one mentioned above. Figures 4.24 and 4.25 show a simple regression tree model for the clustered data previously considered in section 4.5.

4.10.7. Genetic programming models

Evolutionary programming is a population-based approach in which solutions (models) are evolved from groups of individuals by use of Darwinian principles of natural selection. The individuals, which are initially distributed randomly across the search space, are rated in terms of fitness values and the fitter a particular individual, the higher the probability of it surviving to the next generation. The survivors are subsequently allowed to exchange genetic material, and their progeny constitute the succeeding generation.

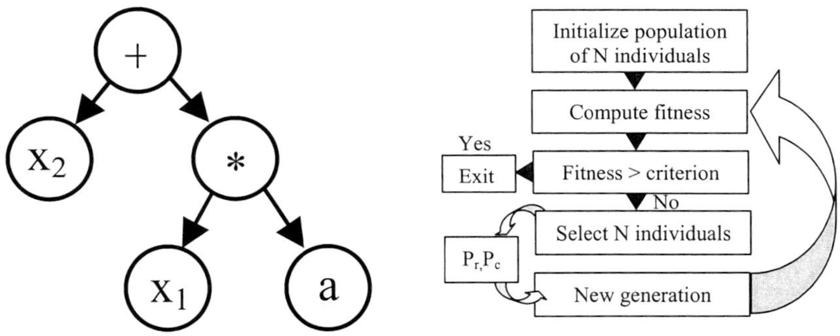

Figure 4.26. Tree-structured representation of the model $y = ax_1 + x_2$, as well as a basic flowchart of a genetic programming algorithm.

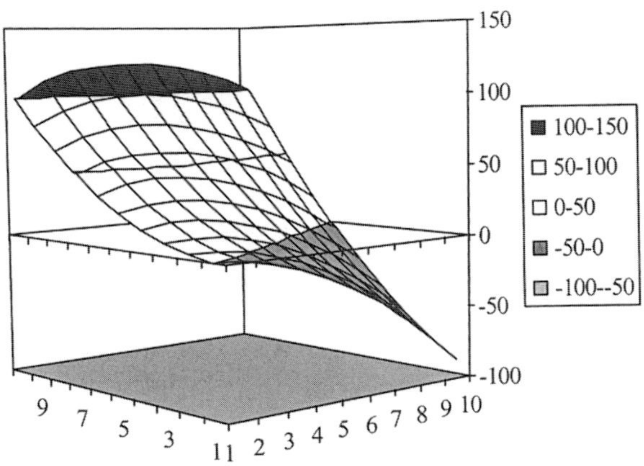

Figure 4.27. Response surface of $f(x_1,x_2) = x_1^2 - x_2^2 + x_1x_2 - x_1/x_2$.

Unlike the chromosomes of genetic algorithms, the individuals in classic evolutionary programming systems typically consist of flexible tree-structures. Each tree is comprised of simple elements or nodes defined by operator and terminal sets. For example, the operator set **O** can be comprised of simple mathematical operators, such as {+, -, *, /}, while the terminal set **T** can be expressed in terms of {**x**, **p**}, where $x = x_1, x_2, \ldots x_m$ constitutes the variable space, and **p** some set of parameters.

An example of such a representation for a linear model of the form $y = ax_1 + x_2$ is shown in Figure 4.18. In this figure, the variable x_1 and the parameter a are combined by a multiplication operator and the result is combined with the variable x_2 with an addition operator. The resultant tree consists of five nodes, viz. two operator nodes **O** = {+, *} and three terminal nodes **T** = $\{x_1, x_2, a\}$. The algorithm is shown diagrammatically in the same figure.

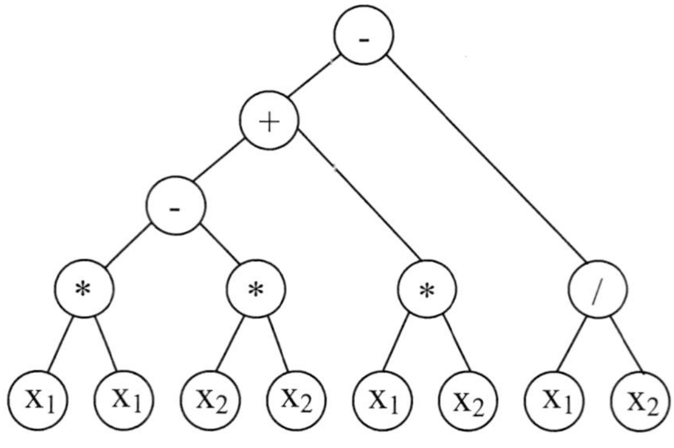

Figure 4.28. Successful models for $f(x_1,x_2) = x_1^2-x_2^2+x_1x_2-x_1/x_2$, developed by genetic programming. The code generated by the algorithm was <x_1 x_1 * x_2 x_1 x_1 - * x_2 x_2 x_1 - * - x_1 x_2 / x_2 x_1 x_1 - * - - +>.

As a simple example, the response surface for a 100 data points $x_i \in [0,100]$, for i = 1,2, $y = f(x_i) = x_1^2-x_2^2+x_1x_2-x_1/x_2$ is shown in Figure 4.27. The terminal set was T = $[x_1, x_2]$ and the operator set O = [+, -, *, /]. The population size was 50, the number of generations 200 and the tree depth restricted to 6. Individuals were selected for crossover via tournaments of size 3. The algorithm could find the solution after 200 generations, 66% of the time, generating tree-structures like the one shown in Figure 4.28.

4.11. CASE STUDY 1: MODELLING OF A SIMPLE BIMODAL FUNCTION

Some of the characteristics of the previously mentioned models can be illustrated by means of simple case studies. For example, consider the set of data generated by $y = 1.1(1-x+2x^2)\exp(-x^2/2)$, to which zero-mean Gaussian with a standard deviation of 0.15 has been added, with $x \in [-2, 5]$. By incrementing x with 0.05 from –2 to 5, a set of 141 data points were generated, to which various models have been fitted for comparison.

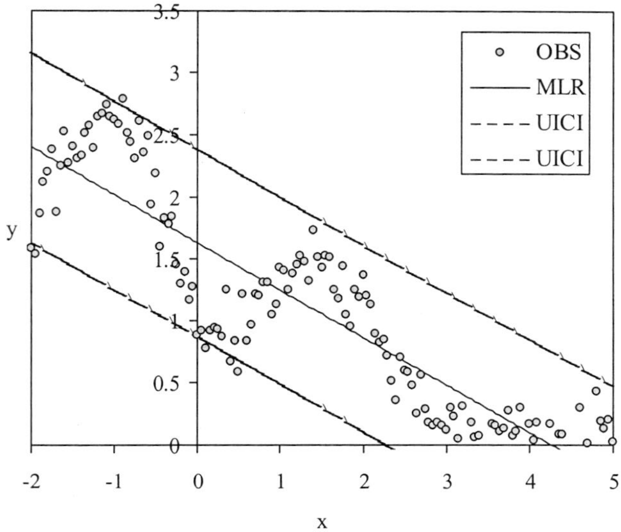

Figure 4.27. Linear regression model, y = 1.633-0.383x (R^2 = 0.811).

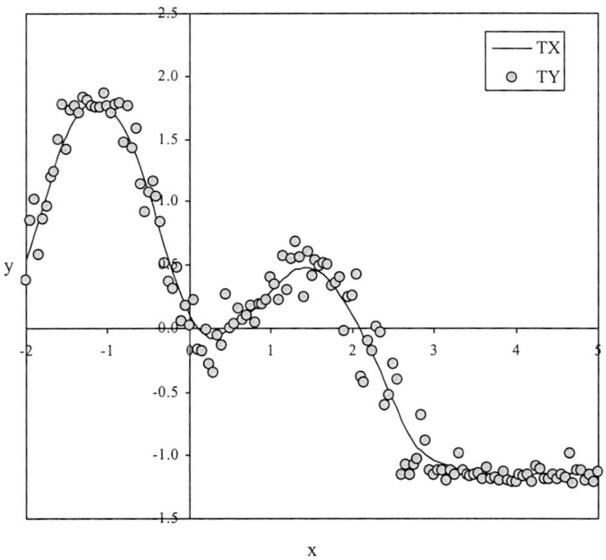

Figure 4.28. Transformed x and y variables (TX and TY) in the ACE model (R^2 = 0.986).

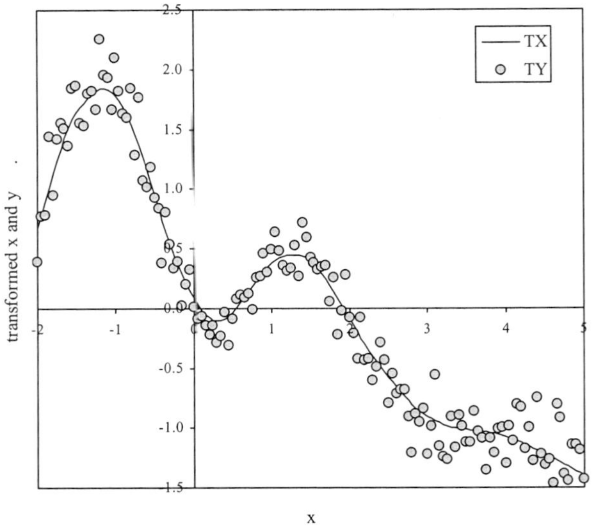

Figure 4.29. Transformed x and y variables (TX and TY) in the AVAS model ($R^2 = 0.975$).

a) Multiple linear regression (MLR)

The fit of the multiple linear regression model is shown in Figure 4.27. Broken lines indicate the lower and upper control limits on individual observations at a 95% confidence level. From the residuals it is clear that although the model accounts for more than 81% of the variance in the data, it does not capture the nonlinear trend. The inadequacy of the model is easy to establish by means of an analysis of the residuals of the model, as the distribution of these residuals deviate markedly from that of a normal distribution.

b) Alternating conditional expectations (ACE) and additive models based on variance stabilizing transformation (AVAS)

As can be expected ACE and AVAS models can explain much more of the variance of the response y. The transformations of the x and y variables for these models are shown in Figures 4.28-4.29 respectively. This can be compared with the best linear regression model in Figure 4.27. The ACE and AVAS models fitted the data with squared correlation coefficients of $R^2 = 0.986$ and $R^2 = 0.975$ respectively.

c) Multilayer perceptron (MLP)

As can be seen from Figure 4.30, a multilayer perceptron with three hyperbolic tangent nodes in the hidden layer and a similar node in the output layer could account for 98.5% of the observed variance in the data. The neural network was trained with a Levenberg-Marquardt algorithm.

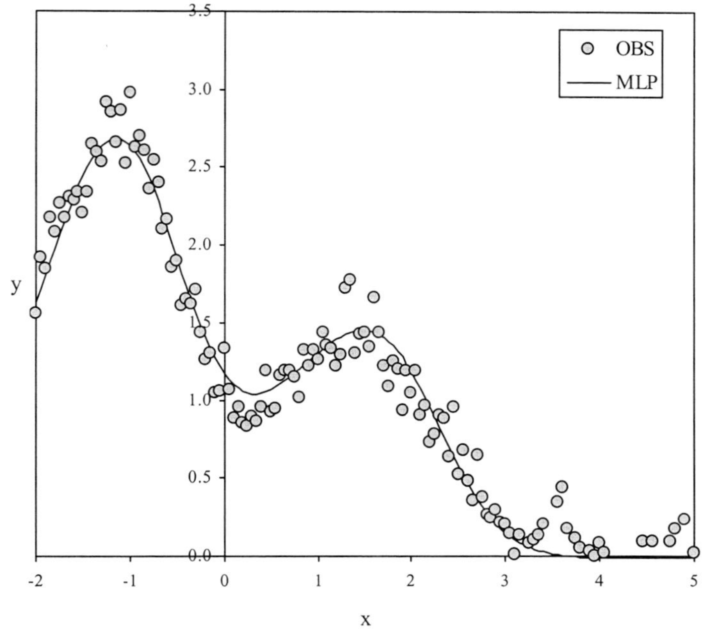

Figure 4.30. Modelling of nonlinear function with a (1-3-1) multilayer perceptron ($R^2 = 0.985$).

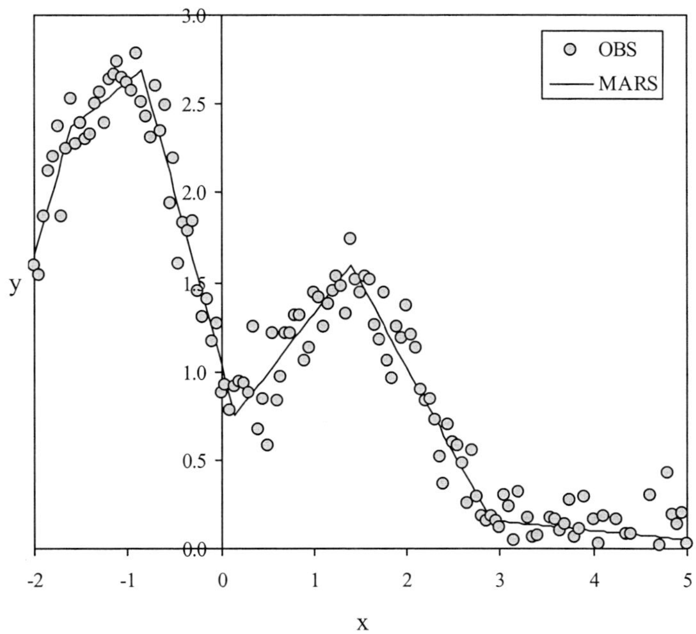

Figure 4.31. Modelling of nonlinear function with a multivariate adaptive regression spline model ($R^2 = 0.970$).

d) Multivariate adaptive regression splines (MARS)

The fit of the multivariate adaptive regression spline model is shown in Figure 4.31. Like the neural network model, the MARS model could also explain 97.0% of the variance of the data, despite the obviously smoother curve fitted by the neural network. The model could be expressed as

$$y = 2.372 + 0.421*\max(0, x+1.6) - 1.780*\max(0, -1.6-x) + 2.608*\max(0, x-0.15) -2.355*\max(0, x+0.85) - 1.633*\max(0, x-1.4) +0.908*\max(0,x-2.9)$$

e) Regression tree (CART)

The data fit and structure of the fitted tree are displayed in Figure 4.32. The regression tree model could explain 98.2% of the variance of the data, which makes it statistically similar to the multilayer perceptron and the multivariate adaptive regression spline models. In rule-based form (rules not simplified), the CART model can be expressed as

IF $x \leq 2.215$ AND $x \leq -0.275$ AND $x \leq -0.575$ AND $x \leq -1.675$ THEN $y = 1.938$
IF $x \leq 2.215$ AND $x \leq -0.275$ AND $x \leq -0.575$ AND $x > -1.675$ THEN $y = 2.495$
IF $x \leq 2.215$ AND $x \leq -0.275$ AND $x > -0.575$ THEN $y = 1.865$
IF $x \leq 2.215$ AND $x > -0.275$ AND $x \leq 0.675$ AND $x \leq -0.025$ THEN $y = 1.319$
IF $x \leq 2.215$ AND $x > -0.275$ AND $x \leq 0.675$ AND $x > -0.025$ THEN $y = 0.9022$
IF $x \leq 2.215$ AND $x > -0.275$ AND $x > 0.675$ THEN $y = 1.320$
IF $x > 2.215$ AND $x \leq 2.725$ THEN $y = 0.6115$
IF $x > 2.725$ THEN $y = 0.6115$

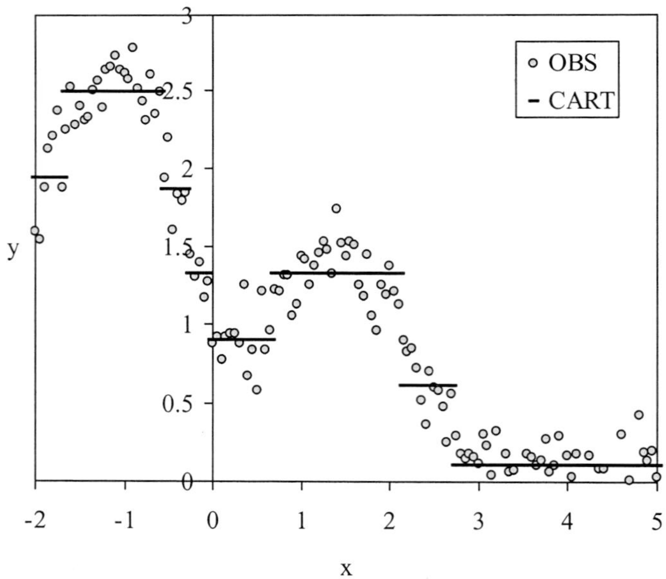

Figure 4.32. Modelling of nonlinear function with a regression tree ($R^2 = 0.982$).

f) Projection pursuit regression (PPR)

The projection pursuit regression model is shown in Figure 4.33. Like the other nonlinear models, the projection pursuit regression model could explain approximately 97.2% of the variance in the response variable. The model shown had a minimum number of 2 terms and smoothing was done. The fraction of observations in the smoothing window (span) was determined by means of cross-validation.

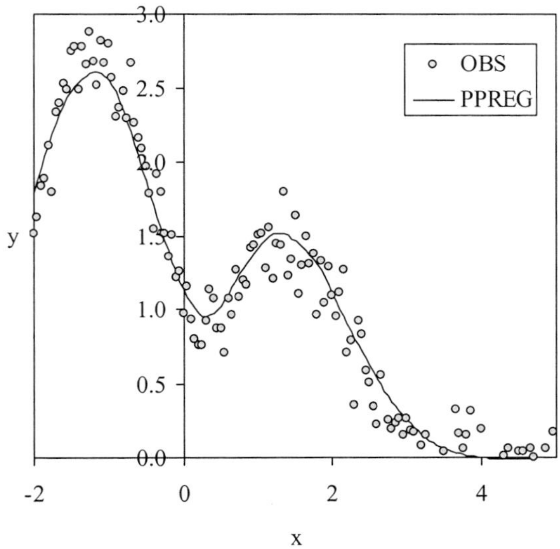

Figure 4.33. Modelling of nonlinear function with a projection pursuit regression ($R^2 = 0.972$).

g) Genetic programming (GP)

A model was fitted to the data by use of genetic programming, as described previously. The terminal set was **T** = [x] and the operator set was **O** = [+, -, *, /, sin, cos, exp]. The fitness function was not based on the errors between actual and predicted values, but rather on the correlation between these values. In each case, the evolved model was scaled to the same range as the observed output and reflected, if the correlation were negative.

After scaling and reflection, the evolved model shown in Figure 4.34 could be expressed as y = 2-sin{exp[sin(x)]}-x-sin{exp[sin(x)]+2x}. The model was evolved by means of 8 populations of 64 individuals each, which could exchange individuals after each generation. Individuals competed for selection by means of 3-way tournaments and reached the solution in less than 20 generations. As can be seen from Figure 4.35, good models were evolved very rapidly and marginal improvements were made at great cost in terms of the complexity of the trees.

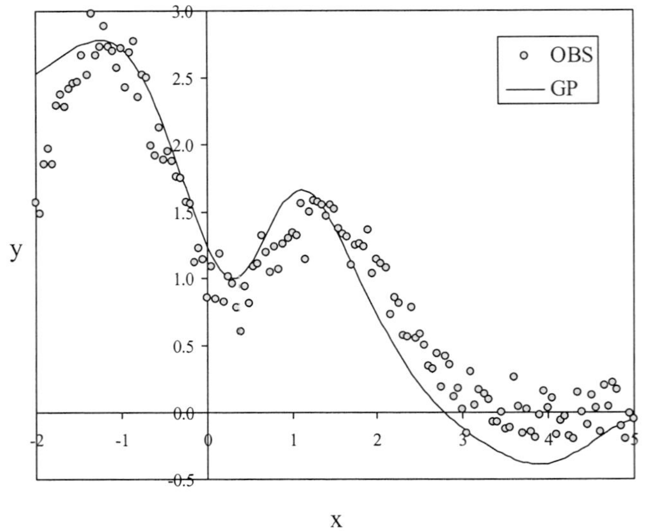

Figure 4.34. Modelling of nonlinear function with genetic programming (R^2 = 0.947). The model generated was y = 2-sin{exp[sin(x)]}-x-sin{exp[sin(x)]+2x}.

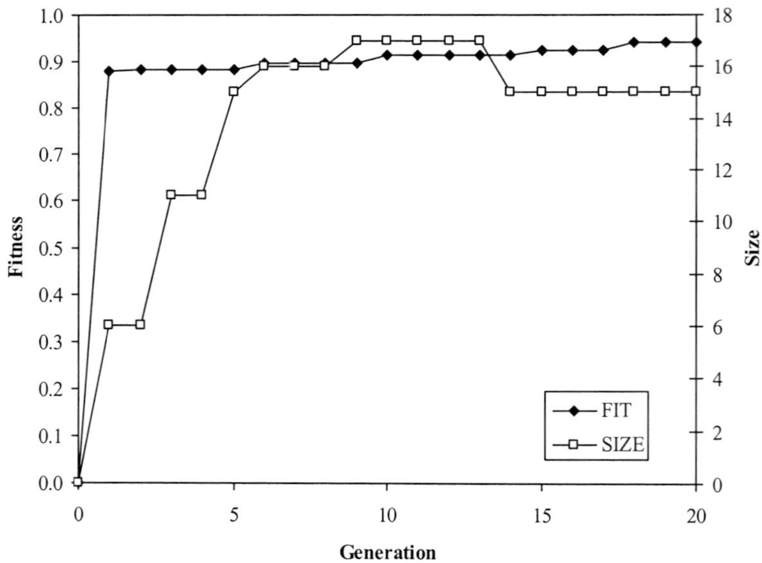

Figure 4.35. Evolution of the genetic programming model over 20 generations. The solid diamonds indicate the fitness of the best individual of each generation, while the empty squares indicated the size of each individual in terms of the total number of nodes in the coded trees.

4.12. NONLINEAR MODELLING OF CONSUMPTION OF AN ADDITIVE IN A GOLD LEACH PLANT

The same case study previously considered with linear regression modeling (section 4.6.2. and 4.9.1) is revisited by means of some of the nonlinear models described above, viz. a general regression neural network (GRNN), regression tree (CART), multivariate adaptive regression splines (MARS), multilayer perceptron (MLP), radial basis function neural network (RBF), as well as genetic programming model (GP). These models were used to represent the relationship between the inputs WESORE, CN, TEMP, AU, RESGRADE, AGITATION, TEMP and EXTRACT to the reagent consumption (SAKRES).

All models were built on a small randomly selected subset of the plant data (44 observations) and tested on an independent test set of data (10 observations). The results of these tests are summarized in Table 4.12. Since the data set is small, the results in this table pertain to the specific training and test sets shown in Figure 4.34.

In this table, MLR denotes multiple linear regression, which had essentially the same form as shown in Table 4.6, since only the EXTRACT and WESORE variables were found to be significant. However, the statistics of the model fit on the test data differ somewhat from that in Table 4.6, since the model was fitted to the training data only in this case. For comparative purposes, the R^2 value of the MLR model is 0.77, as shown in Table 4.12.

The generalized regression neural network (GRNN) consisted of an input layer of 7 nodes, a 44 node layer of Parzen estimators, each of which was positioned on a point in the training data set, a summation layer and an output layer with one node.

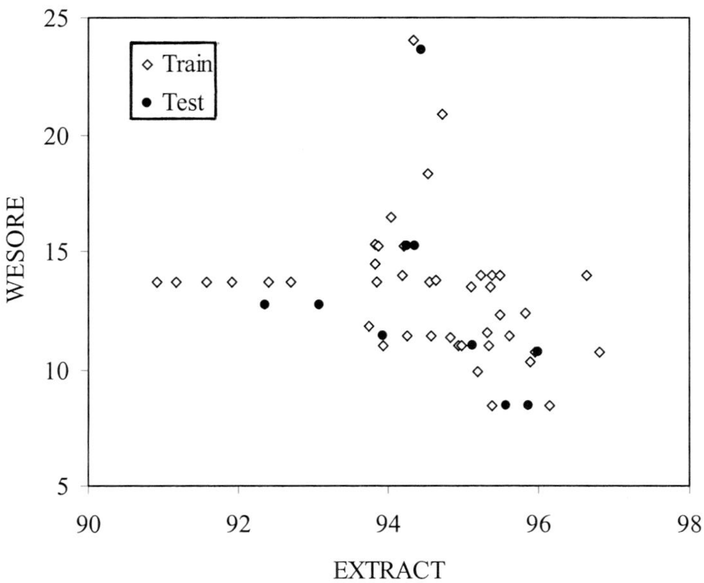

Figure 4.34. Test and training data, as indicated by 10 solid and 44 empty diamonds respectively.

Table 4.12. Comparison of R^2-values of various models on the SAKRES data set. Models: multiple linear regression (MLR), multilayer perceptron (MLP), radial basis function neural network (RBF), generalized regression neural network (GRNN), multivariate adaptive regression splines (MARS), genetic programming (GP) and regression tree (CART).

R^2	0.77	0.94	0.90	0.95	0.92	0.77	0.58	0.78
	MLR	MLP	RBF	GRNN	LS-SVM	MARS	GP	CART

As shown in Figure 4.35, the CART model retained the WESORE and EXTRACT variables only. The model is simple and contains only six terminal nodes. Like CART, the multivariate adaptive regression spline model also contained the WESORE and EXTRACT variables only. The MARS model of the process could be summarized as follows, y = 26.43 + 25.41*max(0,WESORE-13.71) + 63.48*max(0,WESORE-10.99)-78.18*max(0,WESORE-11.41) + 28.11*max(0,EXTRACT-93.82).

The multilayer perceptron consisted of an input layer of 7 nodes, a hidden layer with three bipolar sigmoidal nodes and a single node output layer, also with a hyperbolic tangent activation function. The network was trained by means of the generalized delta rule and could explain 94% of the variance in the data. This and the other neural networks were significantly better than the linear model, since the latter could only explain approximately 77% of the variance of the data, while the neural networks could explain more than 90% of the variance under similar conditions.

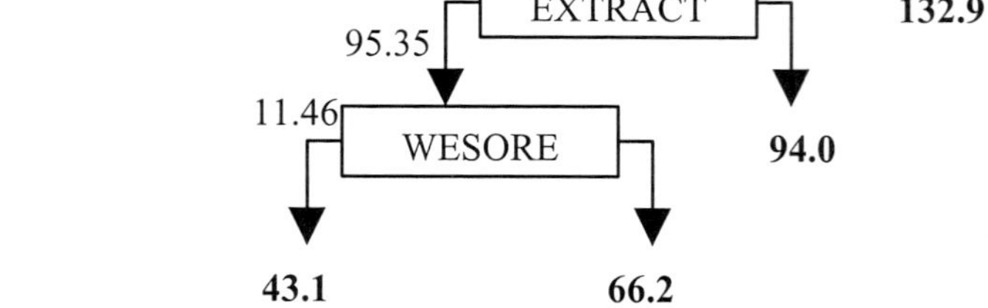

Figure 4.35. CART model of reagent consumption.

The radial basis function neural network consisted of an input layer of 7 nodes, a hidden layer with 44 spherical Gaussian nodes and a single linear output node. The network was trained in two stages. First the hidden layer was optimized by positioning the nodes by use of a k-means clustering algorithm, after which the spreads of the Gaussians were optimized by means of a nearest neighbour heuristic, as described in Chapter 1 (equations 1.70-1.72).

With its emphasis on primal-dual interpretations from convex optimization theory, the least squares support vector machine (LS-SVM) (Suykens et al. 1996; Suykens, 2001) is closely related to Vapnik's standard support vector machine formation, as discussed in Chapter 1 (Vapnik, 1998, Suykens and Vandewalle, 1998).

In feature space, the least squares support vector machine has the following representation:

$$\mathbf{y}(\mathbf{x}) = \mathbf{w}^T\varphi(\mathbf{x}) + \mathbf{b} \tag{4.78}$$

where $\mathbf{x} \in \Re^{nxm}$, $\mathbf{y} \in \Re^{nxp}$. In general, for the nonlinear mapping $\varphi(\cdot)$: $\Re^n \rightarrow \Re^L$, the dimension L can have an infinite dimensionality, which is an important difference with approaches depending on models such as multilayer perceptrons. Given a training set $\{x_k, y_k\}_{k=1}^n$ one considers the optimization problem

$$\min \Im(\mathbf{w},\mathbf{e}) = \tfrac{1}{2}\mathbf{w}^T\mathbf{w} + \tfrac{1}{2}\gamma\Sigma_{k=1}^n e_k^2 \tag{4.79}$$

subject to the equality constraints

$$\mathbf{y}_k(\mathbf{x}) = \mathbf{w}^T\varphi(\mathbf{x}_k) + \mathbf{b} + \mathbf{e}_k, \text{ for all k} \tag{4.80}$$

The cost function with regularization corresponds to a form of ridge regression. The Lagrangian has the form

$$L(\mathbf{w}, \mathbf{b}, \mathbf{e}, \alpha) = \Im(\mathbf{w},\mathbf{e}) - \Sigma_{k=1}^n\alpha_k\{\mathbf{w}^T\varphi(\mathbf{x}_k) + b + e_k - y_k\} \tag{4.81}$$

where α_k are the Lagrangian multipliers or support values. Application of the Mercer conditions gives $\varphi(\mathbf{x}_k)^T\varphi(\mathbf{x}_j) = \mathbf{K}(\mathbf{x}_k,\mathbf{x}_j)$ for k, j = 1, 2, 3, ... n, so that the resulting model for function estimation becomes

$$\mathbf{y}(\mathbf{x}) = \Sigma_{k=1}^n\alpha_k\mathbf{K}(\mathbf{x}_k,\mathbf{x}_j) + \mathbf{b} \tag{4.82}$$

where α_k and $\mathbf{b}$ are the solution to the linear system. In this case Gaussian radial basis function kernels of the form $K(x,x_k) = \exp(-||x-x_k||/\sigma^2)$ was used. The kernel parameters (basis function widths) were optimized from the data via crossvalidation, in conjunction with the γ-values (equation 4.79). The optimal widths for the kernels were found to be between 0.5 and 4. The least squares support vector machine was able to explain 91.8% of the variance of the validation data set.

It was possible to assess the relative importance of the variables in each neural network model, by dithering (approximately 5%) the inputs of a given variable and observing the changes in the output. In each case, the variables EXTRACT and WESORE were found to be the most important. For example, a sensitivity analysis with the radial basis function neural network indicated that if the importance of EXTRACT in explaining model output was 100%, then that of WESORE was 49.0%. This corresponded well with the CART analysis, where EXTRACT was rated as 100% and WESORE as 55.2%, but differed somewhat from the MARS model, which rated both WESORE and EXTRACT as equally important.

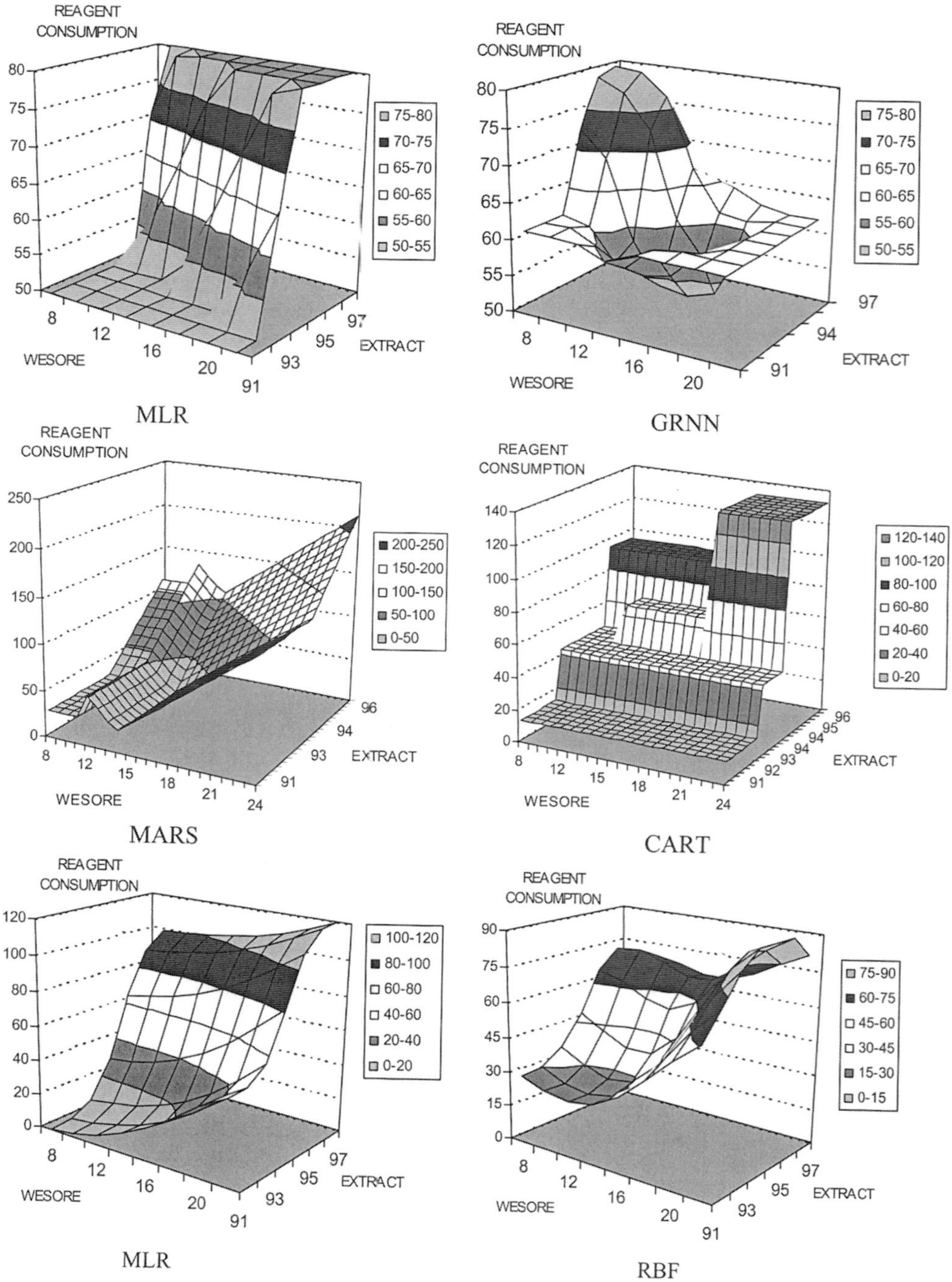

Figure 4.36. Response surfaces of reagent consumption generated with a multiple linear regression model (MLR), generalized regression neural network (GRNN), multivariate adaptive splines (MARS), regression tree (CART), multilayer perceptron (MLP) and radial basis function neural network (RBF).

The genetic programming model was derived with a terminal set consisting of the seven input variables and a basic operator set **O** = {+, -, *, /}. The terminal and operator sets **T** = {**x**} and **O** = {+, -, *, /} were used to generate models relating the input variables $\mathbf{x} \in \Re^m$ to the projected variables $y \in \Re^p$. No constants were used in the terminal sets. Tournament selection in which two or more individuals were randomly selected to compete against each other was used to populate successive generations. This resulted in linear models, which were less effective than the MLR model. The addition of nonlinear operators, such as {exp, log} did not appear to make a material difference.

The response surfaces of some of these models are shown in Figure 4.36. As can be seen from this figure, all the response surfaces have a similar appearance. By its nature, the response surface of the CART model is not smooth, while the interaction term of the WESORE variable can be seen as a ridge in the response surface of the MARS model. With such a small data sets (only 10 points), the statistical differences between the models are negligible, despite the nominal differences in the R^2 values summarized in Table 4.12.

Chapter 5
Topographical Mappings with Neural Networks

5.1. BACKGROUND

Since the emphasis is on the search for interesting structure[34] in the data, graphical display or visualization of these structures is a very important part of exploratory data analysis, as indicated in schematically in Figure 5.1. Principal component analysis is without a doubt the workhorse of the traditional multivariate methods and in the last decade this fast linear technique has been extended in multiple ways to deal with the inevitable non-linearities arising from complex data sets. As discussed previously, neural networks have been especially important in this regard. As indicated in Figure 5.2, topographic projections constitute one of the major classes of techniques that are especially important in data visualization techniques, as well as various forms of feature extraction. Like latent variable projections (Chapter 3), it is an indirect technique, in that the variables are not represented directly by some graphical equivalent, such as is the case with iconographic projections (Pickett and Grinstein, 1988), e.g. Chernoff faces (Chernoff, 1973), Andrews' curves (Andrews, 1972), glyphs and metroglyphs (Anderson, 1954, 1957), hyperboxes (Alpern and Carter, 1991), hierarchical axes (Mihalisin et al., 1991), parallel coordinates (Inselberg and Dimsdale, 1990), etc. These latter methods are not particularly suitable for automated systems or formal decision-making. Wong and Bergeron (1997) gives an overview of these and other techniques developed over the last three decades.

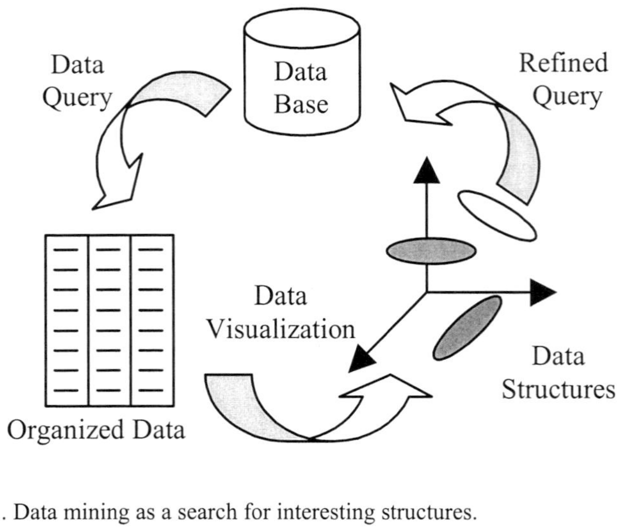

Figure 5.1. Data mining as a search for interesting structures.

Topographic maps are biologically inspired representations of data that combine the characteristics of continuous functions with certain aspects of vector quantization. Individual nodes in topographic or neural maps typically have receptive fields that make them responsive to

[34] In practice this often means the detection of clusters in the data.

subsets of the data. Moreover, these nodes are arranged so that neighbouring nodes respond to subsets of the data with similar distributions or attributes. In the last couple of decades topographic maps have proved to be a popular approach to the exploration of data.

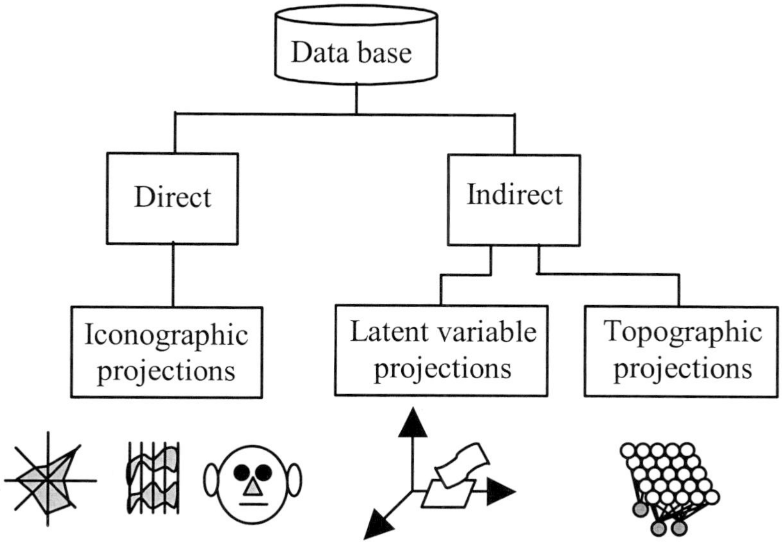

Figure 5.2. Classification of data visualization techniques.

The best-known example is Kohonen's self-organizing feature map (Kohonen, 1982, 1984, 1990) that has been applied to numerous technical problems in various contexts (Kohonen, 1995). Although most people have an intuitive understanding of the topographic characteristics of these maps, it is not easy to define topography rigorously (Bauer et al., 1999). Informally defined, the topography of a map relates to the closeness with which similar data points are located on the map. Topography is thus synonymous with the notion of similarity and the preservation of topology. The preservation of topology between points in the original space and the map space can be characterized by the *continuity* of the mapping, a *ranking* of the distances between pairs of points in the input and the map space, or a direct evaluation of the metric distances between pairs of points in the input and the map space. Some of these concepts are illustrated in Figure 5.3.

These considerations have led to a number of measures by which the preservation of topology can be quantified. Good measures should comply with the following criteria:

- *Uniqueness*: The measure of topology should have a unique, indicative value when applied to a perfect topologically preserving map.
- *Systematic ordering*: The measure should change systematically with systematic changes in the topology of the map, such as when input data from an increasingly higher-dimensional input space are mapped to a fixed coordinate system.
- *Invariance*: The topology measure should not be influenced by scaling of the data or symmetric transformations of the data.

- *Reproducibility*: Quantitatively similar results should be obtained when maps are produced by different instantiations of identical algorithms.

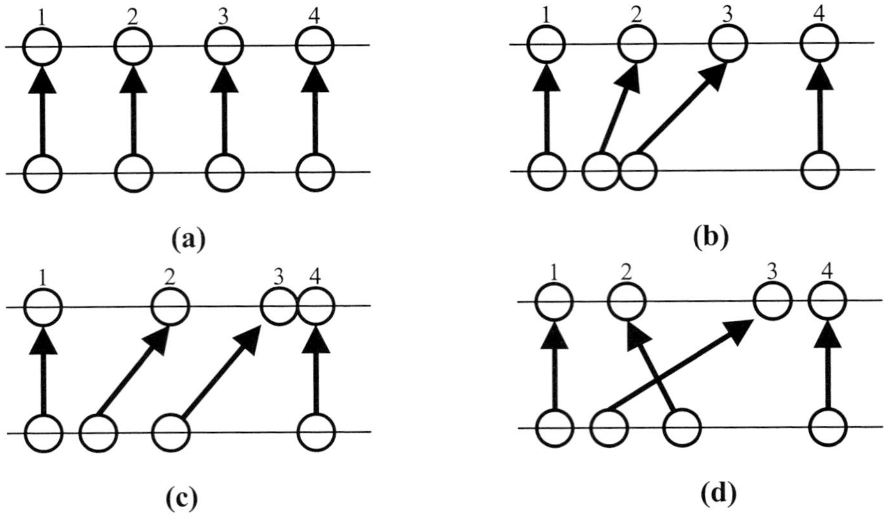

Figure 5.3. Topological, ranking and metric conditions of topographic maps. In (a) a perfect topographic map is shown. In (b) the topological order of the map is preserved, but not the ranking of the neighbours, or the metric distances between corresponding points in the input and output spaces. In (c) the topological order and ranking of neighbours are conserved, but not the metric distances between points in the input and output spaces. In (d) the topological order is violated on a local scale.

5.2. OBJECTIVE FUNCTIONS FOR TOPOGRAPHIC MAPS

Objective functions for the construction of topographic maps can influence the character of the maps significantly. Define the similarities in the input space $\aleph$, as F(i,j) and in the output space $\Im$ as G(p,q), where i and j are points in the input space and p and q are points in the output space, as indicated in Figure 5.4. Assuming that there are n points to be mapped, and that $\wp$ is a one-to-one mapping of points from the input space to the output space, the following metric objective functions can be defined:

- Metric Multidimensional Scaling (Torgerson, 1952)

  Minimize $\Sigma^n_{i=1}\Sigma_{j<i}[F(i,j)-G(\wp(i),\wp(j))]^2$ (5.1)

- Sammon Measure (Sammon, 1969)

  Minimize $[\Sigma^n_{i=1}\Sigma_{j<i}F(i,j)]^{-1}\Sigma^n_{i=1}\Sigma_{j<i}[F(i,j)-G(\wp(i),\wp(j))]^2/F(i,j)$ (5.2)

- Spearman Coefficient (Bezdek and Pal, 1995)

  Maximize $\rho = \Sigma^n_{i=1}(R_i-R_{avg})(S_i-S_{avg})/[(\Sigma^n_{i=1}(R_i-R_{avg}))^{½}(\Sigma^n_{i=1}(S_i-S_{avg}))^{½}]$ (5.3)

These objective functions assume that both F and G are Euclidean distances in the two spaces, with a unity distance between neighbouring points in each space. In the case of the Spearman coefficient ρ (equation 5.3), the linear correlation between the ranks in the input and output spaces is calculated. R_i and S_i are the corresponding rankings in the ordered lists of the F(i,j) and G(p,q) values respectively, R_{avg} and S_{avg} are the corresponding average rankings in the ordered list of the F(i,j) and G(p,q) values respectively. As an example, Figures 5.5 and 5.6 repectively show eight normally distributed clusters (with common standard deviation = 0.02) centred on the vertices of a cube, i.e. (000, 100, 010, 001, 110, 101, 011, 111) projected to two dimensions by means of metric multidimensional scaling and the Sammon criterion. The light triangular markers indicate the clusters on the roof (R) of the cube, while the dark square markers indicate the clusters on the floor (F) of the cube. Intuitively, the Sammon map is better able to capture the structure of the data than the multidimensionally scaled map.

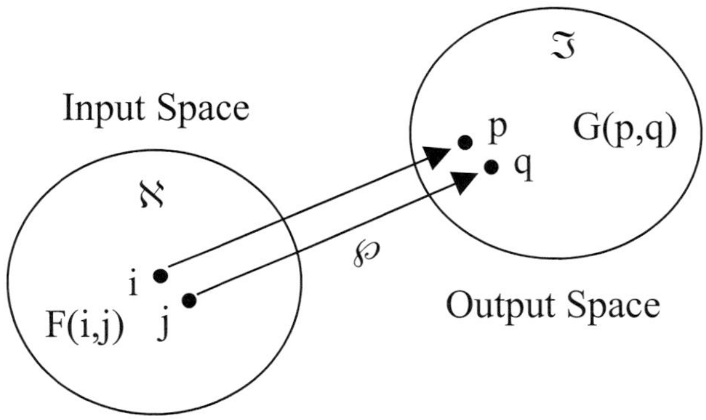

Figure 5.4. Topographic mapping of points from an input space ℵ to an output space ℑ.

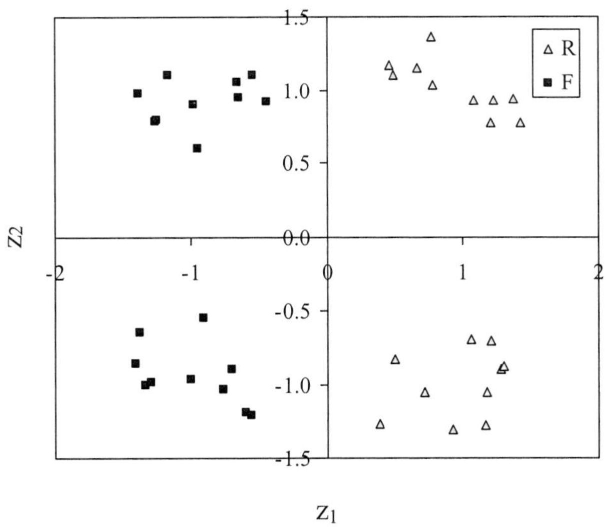

Figure 5.5. Multidimensional scaling map of eight clusters on the vertices of a cube, with R and F indicating points derived from the roof and the floor of the cube respectively.

In so-called C measures, the similarities are expressed as nonlinear functions of the Euclidean distance. That is

$$C = \Sigma^{n}_{i=1}\Sigma_{j<i}F(i,j)G(\wp(i), \wp(j)) \qquad (5.4)$$

Different choices of similarity functions yield different criteria, viz.

- Minimal path length (Durbin and Mitchison, 1990): F(i,j) = Euclidean distance, G(p,q) = 1, if p and q are neighbours on the line, otherwise G(p,q) = 0
- Minimal wiring (Durbin and Mitchison, 1990): G(p,q) = Euclidean distance, F(i,j) = 1, if i and j are neighbours on the square, otherwise F(i,j) = 0
- Minimal distortion (Luttrell, 1990): F(i,j) = squared Euclidean distance, G(p,q) = $\exp[-d^2/\sigma^2]$, where d is the Euclidean distance between p and q, and σ is the length scale in the output space over which nearby output points should represent similar input points. This can be seen as similar to the minimal path length measure, but with a broader neighbourhood function in the output space.
- Inverted Minimal distortion (Mitchison, 1995): G(p,q) = squared Euclidean distance, F(i,j) = $\exp[-d^2/\sigma^2]$, where d is the Euclidean distance between i and j, and σ is the equivalent length scale in the input space over which nearby output points should represent similar input points. This can be seen as similar to the minimal wiring measure, but with a broader neighbourhood function in the input space.

C measures differ from the Spearman coefficient for example, in that the Spearman coefficient is bounded to the range [-1,1] and measures discrepancies in terms of rank order violations, rather than similarities.

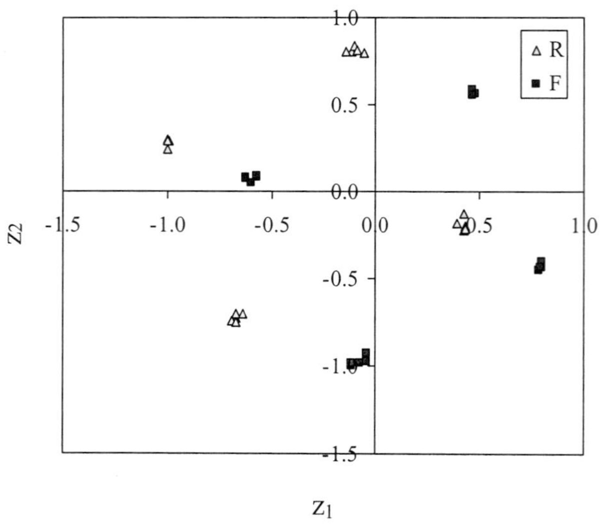

Figure 5.6. Sammon map of eight clusters on the vertices of a cube, with R and F indicating points derived from the roof and the floor of the cube respectively.

5.3. MULTIDIMENSIONAL SCALING

In contrast with other methods of data visualization, where the data are projected from high-dimensional Euclidean data spaces to low-dimensional spaces, multidimensional scaling (MDS) algorithms are designed to visualize proximity[35] or dissimilarity relations between objects by interpoint distances in low-dimensional Euclidean spaces.

Mathematically, the dissimilarity between two objects i and j is defined as a real number δ_{ij}. MDS algorithms determine a spatial representation of the objects, that is each object i is represented by coordinates $x_i \in \Re^p$ in p-dimensional space. The entire embedding configuration of a set of n points can be denoted as $\wp = \{\mathbf{x}_1, \mathbf{x}_2, \dots \mathbf{x}_n\}$. The distance between two points is usually expressed as the Euclidean distance, i.e. $d_{ij} = d(\mathbf{x}_i,\mathbf{x}_j) = ||\mathbf{x}_i\text{-}\mathbf{x}_j||$. In many instances, the raw dissimilarity data are unsuitable for embedding in a Euclidean space, in which case additional processing is required. This is done by transforming the dissimilarities into disparities by means of a monotonic nonlinear transformation $D(\delta_{ij})$. Ideally, the disparities should closely match the distances between embedded points, i.e. $d_{ij} \approx D(\delta_{ij})$.

5.3.1. Metric scaling

Assuming that the transformed dissimilarities match the distances in metric space sufficiently well, and $D_{ij} = D(\delta_{ij})$, then MDS can be formulated as an optimization problem with the cost function

$$\wp_{MDS}(\{x_i\}) = \Sigma_{i=1}{}^{n}\Sigma_{k=1}{}^{n}w_{ik}(||x_i\text{-}x_k||^2 - D_{ij})^2 \qquad (5.5)$$

with squared Euclidean distances used for computational convenience. MDS methods that minimize objective functions of this type are referred to as least squares scaling and belong to the class of metric multidimensional scaling algorithms.[36] The weighting factors w_{ik} are used to weight the disparities individually and can be used to discount disparities with a high degree of uncertainty or to control the scale of the stress function, SSTRESS (equation 5.5). For example, it may be appropriate to use local, intermediate or global normalization

$$w_{ik}{}^{local} = [n(n\text{-}1)D_{ik}{}^2]^{-1}$$

$$w_{ik}{}^{intermed} = [D_{ik}\Sigma_{l,m=1}{}^{n}D_{lm}{}^2]^{-1}$$

$$w_{ik}{}^{global} = [\Sigma_{i,k=1}{}^{n}D_{ik}{}^2]^{-1} \qquad (5.6)$$

5.3.2. Nonmetric scaling and ALSCAL

It is often not possible to construct an explicit function $D(\cdot)$, so that the mapped dissimilarities D_{ij} of a data set match the metric distances adequately. In such a case, the space of the possible transformations $D(\cdot)$ has to be enlarged, incorporating the monotonicity constraint only, so that $\delta_{ij} < \delta_{ik} \Rightarrow D(\delta_{ij}) \leq D(\delta_{ik})$. These order preserving, but otherwise unconstrained

---

[35] Proximity data represent object information by pair-wise dissimilarity values, instead of feature vectors, such a derived by projecting the data from high- to low-dimensional spaces.

[36] The *metric* characterizes the type of transformation $D(\cdot)$ of the dissimilarities and does not refer to the properties of the embedding space.

transformations of the dissimilarities are known as nonmetric MDS algorithms (Shepard, 1962; Kruskal, 1964).

The alternating least squares scaling (ALSCAL) algorithm is widely used to perform multidimensional scaling (Takane and Young, 1977). It minimizes the SSTRESS cost function (equation 5.5) with a weighted Euclidean distance by alternating two phases until convergence. In the model estimation phase, an embedding $\wp$ is derived from the disparities D_{ik}, while the optimal scaling phase transforms the disparities D_{ik} to minimize the SSTRESS for the current configuration of the embedding $\wp$.

5.4. SAMMON PROJECTIONS

As mentioned previously, successful exploratory analysis of process data often depends on the extraction and visualization of compact representative features of the data. With Sammon maps, the objective function is defined as

$$\text{Minimize } S = [\Sigma^{n}_{i=1}\Sigma_{j<i}F(i,j)]^{-1}\Sigma^{n}_{i=1}\Sigma_{j<i}[F(i,j)-G(\wp(i),\wp(j))]^{2}/F(i,j)$$

or

$$\text{Minimize } S = [\Sigma^{n}_{i=1}\Sigma_{j<i}{}^{q}d_{i,j}]^{-1}\Sigma^{n}_{i=1}\Sigma_{j<i}{}^{q}[d_{i,j}-d^{*}_{i,j}]^{2}/d_{i,j} \quad (5.7)$$

where $d_{i,j}$ and $d^{*}_{i,j}$ define the Euclidean distances in the input and output spaces respectively, for a specific pair of observations i and j drawn from a sample of n observations.

Minimization of S (equation 5.7) is an unconstrained optimization problem in the nq variables y_{ij}, with $i = 1, 2, \ldots n$ and $j = 1, 2, \ldots q$. Sammon (1969) used the method of steepest descent for the approximate minimization of S, that is, if $y_i(t)$ is the estimate of y_i at the t'th iteration, then $y_i(t+1)$ is given by

$$y_{ij}(t+1) = y_{ij}(t) - \alpha[A/B] \quad (5.8)$$

where

$$A = \partial S(t)/\partial y_{ij}(t) \text{ and } B = \partial^{2}S(t)/\partial y_{ij}(t)^{2} \quad (5.9)$$

with α the step size for the gradient search, i.e. a nonnegative scalar constant with a recommended value of between 0.3 and 0.4.

With the approach originally formulated by Sammon, as well as other optimization strategies, such as proposed by Chen et al. (1999), it is not possible to obtain an explicit mapping function $\wp$ relating patterns in the input space $\aleph$ with patterns in the output space $\Im$. This means that if new points are to be projected, the optimization procedure has to be repeated. This is a major disadvantage, given that the optimization is computationally expensive, as every step within an iteration nominally requires the calculation of n(n-1)/2 distances. In addition, the error surface is riddled with local minima, and the algorithm is likely to get stuck in one of these.

Various approaches based on cluster analysis (Chang and Lee, 1973; Schachter, 1978; Pykett, 1978) have been proposed to alleviate the computational burden associated with the optimization, but these are only partially successful. More recently, methods based on the use of neural networks (Tattersall and Limb, 1994; Mao and Jain, 1995; De Ridder and Duin, 1997; Lerner et al., 1999; Pal and Eluri, 1998; Lowe and Tipping, 1996) to model the mapping function have removed the need for reoptimization prior to the mapping of new data. These models typically relate the original high-dimensional set of variables to a set of lower-dimensional features. Since no explicit set of exemplars is available to train the neural network, the network has to be trained via so-called hidden target mapping, as explained below.

Development of the neural network models representative of the mapping function $\wp$ entails initialization of the structure and weights of the multilayer perceptron neural network, (random) selection and mapping of a pair of observations (i and j), calculation of the mapping criterion (S) and using the mapping criterion as a basis to update the weight matrix of the neural network, similar to the error criterion that would otherwise have been used in supervised learning. As usual, the dimensionality of the input layer is determined by the number of variables to be mapped, the dimensionality of the output layer is determined by the dimensionality of the projection (smaller than the number of variables to be mapped), and the hidden layer is determined iteratively, depending on the performance of the neural network.

Although calculation of the error criterion (Sammon stress) is computationally intensive, and may lead to excessive training times with large data sets, it is seldom necessary to consider all $n(n-1)/2$ interpoint distances at each iteration. Levine and Kreifeldt (1989, 1994) have considered the unique representation of point patterns in two dimensions with minimal information and based on their work, Tsogo et al. (2001) have shown that only $m(n-2)+1$ distances or similarities are necessary to represent n points uniquely in m-dimensional space. As indicated in Table 5.1, only a small percentage of the interpoint distances therefore needs to be considered in large data sets.

The following case studies illustrate the use of multilayer perceptron type neural networks to visualize data. These networks are compared with a genetic programming algorithm (Chemaly and Aldrich, 2001), briefly outlined below.

Table 5.1 Number of interpoint distances required to uniquely represent data sets with m variables and n observations.

n	100				1000				10000			
m	2	5	10	50	2	5	10	50	2	5	10	50
m(n-2)+1	197	491	981	4901	1997	4991	9981	49901	19997	49991	99981	5E+05
n(n-1)/2	4950	4950	4950	4950	5E+07	5E+07	5E+07	5E+07	5E+07	5E+07	5E+07	5E+07
%	3.98	9.92	19.82	99.01	0.4	1	2	9.99	0.04	0.1	0.2	1

5.5. EXAMPLE 1: ARTIFICIALLY GENERATED AND BENCHMARK DATA SETS

For illustrative purposes, the following simple data sets are considered. Data set BITET consists of an asymmetrically arranged set of four 3-dimensional clusters (A, B, C and D). The clusters are arranged along the vertices of two tetrahedra joined at their bases and with apices pointing in opposite directions. Clusters A, B and C (50 points each) are roughly spherical and arranged along the vertices of the common basis of this bitetrahedron. Cluster D (100 points) is an elongated ellipsoidal cluster that joins the fourth and fifth vertices (apices) of the bitetrahedron.

Mao and Jain (1995) and Pal and Eluri (1998) have previously investigated the SPIRAL data set. It is a data set describing two spirals (500 data points each) in 3D-space, so that $x_1 = ½\cos\theta + \varepsilon$, $x_2 = ½\sin\theta + \varepsilon$ and $x_3 = \sin 2\theta + \cos 2\theta + \varepsilon$, with $\theta \in (-\pi/2, \pi/2)$, and ε a randomly generated noise factor with a normal distribution, $\varepsilon \in (0, 0.25)$.

The third data set SPHERSHELL (Pal and Eluri, 1998) is likewise comprised of three coordinates, and describes a hemisphere with radius 0.6 enclosed in a shell with an inner radius of 2 and outer radius of 2.13. The hemisphere, as well as the shell each contains 500 randomly generated data points.

Although the IRIS data set is neither artificial, nor related to chemical process systems, it has been investigated extensively elsewhere, and serves as a useful benchmark for data mapping algorithms. It consists of 150 data points describing three species of *Iris* (*setosa*, *virginica* and *versicolor*) in terms of sepal length (x_1) and width (x_2) and petal length (x_3) and width (x_4).

5.5.1. Mapping with neural networks

The data were mapped with different neural network models, as well as the original algorithm proposed by Sammon (1969). The results are summarized in Table 5.2. Specifically, SAM refers to the original algorithm proposed by Sammon (1969), based on the use of equation (5.7). SNN1 refers to a multilayer perceptron-type neural network described by among other Tattersall and Limb (1994). This network was trained by means of the generalized delta rule, with the error criterion related to the mismatch between interpoint distances in the input and output spaces of the data.

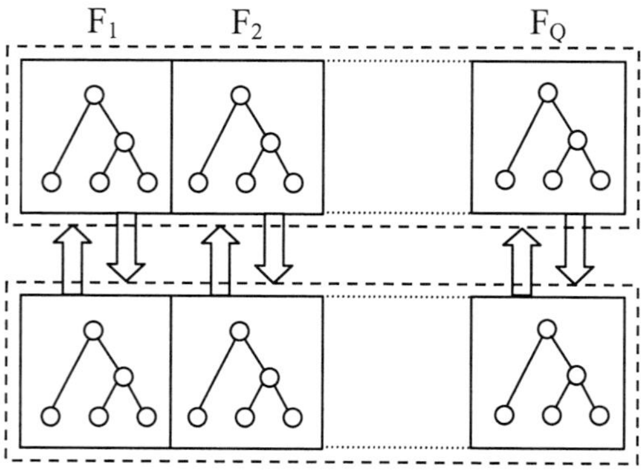

Figure 5.7. Composite tree structures used to project data of the form $\aleph:\Re^p \rightarrow \Re^q$ ($q \leq p$), where the exchange of genetic material is restricted to well-defined regions within the tree structures.

SNN2 refers to the same type of network as SNN1, except that a principal component analysis network (Rubner and Schulten, 1990; Rubner and Tavin, 1989) was first used to project the data, and the weights from this network were consequently used to initialize the Sammon neural network (Mao and Jain, 1995; Pal and Eluri, 1998). SNN3 refers to alternative strategies proposed by Pal and Eluri (1998), making use of statistical sampling and subsets to reduce the n(n-1)/2 number of calculations involved in the computation of the quality of the maps (Sammon stress). With SNN4 (Pal and Eluri, 1998) the idea is the same, except that a Kohonen map is used to extract a small, but adequate representation of the data set prior to generating a Sammon map with a multilayer perceptron.

5.5.2. Evolutionary programming

The terminal set of the evolutionary programming algorithm (similar to the one described in Chapter 4) contained the variables, $\mathbf{T} = \{x_1, x_2, \ldots x_m\}$, while the operator set contained the four basic arithmetic operators, that is $\mathbf{O} = \{+, -, *, /\}$. In all cases the size of the chromosomal population was 100, and the reproduction, crossover and mutation probabilities were 0.10, 0.60 and 0.04 respectively.

Table 5.2. Comparison of the genetic programming algorithm with various methods in terms of the Sammon stress (equation 5.5).

Data set	**EP[a]**	**SAM**	**SNN1**	**SNN2**	**SNN3**	**SNN4**
BITET[b]	0.0513	-	0.0597	-	-	-
SPIRAL[c]	0.00404	0.02931	0.02428	0.34153	0.34435	0.34579
SPHERSHELL[d]	0.0536	0.000889	0.05036	0.09	0.03	0.05
IRIS[e]	0.00916	0.00659	0.00497	0.01252	0.0617	0.0424

[a]Average of three runs for each data set (b)-(e) below.
[b]$(F_1, F_2) = (x_1, x_1+x_2)$; (x_2, x_3-x_1) and (x_3-x_2, x_1); [c]$(F_1, F_2) = (x_2, x_3)$; (x_3, x_2) and $(x^2/x_1^2, x_3)$;
[d]$(F_1, F_2) = (x_1, x_2)$; (x_2, x_1) and (x_1, x_2); [e]$(F_1, F_2) = (x_1-x_2, x_3^2/x_4+x_2x_3)$; (x_2, x_1+x_4) and $(-x_1-x_4, x_2)$;
[e]$(F_1, F_2) = (x_3 + x_4/[1+\exp(x_1)], x_2)$; (x_1+x_4, x_2) and $(x_2, [1/(1+\exp(x_4)][1/(1+\exp(x_3+x_4) + x_1+x_3])$.

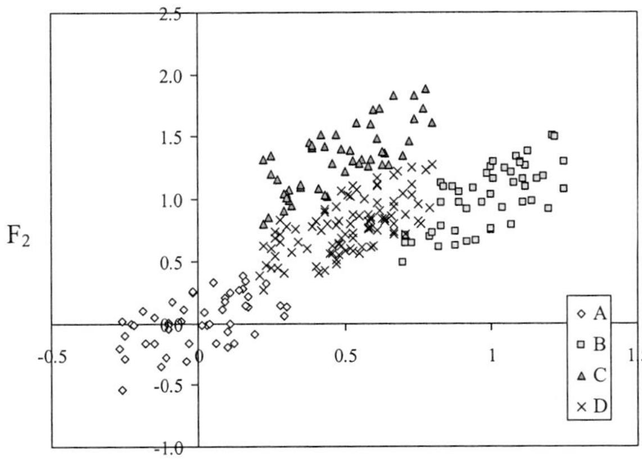

Figure 5.8. Typical Sammon map of the BITET data set (Example 1) generated by the evolutionary programming algorithm, S = 0.0472, $F_1 = x_1$ and $F_2 = x_1 + x_2$. The clusters are indicated by different labels, as shown in the legend.

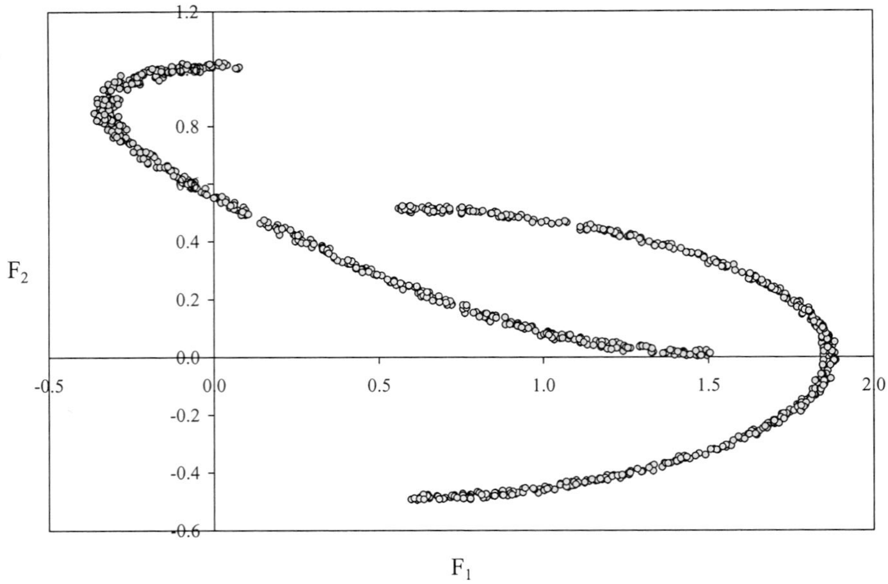

Figure 5.9. Typical Sammon map of the SPIRAL data set generated by the evolutionary programming algorithm, S = 0.00403, $F_1 = x_3$ and $F_2 = x_2$.

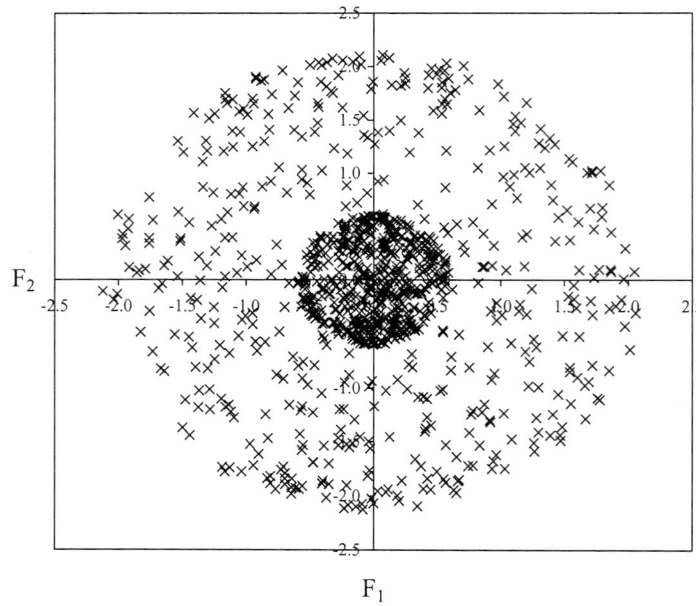

Figure 5.10. Typical Sammon map of the SPHERSHELL data set generated by the evolutionary programming algorithm, S = 0.0531, $F_1 = x_2$ and $F_2 = x_1$.

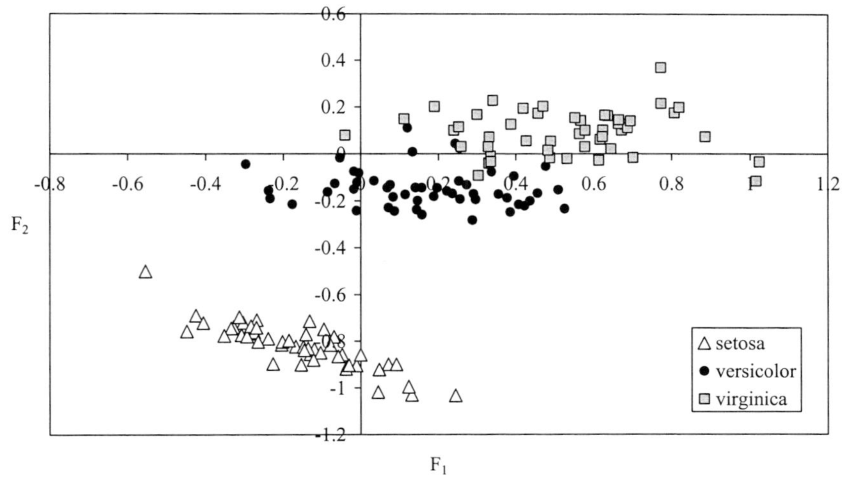

Figure 5.11. Typical Sammon map of the IRIS data set generated by the evolutionary programming algorithm, $S = 0.00657$, $F_1 = x_3 + x_4/[1+\exp(x_1)]$; $F_2 = x_2$.

An elitist strategy was followed, in that the best individual was automatically retained in successive generations. The trees were constrained to a maximum depth of 7, which is not particularly restrictive, given the simplicity of the data sets. Individuals were selected by means of a 3-way tournament method, while the fitness of each individual was defined as the inverse of the Sammon stress (equation 5.5). Since the algorithm had to extract more than one feature, the trees in the population had composite structures, as shown in Figure 5.7. This meant that for each feature (typically two or three), the exchange of genetic material was confined to the subpart of the tree representing the feature. The co-ordinates of the individual points in each data set were presented to the genetic programming algorithm, which projected the data to two-dimensional feature spaces with co-ordinates y_1 and y_2.

From the overall results in Table 5.2, the evolutionary programming (EP) algorithm appears to map the data at least as accurately as the neural networks. The models generated by the evolutionary programming algorithm are shown by means of footnotes in Table 5.2. As can be seen from this table, the models are fairly simple, in fact, adequate representations can be constructed by simple linear combinations of the variables. Typical maps generated by these algorithms are shown in Figures 5.8-5.11.

5.6. EXAMPLE 2: VISUALIZATION OF FLOTATION DATA FROM A BASE METAL FLOTATION PLANT

The following data set was collected from a base metal flotation plant. It consisted of approximately 1500 observations on 13 variables that described the ore and reagent feed rates to the plant, as well as other operating conditions. The variables were denoted by x_1, x_2, ... x_{13}. The plant data exhibited a clustered structure, owing to the way in which the plant was operated. This is shown in a principal component map of the data in Figure 5.7, where the concentration of one of the valuable metals has been superimposed on the data in a discretized form as "high", "medium" and "low".

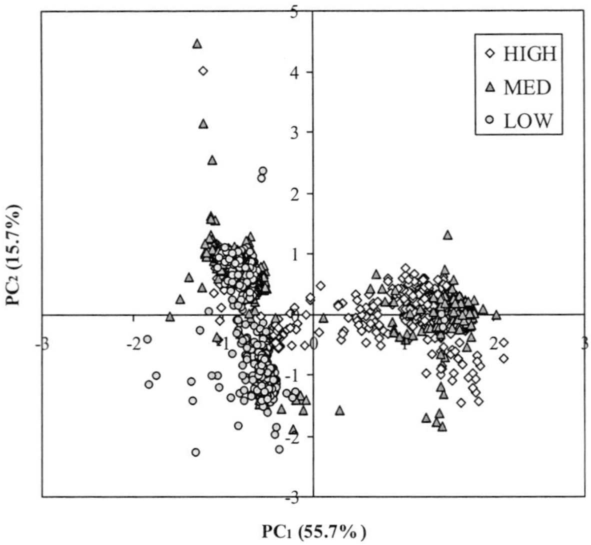

Figure 5.12. Principal component map of 13 plant variables on a base metal flotation plant. The first two principal components (PC_1 and PC_2) respectively explain 55.9% and 14.1% of the variation in the data. The discretized values of the concentration of the valuable metal (not part of the mapped data set) are superimposed on the map.

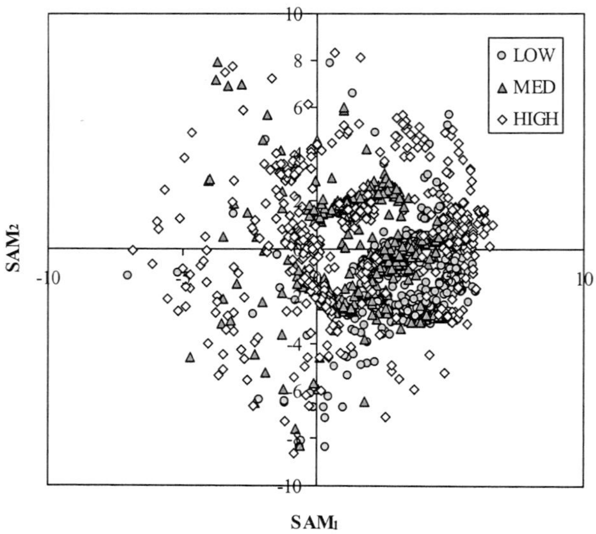

Figure 5.13. Sammon map with Sammon stress = 0.1032 of the base metal flotation data. The discretized values of the concentration of the valuable metal (not part of the mapped data set) are superimposed on the map.

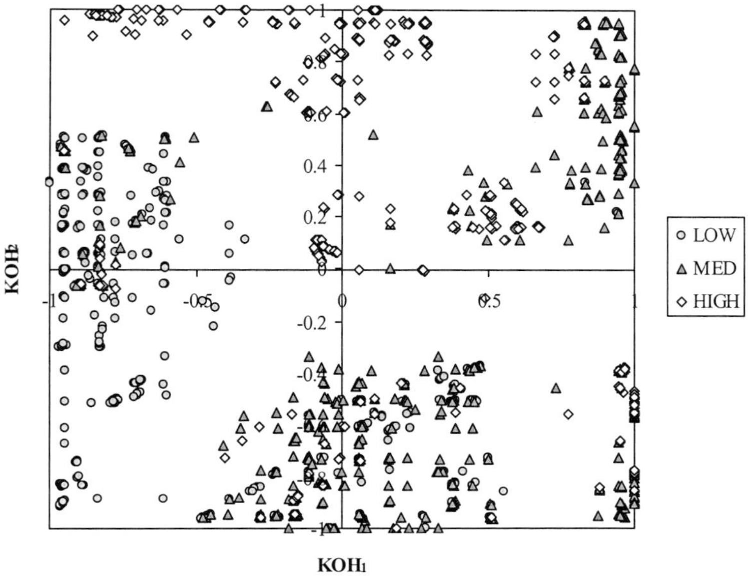

Figure 5.14. Kohonen map of the base metal flotation data shown in Figure 5.12.

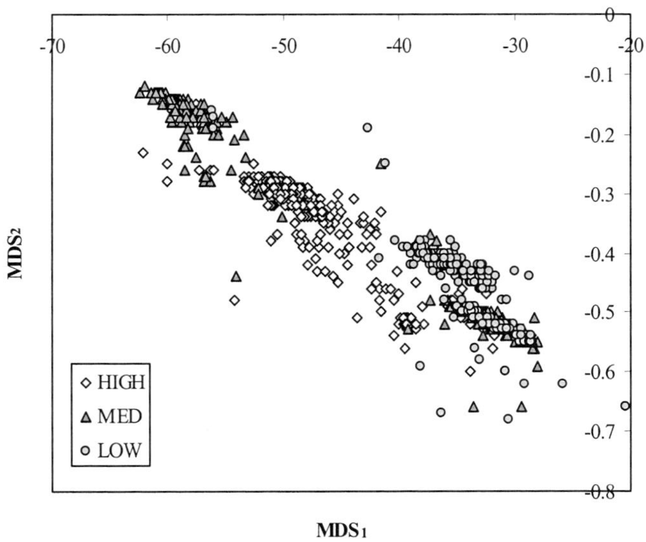

Figure 5.15. Multidimensional scaling (ALSCAL) map of the base metal flotation data shown in Figure 5.12. The ALSCAL stress is 0.00092.

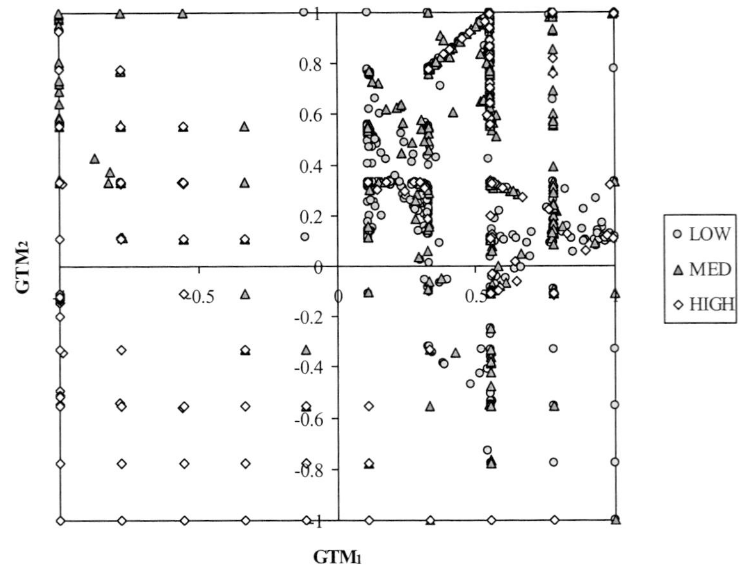

Figure 5.16. Generative topographic map of the base metal flotation data shown in Figure 5.12.

The data were subsequently mapped by use of four topographical models, viz. a Sammon neural network, multidimensional scaling, Kohonen neural network and a generative topographic mapping neural network (Bishop et al., 1997, 1998). Some difficulties were experienced with the training of the Sammon neural network, but these could be alleviated by orthogonalization of the 13 variables by means of principal component analysis, i.e. the network was trained on the scores of the 13 principal components, rather than the original variables themselves. In all cases a random selection of only 2% of the data were used for the generation of the mappings.

Table 5.3. Classification with linear discriminant analysis.

Features	SAM	MDS	KOH	GTM	PCA	Original
%Misclassification	48.5%	40.7%	36.4%	34.6%	37.4%	22.9%

Table 5.4. Classification with classification trees.

Features	SAM	MDS	KOH	GTM	PCA	Original
%Misclassification	21.0%	19.7%	13.1%	18.8%	16.9%	6.6%
Terminal nodes	84	71	59	90	73	40

As a criterion of the quality of the maps, the features extracted from the data were used to classify the data into the three classes indicated in the Figures 5.12-5.16. Classification trees, as well as linear discriminant analysis were used to this effect and the results are summarized in Tables 5.3-5.4.

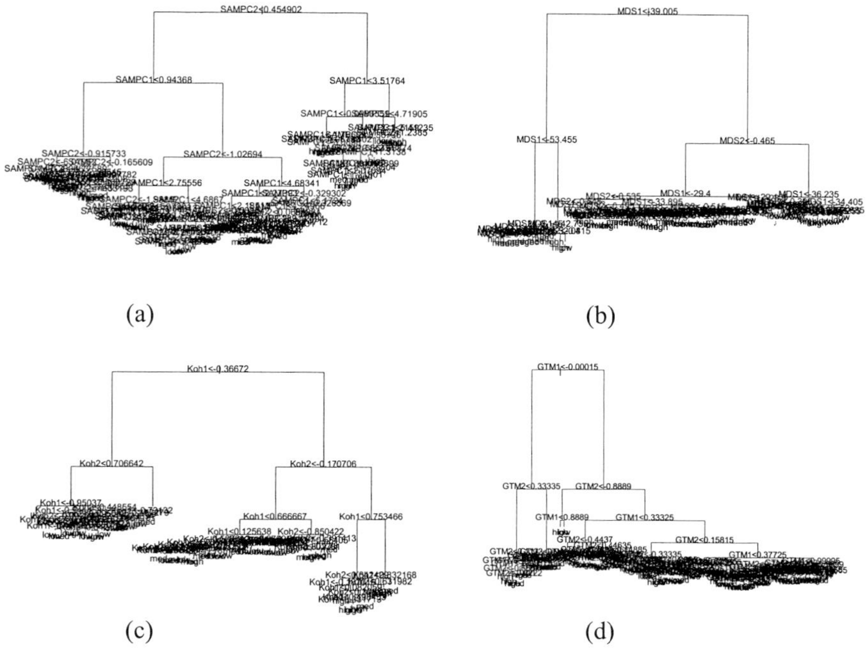

Figure 5.17. Classification tree based on (a) SAM, (b) MDS, (c) Kohonen and (d) GTM features.

With linear discriminant analysis based on the 13 original variables, the rate of misclassification was approximately 22.9%, which was lower than the rates obtained with the models based on the 2 features only. The best models were built with features obtained with the generative topographic mapping (34.6%) and the closely related Kohonen neural network (36.4%), while the least successful model was based on the features extracted with the Sammon neural network (48.5%). When the regression trees were used to classify the data, the Kohonen neural network (13.1%) and the generative topographic mapping (18.8%) performed better than the Sammon map and the multi-dimensional scaling maps again. This should not necessarily be seen as a reflection on the quality of the maps derived with the different techniques, since the different mappings were produced by means of different objective functions.

If anything, it tends to favour the projections where the emphasis has been placed on the retention of the global structure of the data. For example, the maps obtained with the Kohonen and generative topographic mapping neural networks appear to contain more distinct clusters than the maps derived with the other two techniques, which favours separability of the elements of the map.

The structures of the tree classifiers are shown in Figure 5.17(a)-(d). The tree based on the features extracted by the Kohonen neural network was the smallest, containing only 40 terminal nodes. Interestingly, the largest tree was based on the features extracted with the generative topographic map and contained 90 terminal nodes. The posterior probability distributions from typical observations of the classes obtained with the generative topographic mapping are shown in Figure 5.18.

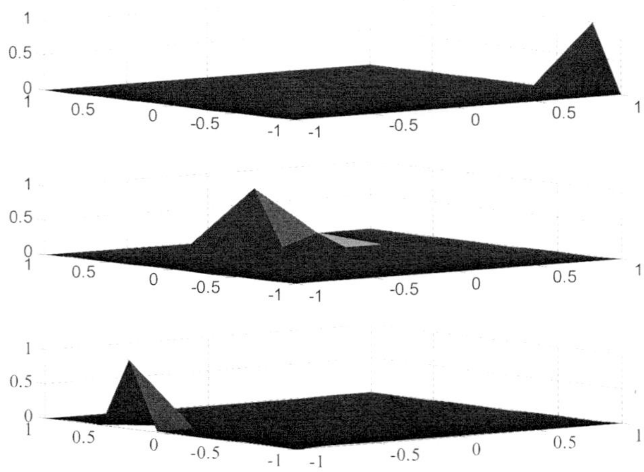

Figure 5.18. Posterior probability distributions from typical observations of the classes "low" (top), "medium" (middle) and "high" (bottom).

5.7. EXAMPLE 3: MONITORING OF A FROTH FLOTATION PLANT

With topological maps of froth structures, process changes can be tracked continuously, as opposed to the identification of discrete classes (Aldrich et al., 1995). The methodology of the control of a flotation cell by means of topological process maps is shown schematically in Figure 5.19.

The basic approach is similar to that discussed for multivariate statistical process control in Chapter 3. Process and image variables that are measured on-line are stored in a data base, from where a predefined set of process variables are fed to a neural network, which generates a topological process map. In practice, Kohonen's self-organizing feature maps have been used extensively, owing to their robustness and convenience, although these are by no means the only suitable approach (Aldrich et al., 1995). The map itself can be interpreted directly by an experienced process operator. Alternatively, the map is used by some form of a knowledge-base or decision support system to assist the operator in the final control decision. Although some work is currently being done to assess the potential of closed-loop model-based control systems, such systems are not presently used in industry, especially since flotation processes are characterized by considerable disturbances associated with variations in the ore feed.

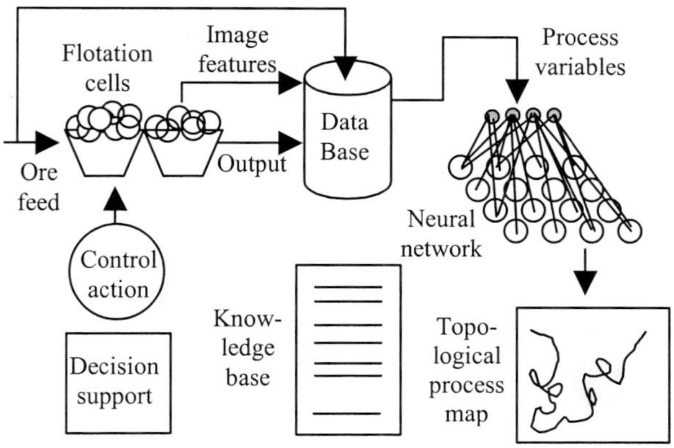

Figure 5.19. Monitoring of flotation cells with topological process maps.

In this example, five image features are projected to a process chart created by a generative topographic map. These neighbouring grey level dependence matrix[37] features were based on those originally proposed by Haralick et al. (1973) and Haralick (1979), for the characterization of textures (Siew et al., 1988; Muhamad and Deravi, 1992). Moolman (1995) and Moolman et al. (1998) give an extended review of these parameters and their use in the characterization of flotation froth images.

Small Number Emphasis: $SNE = \Sigma_r\Sigma_s[Q(r,s)/s^2]/R$ (5.10)

A rough measure of the average bubble size of the froth. The higher the SNE, the smaller the bubbles.

Entropy: $\varepsilon_S = -\Sigma_i\Sigma_j[f(i,j,d,a)\log\{f(i,j,d,a)\}]$ (5.11)

The entropy is a measure of the complexity of the image, i.e. a complex image tends to have a higher entropy than a simple one.

Inertia: $I = \Sigma_i\Sigma_j[(i-j)^2f(i,j,d,a)]$ (5.12)

The inertia is a measure of the number of local variations in the image. Therefore an image with a large number of local variations will have a larger inertia value.

Local Homogeneity: $LH = \Sigma_i\Sigma_j[f(i,j,d,a)/\{1-(i+j)\}^2]$ (5.13)

This parameter is a measure of the tendency of similar grey levels to be neighbours.

[37] This matrix takes the form of a two-dimensional array *Q*, where *Q(r,s)* can be considered as frequency counts of greyness variation of a processed image. The dimensions of the array are *r* x *s* where *r* represents the number of grey levels and *s* the number of possible neighbours to *a* pixels in an image. The *Q* matrix can be computed (for positive integer *d*, *a*) by counting the number of times that the difference between each element in the image function *f(i,j)* and its neighbours is equal to or less than *a* at a certain distance *d*.

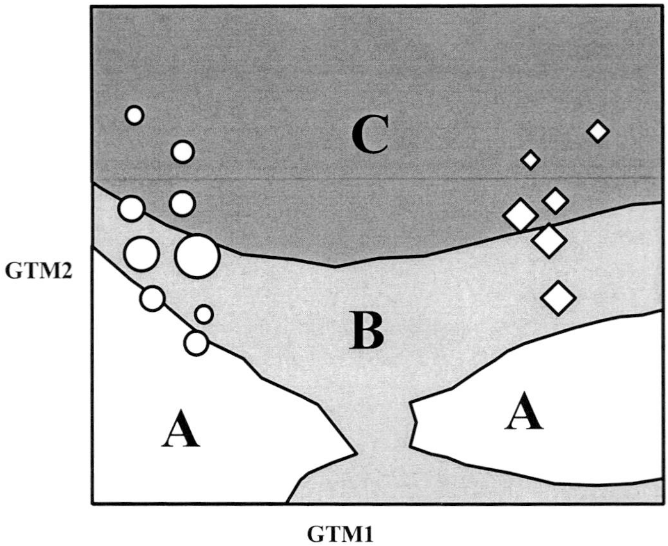

Figure 5.20. Process map of GTM features of froth flotation, with contours indicating the stability of the froth. The circular markers indicate depressant-rich conditions, while the diamond-shaped markers indicate conditions of relatively low depressant concentrations. Larger markers indicate more recent observations than smaller markers. Regions denoted by A indicate highly stable, immobile, sticky froths, region B represents an ideal froth, highly mineralized, with intermediate bubble size, mobility and stability, characterized by a closely packed polyhedral structure, while the shaded regions above B indicate highly mobile, runny froths with small bubbles, low stability and low mineralization.

$$\text{Correlation: } R = \sum_i\sum_j[(i-\mu_x)(j-\mu_y)\{f(i,j,d,a)/\sigma_x\sigma_y\}] \qquad (5.14)$$

where μ_x and σ_x are respectively the mean and standard deviation of the row sums of the matrix, and μ_y and σ_y are the mean and standard deviation of the column sums. This is a measure of the grey level linear dependencies in the image.

The distance metric d in equation (5.14) can be defined by as $d\{(k,l),(m,n)\} = \max\{|k-m|,|l-n|\}$, where (k,l) and (m,n) are the coordinates of two pixels separated by distance d (Haralick, 1973). A normalization factor may be applied to either the feature matrix or to each of the feature parameters. The features (equations 5.10-5.14) characterized the texture and mineral loading of the froths and were mapped to two-dimensional process map by means of a GTM neural network.

An example of such a map is shown in Figure 5.20. The stability of the froth was measured via comparison of a series of digitised images of the froth taken in rapid succession at regular intervals (Moolman, 1995) and this measure was superimposed on the features extracted with the GTM network, as indicated in Figure 5.20. The region indicated by B represents conditions of optimal stability. Too stable a froth (two regions A) prevents optimal recovery of the concentrate, while too unstable a froth (region C) has a similar net effect.

5.8. EXAMPLE 4: ANALYSIS OF THE LIBERATION OF GOLD WITH MULTI-DIMENSIONALLY SCALED MAPS

Comminution processes are used to liberate minerals prior to further processing and it is well-known that they play an important role in the overall profitability of mineral processing operations. Although high-grade concentrates can have a considerable impact on the economics of downstream processing, the cost of grinding itself is high and hence the optimal particle size has to be considered carefully.

Not surprisingly, models that can be used for the quantitative estimation of mineral liberation or the leachability of an ore as related to mineralogy and particle size are of great practical use in the design and simulation of plants, especially when these models can be combined with other quantitative models (Bonifazi and Massacci, 1995). Unfortunately, most liberation models depend on parameters that are difficult or tedious to determine beforehand, such as intercept lengths, diameter of the valuable metal particulates, particle size at which liberation reaches a particular level, etc. Although some progress has been made with regard to these difficulties, e.g. by making use of diagnostic leaching to determine the mineralogy of the gold ores (Lorenzen and Van Deventer, 1994; Lorenzen, 1995), these advances have not been sufficient to date. In this case study, the liberation of gold from various South African gold ores is evaluated by use of topographic mapping with Kohonen neural networks and the topological features extracted from the data are used to develop liberation models.

5.8.1. Experimental data

The experiments were based on nine different gold ores obtained from South African mines, viz. ores from the St Helena, Beatrix, Western Deep Levels (WDL), Free State Geduld (FSG), Unisel, Harmony, Kinross, Leslie and Barberton mines. Except for the refractory Barberton ore, all ores were fed directly to an autogenous grinding mill. The Barberton ore was crushed to -2000 mm before milling. Diagnostic leaching was used to determine the deportment of the gold in the different ores, as summarized below.

a) St Helena and Unisel gold ores

Approximately two thirds of the Unisel gold ore with an average head grade of 5.5 g/ton are sourced from the Basal reef and one third from the Leader reef. The St Helena ore has an average head grade of 4-5 g/ton and is derived from the Basal reef only. In the Basal reef, approximately 50% of the gold occurred as free gold, while the balance consisted of gold included in the pyrite (approximately 35%), gold attached to pyrite (approximately 10%), and gold associated with carbonaceous matter (approximately 5%). The sizes of the hackly (sharp angular) gold grains varied considerably, ranging from a few microns up to a few millimetres in diameter.

The gold attached to the pyrites could be extracted completely via cyanidation. The pyrite grains in which the gold were included had an average size of approximately 10 μm. The estimated recoverability of these gold grains by means of cyanidation depended on the size of the particles. Approximately 80% of the gold could be recovered from particles larger than 20 μm, 50% from particles ranging from 5-20 μm and 10% from particles smaller than 5 μm. The gold grains associated with the hydrocarbons were very small, averaging roughly 2 μm in diameter.

b) Beatrix gold ore

The Beatrix reef is located in the Orange Free State gold fields and has an average head grade of 6 g/ton (1994). According to mineralogical analyses reported by the mine, gold occurs in four main forms in the reef, viz. free gold, gold associated with quartz and silicates, gold associated with pyrite and gold associated with carbonaceous materials.

Approximately 70% of the gold occurred as free gold interstitially dispersed in the quartz. These gold grains varied in size from 210 x 75 μm^2 to 5 x 2 μm^2, with a median size of 38 x 14 μm^2. Gold locked up in quartz and other silicates, such as chlorite, had a median size of 32 x 10 μm^2 and comprised 10% of the Beatrix ore. In contrast, the gold associated with the pyrite had a smaller median size of approximately 15 x 6 μm^2 and comprised roughly 5% of the total gold in the ore. The balance of the gold was associated with the hydrocarbons in the ore and had a median size of approximately 8 x 3 μm^2.

c) Kinross and Leslie gold ores

The Kinross and Leslie gold fields are located in the eastern and western parts of the Evander gold fields in Mpumalanga, respectively. Their gold mineralogies differ significantly from each other and the other gold fields, with an average head grade of 7 g/ton for the Kinross and 5-6 g/ton for the Leslie gold ore (1994). Owing to the occurrence of shale, which interfered with the extraction of the gold by use of cyanidation, the Leslie ore was milled to 90% -75 μm, as opposed to 70% -75 μm for the other ores.

d) Barberton gold ore

The Barberton gold mine is located in Mpumalanga as well, and owing to the presence of large amounts of pyrite and arsenopyrite in the gold ore, extraction by means of cyanidation is not feasible. Instead, the gold from this ore is removed by use of bacterial leaching.

e) Western Deep Level, Free State Geduld and Harmony gold ores

The Free State Geduld (FSG) and Harmony gold mines are located in the Free State gold fields, while the Western Deep Levels (WDL) gold mine is situated in the West Wits Line.

5.8.2. Milled and unmilled ores

The unmilled ore consisted of eight particle size fractions, namely +6700 μm, -6700 to +1500 μm, -1500 to +300 μm, -300 to +150 μm, -150 to +106 μm, -106 to + 75 μm, -75 to +53 μm and -53 μm. A representative sample of the unmilled ore was consequently reduced by crushing and milling (depending on the particle size) to produce a sample of 70% -75 μm. The milled sample was also screened into the abovementioned eight size fractions.

After pretreatment of the ores, diagnostic leaching was performed once on each of the particle size fractions of the milled and unmilled or crushed ores. The results are summarized in Tables 5.5. The mineralogical data in Tables 5.5 can be visualized by means of hierarchical cluster analysis, based on Euclidean distances and the single link method of agglomeration, as the cluster shapes probably deviate substantially from spherical geometries. As can be seen from the dendrogram in Figure 5.21, the deportment of the gold in the FSG, WDL and

Barberton ores are different from that in the Beatrix, Helena, Kinross, Leslie, Unisel and Harmony ores before milling. After milling, the deportment of the gold in the Beatrix and St Helena ores are still very similar (Figure 5.22), as it is in the Kinross and Leslie ores.

```
                       Rescaled Distance Cluster Combine

  C A S E      0         5        10        15        20        25
Label     Num  +---------+---------+---------+---------+---------+

BEATRIX    2   -+---+
HELENA     5   -+   +---------+
KINROSS    6   -+---+         +-------------+
LESLIE     7   -+             I             +-------------------+
UNISEL     8   ---------------+             I                   I
HARMONY    4   -----------------------------+                   I
FSG        3   ---------------------+-------+                   I
WDL        9   ---------------------+       +-------------------+
BARBER     1   -----------------------------+
```

Figure 5.21. Dendrogram using single linkage clustering for unmilled ores, based on the five variables Free Gold, Base Metal Sulphides (BMS), Pyrite, Silicates and Carbon.

In contrast, as far as the deportment of the gold is concerned, the Barberton and Harmony gold ores are relatively dissimilar from each other and the other ores prior to milling. These relationships are changed by the milling process, although the main groupings of the ores remain the same, with Barberton, WDL and FSG in the one group and the remainder of the ores in the other group. These changes can be visualized by use of the topological feature map in Figure 5.23, based on multidimensional scaling of the variables characterizing the ores.

```
                 Rescaled Distance Cluster Combine

  C A S E      0         5        10        15        20        25
Label     Num  +---------+---------+---------+---------+---------+

HARMONY    4   -+-----+
KINROS     6   -+     I
BEATRIX    2   -+     +-----------------------------------------+
HELENA     5   -+---+ I                                         I
UNISEL     8   -+   +-+                                         I
LESLIE     7   -----+                                           I
BARBER     1   ---+-------------------------------------+       I
WDL        9   ---+                                     +-------+
FSG        3   -----------------------------------------+
```

Figure 5.22. Dendrogram using single linkage clustering for milled ores, based on the five variables Free Gold, Base Metal Sulphides (BMS), Pyrite, Silicates and Carbon.

Table 5.5. Approximate analysis (%) of gold in the various ores.

			Kinross		
	Free Au	BMS	Pyrite	Silicates	Carbon
Unmilled	36.49	13.31	6.93	38.87	4.4
Crushed	36.77	13.42	7.84	38.33	4.4
Milled	84.91	3.64	3.84	5.24	2.38
			Leslie		
	Free Au	BMS	Pyrite	Silicates	Carbon
Unmilled	32.08	8.25	11.02	46.03	2.62
Milled	75.82	5.08	3.23	13.83	2.04
			Unisel		
	Free Au	BMS	Pyrite	Silicates	Carbon
Unmilled	40.73	9.14	16.40	22.41	11.32
Milled	74.09	8.62	10.18	3.65	3.47
			Free State Geduld		
	Free Au	BMS	Pyrite	Silicates	Carbon
Unmilled	25.2	35.3	29.4	10.1	-
Milled	47.4	23.8	24.8	4.0	-
			Western Deep Levels		
	Free Au	BMS	Pyrite	Silicates	Carbon
Unmilled	8.3	29.9	46.1	15.7	-
Milled	20.2	30.1	42.1	7.5	-
			Barberton		
	Free Au	BMS	Pyrite	Silicates	Carbon
Unmilled	7.36	14.51	65.65	4.45	8.04
Milled	25.9	28.96	40.24	1.2	3.7
			Beatrix		
	Free Au	BMS	Pyrite	Silicates	Carbon
Unmilled	17.4	10.22	9.93	55.3	7.14
Milled	64.05	8.03	8.58	11.45	7.89
			St Helena		
	Free Au	BMS	Pyrite	Silicates	Carbon
Unmilled	23.21	6.31	6.44	53.72	10.31
Crushed	17.42	6.4	6.61	58.41	11.13
Milled	68.76	9.97	7.36	8.04	5.88
			Harmony		
	Free Au	BMS	Pyrite	Silicates	Carbon
Unmilled	18.58	6.47	4.56	36.94	33.45
Milled	89.43	4.81	1.3	1.98	0.54

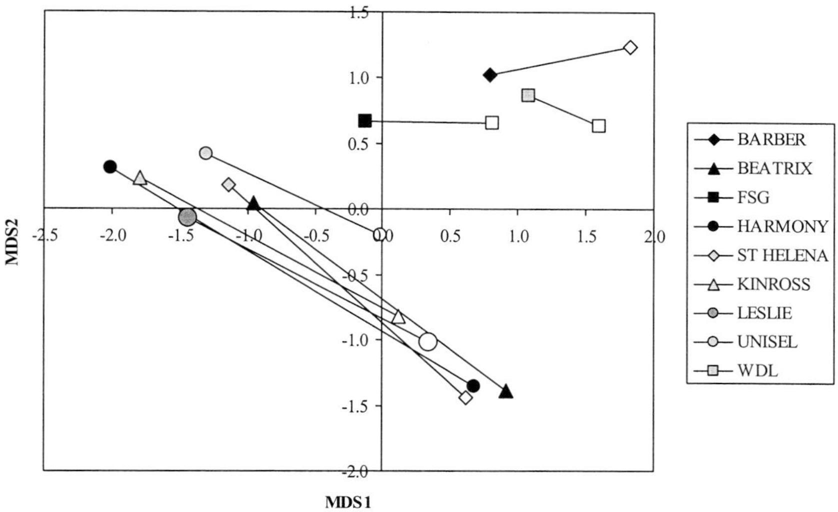

Figure 5.23. Topological feature map of gold deportment in the milled and unmilled gold ores. Solid markers show the ores before milling, while ores before and after milling are connected by solid lines.

5.9. EXAMPLE 5. MONITORING OF METALLURGICAL FURNACES BY USE OF TOPOGRAPHIC PROCESS MAPS

Optimal control of submerged arc furnaces requires that the electrical and metallurgical aspects of the process, as well as concomitant interrelationships among these phenomena be accounted for. Although the electrical operation of submerged arc furnaces tends to be controlled comparatively accurately (Reuter, 1997), it is difficult to characterize the particulate feed or to define the process metallurgy. Moreover, it is often not realized that whereas the dynamics of the electrical operation allows feedback control, the metallurgy can only really be controlled in a feed-forward mode. This can be attributed to the large temperature gradients associated with the non-steady state (tap-to-tap) operation of the furnaces, which can result in complex dynamics which can be very difficult to identify and control.

Aldrich and Reuter (1999) have discussed the use of topographic maps for the monitoring of metallurgical furnaces and have shown that these maps can work where process control charts based on the principal components of process data may be inadequate. As with latent variable methods, the process data are projected to a process map, which can give the operator an indication of the state of the process, as well as a means of fault diagnosis in the case of process deviations. In the latter case, the features of the process map have to be mapped back to the original variables.

Aldrich and Reuter (1999) have considered the operation of a carbide furnace that could be represented by a set of seven process variables, (x_1-x_7). x_1 represented the furnace load, x_2 and x_3 represented the electrical variables, x_4 represented a raw material additive, and x_5, x_6 and x_7 represented the reductants, all of which were equated to daily averages. A map based on topological features (SomX and SomY) of the variables extracted with a Kohonen neural network are shown in Figure 5.24. The process quality variable (the product of the grade and production rate) is superimposed on the map. As can be seen from this figure, the NOR (where process quality is maximal) lies in the upper right-hand corner of the map (based on historic process data). It is also clear that the map is not smooth, but that two zones of near-maximal operation are separated by a valley of poor performance, making control difficult.

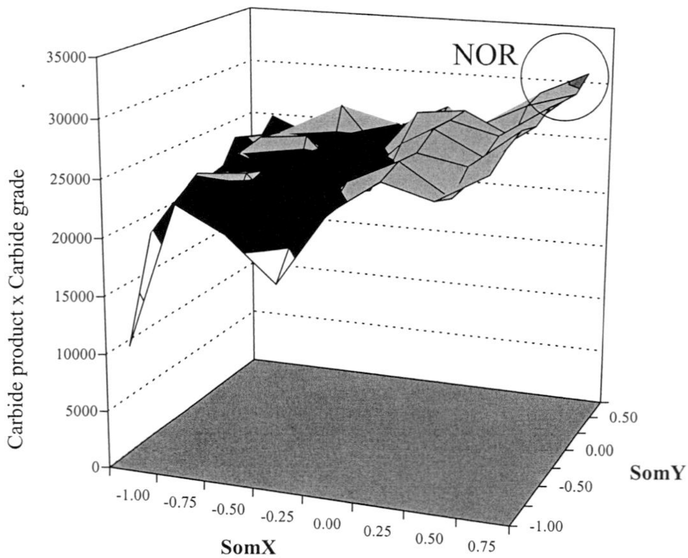

Figure 5.24. The NOR in the projected feature space (grey areas on the surface chart).

Reconstruction of the original variables from the features could be accomplished by means of a second multilayer perceptron neural network. In this case features derived by means of a Sammon map were used. The unipolar sigmoidal neural consisted of a 3-node input layer, a 4-node hidden layer and a 1-node output layer, and was trained by the generalized delta rule on exemplars of the form (SAMX1, SAMX2, SAMX3 | x_1, x_2, ... x_7).

The exemplars themselves were derived from the small region (NOR) in the top right-hand corner of the surface map shown in Figure 5.24. Once the first neural network had mapped new data, the second neural network could be used to reconstruct process variables from the mapped features scores. Comparison of the reconstructed variables with the actual variables could be used to derive SPE charts, similar to those described in Chapter 3. These results could be used to analyze changes in the structure of the data, which could in turn be used to identify variables contributing to process deviations.

Application of the methodology to the monitoring of the furnace can be illustrated by considering two cases. The first is based on actual plant data, in which the consumption of one of the reductants (x_7) fell below a critical level. Under these circumstances models for production are not affected significantly, but those for the grade tend to break down completely. In the second case, the breakdown of a resistance sensor (x_3) was simulated by incorporating a

100% upward bias in all the electrode resistance measurements. These effects are clearly visible on the process map in Figure 5.25, which shows four scenarios.

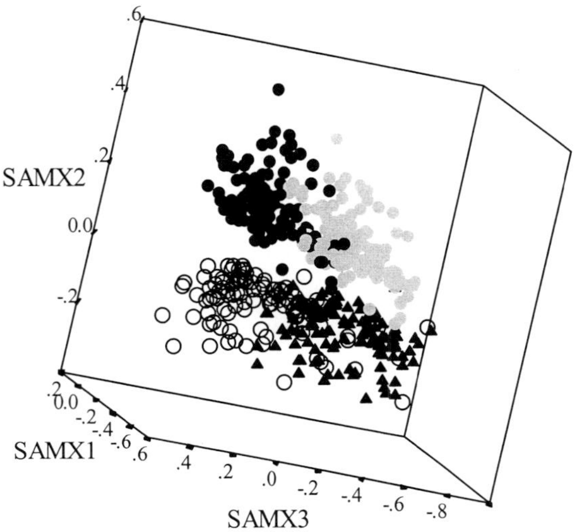

Figure 5.25. Multivariate map of process faults. Normal operation is indicated by the solid black circles, while a 100% bias in the resistance measurements (x_3) is indicated by the solid grey circles, low reductant (x_7) consumption is indicated by empty circles, and both low reductant consumption (x_7) and resistance measurement (x_3) bias are indicted by solid triangles.

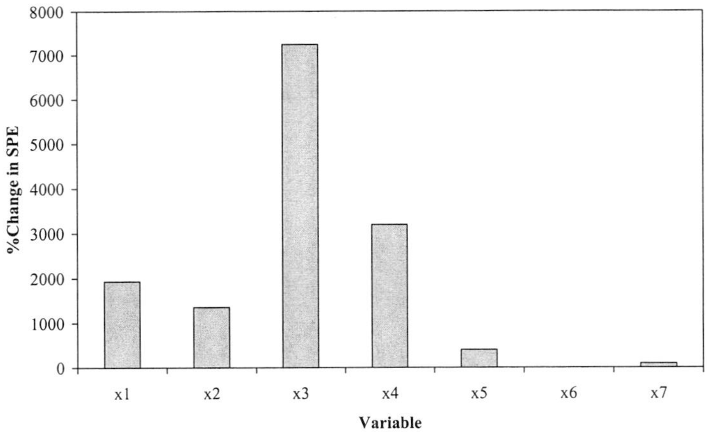

Figure 5.26. Identification of variable(s) contributing to process deviation associated with a bias in resistance measurements. The relatively high value of the raw material additive (x_4) and furnace loads (x_1) can be related to the relatively high correlation between the resistance and these two variables, viz. $R^2 = -0.514$ (x_1) and $R^2 = -0.502$ (x_4).

The solid black circles in Figure 5.25 show the NOR, the empty circles show the deviation associated with the lowered reductant consumption, the solid grey circles show the deviation associated with the sensor failure, while the solid triangular markers show deviations associated with both the lower reductant consumption and sensor failure combined. As can be seen from this map, the scores of the projected process conditions form relatively well-defined clusters, and are readily distinguishable from the map.

In Figure 5.26, the change in the values of the squared prediction errors (SPE) for all the variables are shown for the deviation associated with a bias in the resistance measurements. As can be seen from this map, the change in the SPE of variable x_3 is significantly larger than in the other variables.

A similar map can be shown for the sensor failure (Figure 5.27), and again it can be seen that the cause of the deviation is readily identifiable from the relatively large value of the percentage change in the SPE-value of the reductant consumption (x_7).

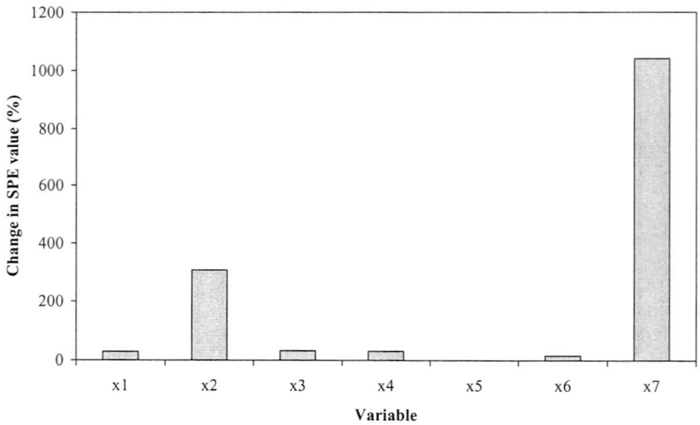

Figure 5.27. Identification of variable(s) contributing to process deviation associated with lower reductant consumption (x_7).

Chapter 6
Cluster Analysis

Data analysts often encounter situations best resolved by the definition of groups of homogeneous objects, whether they are individuals, products or even process system behaviours. In all instances, the analysts are looking for a natural multivariate structure among the observations. This is especially important where little prior information on the data is available and the analyst has to make as few assumptions as possible. The method most commonly used to accomplish this goal is cluster analysis, which aims to group individuals or objects that are more homogeneous than objects in other groups. The clustering task typically consists of the following steps, as indicated in Figure 6.1.

i. Pattern representation (including feature extraction and/or selection)

ii. Definition of a pattern similarity measure

iii. Grouping or clustering of the patterns or observations

These steps can be followed by data abstraction (interpretation of the clusters) and assessment of the quality of the output, where required.

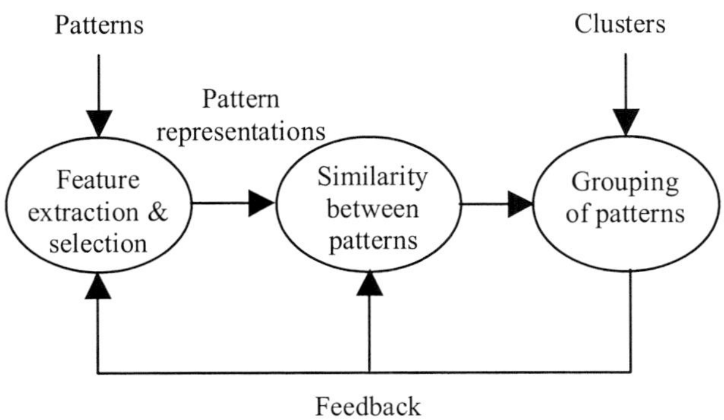

Figure 6.1. Various steps in cluster analysis.

Feature extraction in step (i) is based on the transformation of the input variables to produce features that can be used for clustering and among other, the techniques discussed in Chapters 3 and 5 can be used for this purpose. Determination of the interpattern similarity is fundamental to the analysis and is therefore considered in more detail below.

6.1. SIMILARITY MEASURES

Measures can be separated into two broad classes, namely distance-type measures, and matching-type measures. The former is used with data having metric properties, while the latter is applicable to data with qualitative components.

6.1.1. Distance-type measures

Given the space (the original variable space or a derived discriminant space) in which observations or objects are represented, the fundamental problem of classification is reduced to the selection of a metric or distance measure. The reason is that an observation that needs to be assigned to a particular class can be assigned to the group to which it is judged to be the closest, as indicated by the metric.

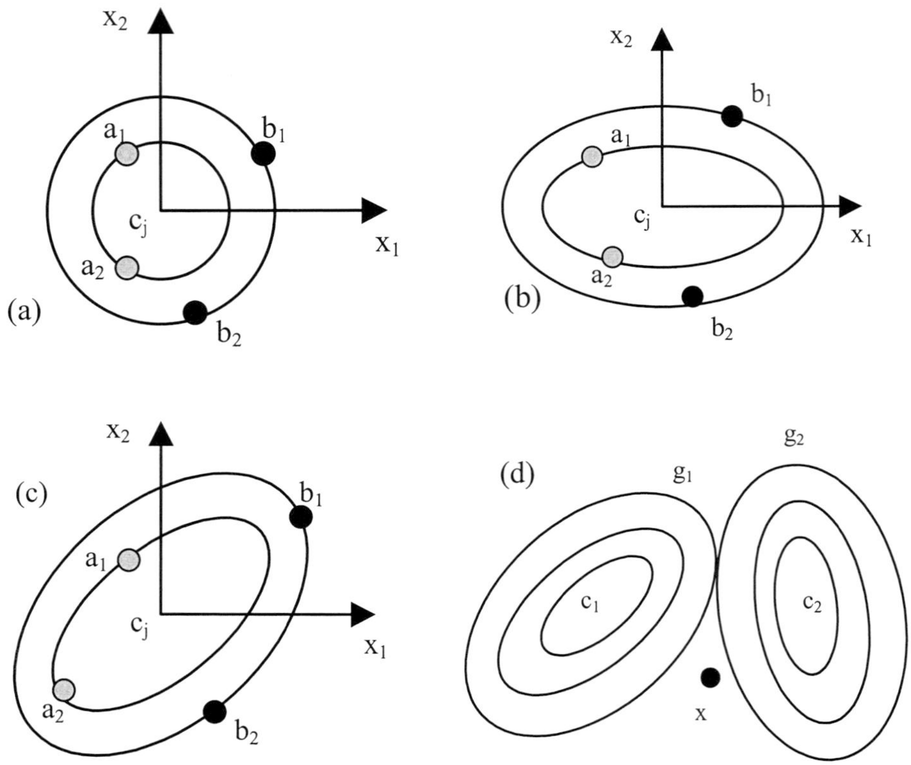

Figure 6.2(a) Squared Euclidean distance measure, (b) squared Euclidean distance with differently weighted variables, (c) generalized squared distance measure and (d) classification based on different within-group dispersions.

Although a large number of different metrics have been proposed previously, most of these distance measures satisfy narrowly defined optimal criteria based on a set of specific assumptions that usually cannot be validated by the available data. As far as the empirical analysis of data is concerned, the prescription of a distance measure is therefore largely a trial-and-error procedure, based on intuition and prior knowledge of the data (and sometimes good luck!).

A class of positive semiquadratic forms provides a useful general set of squared distance measures. Specifically, if $\mathbf{x}^T = (x_1, x_2, \ldots x_m)$ is a vector of m variables that needs to be assigned to one of g prespecified groups, then the distance between the observation x and the centroid c_j of the j'th group can be expressed by

$$d^2(j) = (\mathbf{x} - \mathbf{c}_j)^T\mathbf{M}(\mathbf{x} - \mathbf{c}_j) \quad (6.3)$$

In this case **M** is a positive semidefinite matrix to ensure that $d^2(j) \geq 0$ and the observation is assigned to the group for which $d^2(j)$ is the smallest for j = 1, 2, ... g. Different choices of **M** leads to different metrics, as indicated in Figure 6.2.

The use of an unweighted squared Euclidean distance may be inappropriate when the observations are noncommensurable and are likely to have appreciably different variances. In this case elliptic or ellipsoidal distance measures can be used to compensate for the differences in the variances of the variables. As before, in Figure 6.2(b), observations a_1 and a_2 would be the same distance from the centroid of the class c_j, and closer than b_1 and b_2 that would also be the same distance from c_j. This is accomplished by setting M to a diagonal matrix with elements equal to the reciprocals of the variances of the different variables.

Yet another distance measure well suited to correlated variables (as well as differences in the variances of the variables) can be obtained by setting **M** to the inverse of the covariance matrix S of the variables (i.e. the Mahalanobis distance). Geometrically this means that the ellipse or ellipsoid is tilted so that the major axis is oriented in a direction reflecting the correlation between the variables, as indicated in Figures 6.2(c) and 6.2(d). As before, points lying on a particular ellipse are considered to be equidistant from the centroid of the ellipse (the class centroid). In Figure 6.2(d), the two groups g_1 and g_2 have different variable structures or within-group correlations. As an example, consider the fluids in Table 6.1.

Table 6.1. Physical properties of fluids, viscosity (μ), density (ρ) and surface tension (σ).

Fluid	**μ (cP)**	**ρ (kg/m^3)**	**σ (mN/m)**
Water (W)	0.98	997	72.0
Ethyl alcohol (E)	1.09	790	21.3
Aniline (A)	3.34	1020	36.7

These similarities of these fluids can be expressed in terms of their Euclidean distances from each other, as follows.

$$d_{WE} = \{(0.98\text{-}1.09)^2 + (997\text{-}790)^2 + (72.0\text{-}21.3)^2\}^{1/2} = 213.11$$
$$d_{WA} = \{(0.98\text{-}3.34)^2 + (997\text{-}1020)^2 + (72.0\text{-}36.7)^2\}^{1/2} = 230.53$$
$$d_{EA} = \{(1.09\text{-}3.34)^2 + (790\text{-}1020)^2 + (21.3\text{-}36.7)^2\}^{1/2} = 42.20$$

Water and aniline are therefore relatively similar, while ethyl alcohol is relatively different from water and aniline. However, if the densities are expressed in different units, such as g/cm^3, the relative similarities between the fluids will change significantly, i.e.

$$d'_{WE} = \{(0.98\text{-}1.09)^2 + (0.997\text{-}0.790)^2 + (72.0\text{-}21.3)^2\}^{1/2} = 50.70$$
$$d'_{WA} = \{(0.98\text{-}3.34)^2 + (0.997\text{-}1.020)^2 + (72.0\text{-}36.7)^2\}^{1/2} = 15.57$$
$$d'_{EA} = \{(1.09\text{-}3.34)^2 + (0.790\text{-}1.020)^2 + (21.3\text{-}36.7)^2\}^{1/2} = 35.38$$

In this case we would conclude that all fluids are more or less equally similar (or dissimilar). (Clearly the scaling (weighting) of the variables is an important consideration and in the absence of any prior information it would be sensible to weight all the variables equally).

6.1.2. Matching type measures

Matching type measures, or association coefficients are appropriate where the data are nominally scaled. These measures usually range from 0 to 1, and are based on the view that objects with many attributes in common are more similar than those with relatively few attributes in common. For example, consider the following table of minerals.

Mineral	**System**	**Colour**	**Lustre**	**Streak**
Cu	Cubic	Red	Metallic	Red
$FeCr_2O_4$	Cubic	Black	Metallic	Brown
Fe_2O_3	Hexagonal	Black	Dull	Red
$FeTiO_3$	Hexagonal	Black	Metallic	Black

The attributes of the minerals can be expressed as follows, with (system, colour, lustre, streak) expressed as (hexagonal = 1, red = 1, metallic =1, red = 1), if true, otherwise 0. That is

Mineral	**System**	**Colour**	**Lustre**	**Streak**
Cu	0	1	1	1
$FeCr_2O_4$	0	0	1	0
Fe_2O_3	1	0	0	1
$FeTiO_3$	1	0	1	0

The similarities of the minerals can be calculated as follows, by means of an association table. For each pair of minerals, there are four values, namely the number of positive (1 or +) attributes that the two individuals share (+/+ comparison), the number of positive-negative (+/-) attribute pairs, the number of negative-positive (-/+) attribute pairs, as well as the number of negative attributes that the two individuals have in common (-/-). Table 6.2 shows these values for the Cu and $FeCr_2O_4$ minerals.

Table 6.2. Similarities of the Cu and $FeCr_2O_4$ minerals.

		Cu	
		+	-
$FeCr_2O_4$	+	**1**	**0**
	-	**2**	**1**

The four criteria by which individuals are compared, can be denoted by a, b, c and d respectively, and similarity coefficients can be defined on this basis. For example (a+d)/(a+b+c+d), a/(a+b+c), 2a/(2a+b+c), a/(a+b+c+d), etc.

With a similarity coefficient defined as (a+d)/(a+b+c+d), the similarity coefficient of the Cu and $FeCr_2O_4$ would be S(Cu-$FeCr_2O_4$) = (1+1)/(1+2+0+1) = 0.5. This would make the Cu more similar to $FeCr_2O_4$ than to Fe_2O_3, with S(Cu-Fe_2O_3) = (1+0)/(1+2+1+0) = 0.25.

Apart from the choice of association measure, the assignment of binary codes is also crucial. For example, if the binary codes are switched, as shown in Table 6.3, the results of the above calculations could be affected, depending on the definition of the similarity coefficient.

Table 6.3. Similarities of the Cu and $FeCr_2O_4$ minerals after switching of binary codes.

Mineral	System	Colour	Lustre	Streak
Cu	1	0	0	0
$FeCr_2O_4$	1	1	0	1
Fe_2O_3	0	1	1	0
$FeTiO_3$	0	1	0	1

Note that it is easy to transform distance measures into association measures, but not vice versa, owing to the necessity for distances to satisfy certain conditions. In particular, a function d(x,y) of pairs of points of a set C, is said to be a metric for C, if

$$d(x,y) \geq 0;\ d(x,y) = 0, \text{ iff } x = y \qquad \text{positivity} \qquad (6.1)$$

$$d(x,y) = d(y,x) \qquad \text{symmetry} \qquad (6.2)$$

$$d(x,z) + d(y,z) \geq d(x,y) \qquad \text{triangle inequality} \qquad (6.3)$$

6.1.3. Contextual and conceptual similarity measures

Apart from the previously considered and widely used similarity measures, some distance measures also account for the context of objects (Michalski et al., 1983). In this case the similarity between points can be defined as

$$d(i,j) = f(\mathbf{x}_i,\mathbf{x}_j,\xi) \qquad (6.4)$$

where ξ defines the context (set of surrounding points) of an observation. An example of such a metric is the mutual neighbour distance[38] (Gowda and Krishna, 1977), defined as

$$d_{MN}(i,j) = NN(i,j) + NN(j,i) \qquad (6.5)$$

where NN(i,j) is the neighbour number of $\mathbf{x}_j$ with respect to $\mathbf{x}_i$. This concept is illustrated in Figures 6.3. In Figure 6.3(a), the nearest neighbour of B is C and C's nearest neighbour is B. Thus NN(B,C) = NN(C,B) = 1 and $d_{MN}(B,C) = 2$. By adding points D, E and F to points A, B and C, as in Figure 6.3(b), point C's context is changed and B is no longer closer to C than to A. This can be verified by the fact that NN(B,C) = 4 and NN(C,B) = 1, thus $d_{MN}(B,C) = 5$. In contrast, NN(A,B) = 2, NN(B,A) = 1 and $d_{MN}(A,B) = 3$. Thus, the distance between B and C has increased, although the points have not moved.

In the case of conceptual clustering (Michalski et al., 1983; Wang, 2001), the similarity between $\mathbf{x}_i$ and $\mathbf{x}_j$ is defined as

$$d_{CC}(i,j) = f(\mathbf{x}_i,\mathbf{x}_j,\zeta,\xi) \qquad (6.6)$$

[38] The mutual neighbour distance is not a metric, as it violates the triangular inequality (equation 6.3).

where ζ is a set of predefined concepts and ξ represents the context, as before. Figure 6.4 illustrates this notion. In this figure, points B and C belong to the same concept (rectangle) and can therefore can be viewed as more similar than points A and B, since A belongs to a different concept (circle).

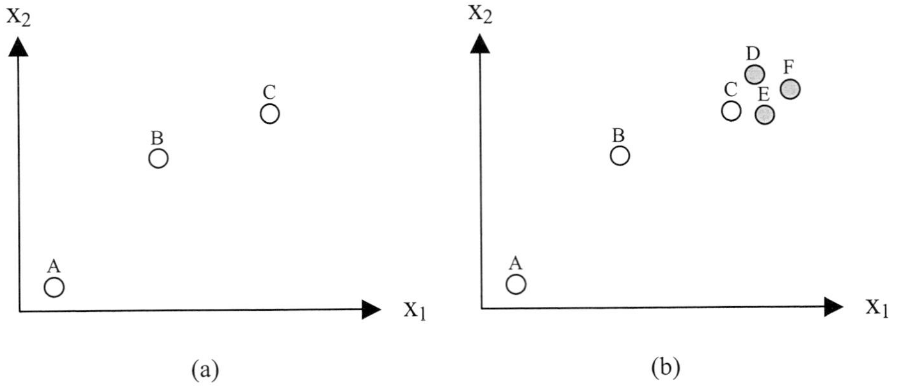

Figure 6.3. Contextual clustering of data, where the context of point C is changed by the absence (a) or presence (b) of neighbouring points D, E and F.

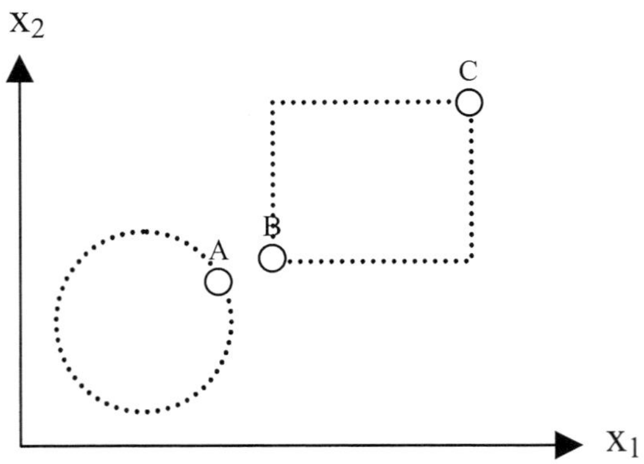

Figure 6.4. Conceptual similarity between points A, B and C. Points B and C can be viewed as more similar than points A and B, because they belong to the same concept (rectangle), while A belongs to a different concept (circle).

6.2. GROUPING OF DATA

Once a satisfactory distance measure has been selected, the objects (patterns, exemplars or observations) can be grouped. Generally speaking, the cluster analysis problem involves partitioning of a set of n objects into $k \leq n$ disjoint subgroups or clusters. The effectiveness of clustering can be measured by various criteria, e.g. minimization of the ratio of the within-group variance to between-group variance. Also, additional criteria may be applicable, such as the specification of a set number of clusters k. The general class of clustering problems is computationally difficult, and generally requires the examination of a number of (N_p) partitions, viz.

$$N_p = \Sigma^n_{g=1}(1/g!)\{\Sigma^n_{g=1}(-1)^{g-i}[(g!/(i!(g-i)!)](i^n)\} \quad (6.7)$$

growing exponentially as the number of objects increases (Joseph and Bryson, 1998). By sequentially ordering the objects prior to clustering, the number of possible partitions decreases to

$$N_p' = \Sigma^n_{g=1}(n-1)!/[(g-1)!(n-g)!] = 2^{n-1} \quad (6.8)$$

Although this is still an exponential number, it can be solved in polynomial time (Joseph and Bryson, 1998).

The large number of methods that have been proposed for cluster analysis, can be divided into four categories.

i. Hierarchical methods: Strictly hierarchical methods depend on similarity or dissimilarity measures as discussed above, and provide a complete scheme of division into all possible numbers of groups from 1 to n. These methods were first used in the fields of ecology and taxonomy in the 1950s and remain the most popular techniques today. The k-group partition is obtained either by splitting one group in the (k-1) division, or by fusing two of the groups in the (k-1) division. Agglomerative methods are used predominantly, i.e. starting from n individuals or objects, they build up groups, until all individuals are assigned to a single group. In contrast, divisible methods start with a single group, which is progressively fragmented. In either case, a tree diagram or dendrogram, is usually produced, showing how the groups are built up or divided. Partitioning into a given number of groups is found by simply cutting the dendrogram at an appropriate level.

ii. Optimal partitioning: The individuals are sub-divided into k groups, in order to optimize some criterion, such as the within-group covariance matrix. The k-means clustering method introduced in 1976, is a well-known example of this approach to cluster analysis. This is not a strictly hierarchical method, in that the optimal division into k + 1 groups cannot be achieved by splitting one of the k groups.

iii. Distribution mixtures: In practice, this amounts to fitting mixtures of multivariate normal distributions to the data. The means of the distributions differ, or are assumed to be the same. The final result is not a partitioning of the data, but rather a set of posterior probabilities of group membership, which can be used to assign individuals to the groups for which probabilities are the highest. Pearson first raised the concept in 1894, and practical algorithms have been around since 1965.

iv. Non-parametric estimation of local densities: These methods do not make any assumptions with regard to the distribution of the data, and do not use distance measures determined prior to cluster formation. They have the drawback that the data structure is not always readily apparent from the resulting analysis.

Although most of these methods can be implemented in numerical algorithms, use of a particular algorithm requires at least some a priori knowledge of the structure of the data (which is often not available in the first place). For example, complete link, group average and centroid hierarchical methods tend to favour spherical clusters with high internal affinity. In contrast, the single link hierarchical method is predisposed towards elongated clusters, which could be comprised of very dissimilar members. This is referred to as chaining on account of the fact that a unit may join a group on the basis of the similarity of just one of its members to that group. Likewise, the application of k-means algorithms either requires specification of the number of clusters to be identified, or the distances between clusters. Again, this information is often not available in practice. Since hierarchical and k-means cluster analysis are used in most applications, these methods are considered in more detail below.

6.3. HIERARCHICAL CLUSTER ANALYSIS

The variants of hierarchical cluster analysis are distinguished by the way in which the groups are formed. The so-called single link, and complete link agglomeration of the data illustrate these concepts.

6.3.1. Single link or nearest neighbour method

The single link method uses a nearest neighbour rule, and starts out by finding the two objects separated by the shortest distance, as indicated in Figure 6.5. These two objects constitute the first cluster. In the next step, there are two possibilities. Either a third object can join the first two, or the two closest unclustered objects can join to form a second cluster. This depends on whether the distance between the clustered objects and one of the unclustered objects is smaller than the distance between the two closest unclustered objects. This agglomerative process continues until only a single cluster remains.

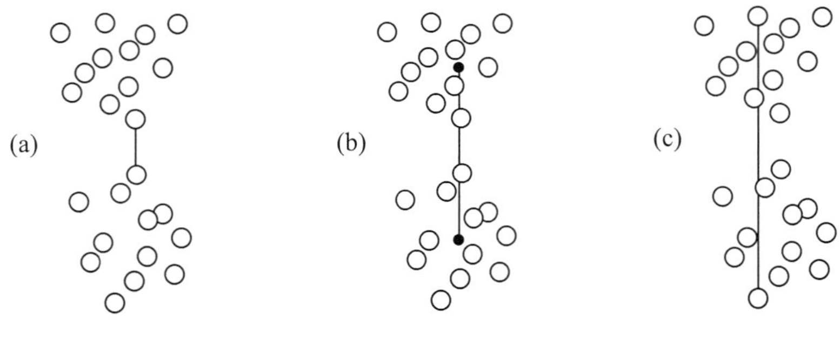

Figure 6.5. Inter-cluster distances defined by (a) nearest neighbour, (b) average link (similar to centroid), and (c) furthest neighbour methods.

For example, consider the set of data (5 observations on 4 variables) shown below. The 5 x 5 Euclidean distance matrix is derived by calculating $d_{ij} = \{\Sigma_{i=1}{}^{N}\Sigma_{j=1}{}^{N}(x_i - x_j)^2\}^{½}$ for all the observations, e.g. $d_{12} = \{(1.2\text{-}2.5)^2 + (2.2\text{-}1.5)^2 + (2.0\text{-}1.0)^2 + (1.5\text{-}2.0)^2\}^{½} = d_{12} = 1.852$

x_1	x_2	x_3	x_4
1.2	2.2	2.0	1.5
2.5	1.5	1.0	2.0
2.2	1.2	1.5	2.5
2.5	2.5	1.2	1.5
1.5	1.3	2.3	2.2

The similarity matrix (D_1) is shown below.

	1	**2**	**3**	**4**	**5**
1	0.000	1.852	1.803	1.556	1.217
2	1.852	0.000	0.825	1.136	1.664
3	1.803	0.825	0.000	1.694	1.109
4	1.556	1.136	1.694	0.000	2.035
5	1.217	1.664	1.109	2.035	0.000

The minimum distance is d_{23} = 0.825, and these two objects are therefore joined first. The distances between the cluster (23) and the remaining objects are consequently calculated.

$$d_{(23)1} = \min\{d_{21}, d_{31}\} = d_{31} = 1.803$$

$$d_{(23)4} = \min\{d_{24}, d_{34}\} = d_{24} = 1.136$$

$$d_{(23)5} = \min\{d_{25}, d_{35}\} = d_{35} = 1.109$$

```
                      Rescaled Distance Cluster Combine

             0         5        10        15        20        25
             +---------+---------+---------+---------+---------+

Object 2      -+-----------------------------------+
Object 3      -+                                   +-+
Object 5      -------------------------------------+ +---------+
Object 4      ---------------------------------------+         I
Object 1      -------------------------------------------------+
```

Figure 6.6. Dendrogram of data using single linkage based on Euclidean distance.

A new matrix of distances (D_2) can therefore be formed

	1	**(23)**	**4**	**5**
1	0.000	1.803	1.556	1.217
(23)	1.803	0.000	1.136	1.664
4	1.556	1.136	0.000	2.035
5	1.217	1.109	2.035	0.000

The smallest distance in this matrix is $d_{(23)5}$ = 1.109. This means that object number 5 will join the cluster formed by objects number 2 and 3. A new distance matrix, with elements 1, (235) and 4 can thus be formed, and the whole process can be repeated, until a single cluster containing all the elements remains. The results are usually summarized in the form of a dendrogram, as shown below in Figure 6.6.

6.3.2. Complete link or furthest neighbour method

The complete link method is the opposite of the single link method, in the sense that distance is now defined as the distance between the most distant pair of individuals. The approach is best considered by way of illustration, as before. Considering the same data set (x_1, x_2, x_3, x_4, x_5) as above, the first step again results in the fusion of individuals 2 and 3, as was the case with the single link method. However, the distance between this cluster, and the remaining individuals is computed differently, as follows.

$$d_{(23)1} = \max\{d_{21}, d_{31}\} = d_{21} = 1.852$$

$$d_{(23)4} = \max\{d_{24}, d_{34}\} = d_{34} = 1.694$$

$$d_{(23)5} = \max\{d_{25}, d_{35}\} = d_{25} = 1.664$$

This gives D_2

	1	**(23)**	**4**	**5**
1	0.000	1.852	1.556	1.217
(23)	1.852	0.000	1.694	1.664
4	1.556	1.694	0.000	2.035
5	1.217	1.664	2.035	0.000

The smallest distance in this matrix is $d_{15} = 1.217$. This means that object number 1 will join object number 5 to form a new cluster (15). A new distance matrix, with elements (15), (23) and 4 can thus be formed, and the whole process can be repeated, until a single cluster containing all the elements remains. The resulting dendrogram is shown in Figure 6.7. Variants such as average linkage[39] and centroid[40] linkage of the two extremes (nearest and furthest neighbours) are handled in a similar way.

```
                        Rescaled Distance Cluster Combine

             0         5        10        15        20        25
             +---------+---------+---------+---------+---------+

Object 2     -+----------------------------------+
Object 3     -+                                  +-------------+
Object 4     ------------------------------------+             I
Object 1     -----------------+-----------------------------+
Object 5     -----------------+
```

Figure 6.4. Dendrogram of data using complete linkage based on Euclidean distance.

The great advantage of the hierarchical distance class of methods is their flexibility in terms of the use of similarities[41] (which could be based on combinations of continuous, binary, or multicategory variables). The disadvantage of the approach is that the (dis)similarities have to be determined prior to grouping of the data, although these (dis)similarities depend on the scale of the original variables, as well as correlation between the variables. For example, if the data were continuous, and more or less normally distributed, the Mahalanobis distance might

[39] Average distances between all pairs of points or clusters are considered.
[40] Based on distances between the centroids of points or clusters.
[41] Note that once the dissimilarity matrix has been calculated, the original features of the data are forgotten.

be a suitable dissimilarity measure. Unfortunately this distance cannot be calculated if the groups are unknown, and standardizing on the basis of the whole data set may lead to long thin clusters, and inefficient clustering.

6.4. OPTIMAL PARTITIONING (K-MEANS CLUSTERING)

k-Means clustering is an optimization or partitioning technique in which n objects on m measurements are partitioned, based on the similarities (Euclidean distances) of the objects. It is a simple algorithm in which a predetermined number of cluster centres have to be specified prior to running the algorithm. The k starting points can be formed by a variety of ways, including random selection of k objects from the sample, or selection of k objects that are furthest apart in the sample. The algorithm then proceeds to assign each object to the nearest cluster. Cluster centres are subsequently recomputed, objects reassigned, etc., in an iterative fashion, until the clusters are all stable.

More formally, assume partitions are defined by P(n,k), i.e. n objects, each of which assigned to 1 of k clusters, the mean of the j'th variable in the L'th cluster is denoted by $x_{avg}(L,j)$, while the number of objects assigned to the L'th cluster is N(L). The distance between the i'th object and the L'th cluster can then be expressed as

$$d(i,L) = \{\Sigma_{j=1}^{m}\{x(i,j)-x_{avg}(L,j)]^2\}^{1/2} \qquad (6.9)$$

Moreover, the error component of the partition can be defined as

$$E[P(n,k)] = \Sigma_{i=1}^{m}d[i,L(i)]^{2} \qquad (6.10)$$

L(i) is the cluster that contains the i'th object, and d[i,L(i)] is the Euclidean distance between the i'th object and the mean of the cluster containing the object. A search is conducted to minimize E, by moving objects from one cluster to another, until no further reduction in E can be attained. In addition, unlike the hierarchical methods discussed previously, assignment of objects to clusters is not irrevocable.

6.5. SIMPLE EXAMPLES OF HIERARCHICAL AND k-MEANS CLUSTER ANALYSIS

Some of the characteristics of hierarchical and k-means cluster analysis methods can be highlighted by way of three simple examples. All data sets consist of two variables (x_1 and x_2) only, for easy visualization. The first set (*Strips*) can be seen as two elongated obliquely located clusters in Figures 6.5-6.6, the second (*Sphere*) as a circle enclosed by a ring (Figures 6.7-6.8), and the third (*Oscul*) as two osculating triangle-shape clusters, as shown in Figures 6.9-6.10.

The first set of clusters are difficult to separate with average, centroid or furthest link methods, as well as k-means algorithms that tend to favour spherical clusters. In contrast, these clusters are easily identified by use of single link hierarchical agglomeration algorithms. The second set of clusters containing both a spherical and a non-spherical cluster again presents problems for average, centroid or furthest link methods, but does not pose a problem for single link methods.

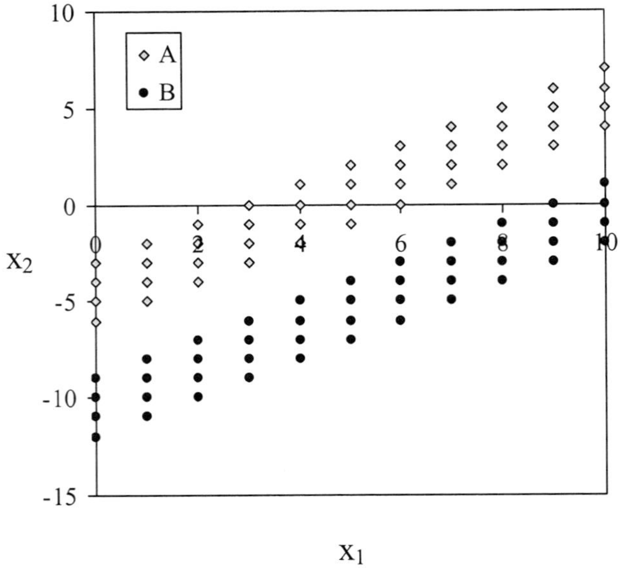

Figure 6.5. Single linkage clustering of *Strips* data set.

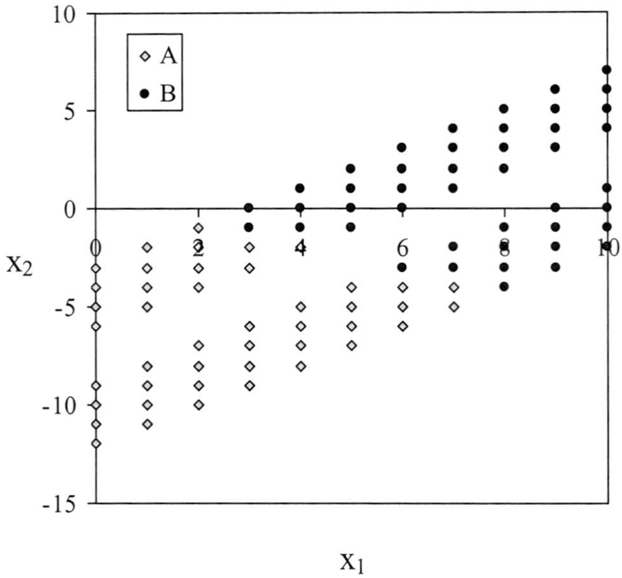

Figure 6.6. 2-Means clustering of *Strips* data set. Similar results are obtained with average, centroid or furthest link agglomerative hierarchical clustering algorithms.

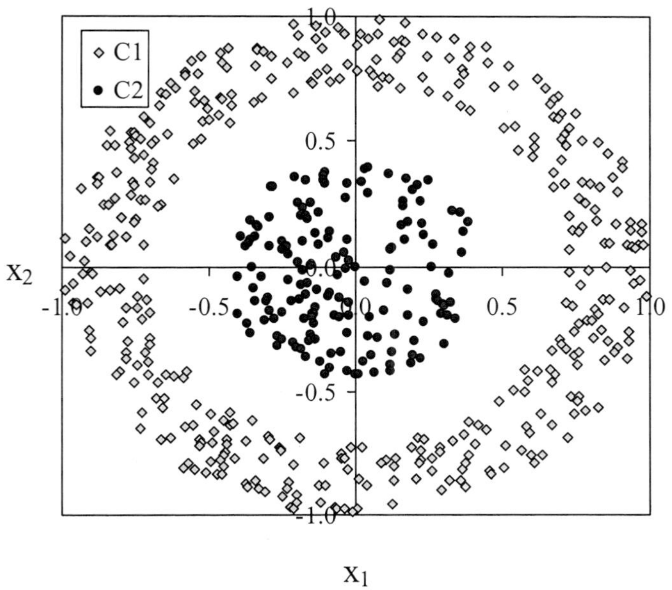

Figure 6.7. Single linkage clustering of *Sphere* data set.

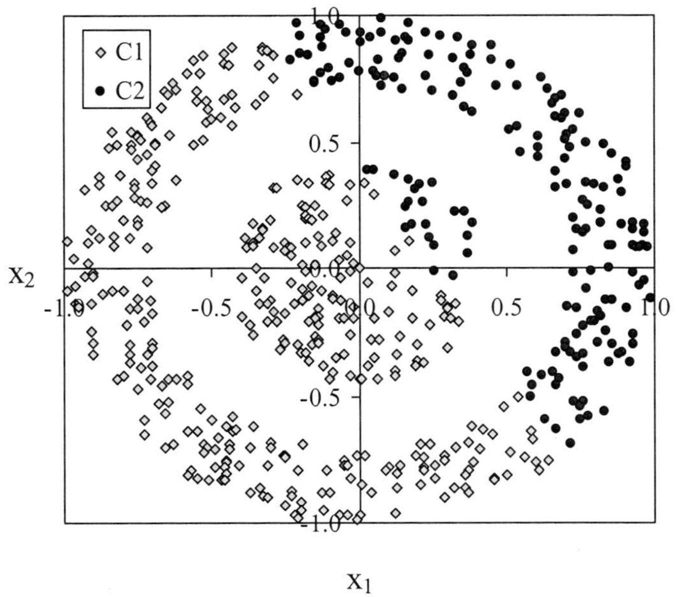

Figure 6.8. 2-Means clustering of *Sphere* data set. Similar results are obtained with average, centroid or furthest link agglomerative hierarchical clustering algorithms.

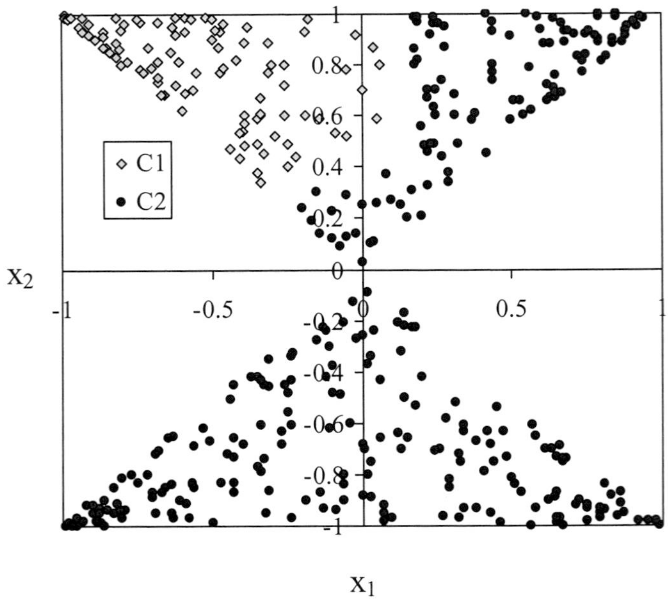

Figure 6.9: Single linkage clustering of *Oscul* data set.

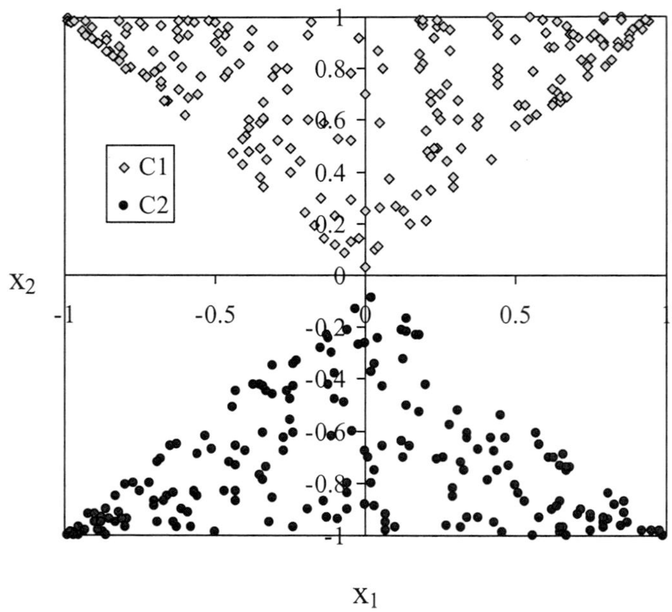

Figure 6.10. 2-Means clustering of *Oscul* data set. Similar results are obtained with average, centroid or furthest link agglomerative hierarchical clustering algorithms.

On the other hand, the osculating clusters are difficult to separate with single link methods, owing to the tendency of these techniques to chain. In fact, the outcome of the single link cluster analysis is very sensitive to the locations of points near the adjoining tips of the clusters. Slight changes in the coordinates of just a few of these points could result in perfect separation of the two clusters, or no separation of the clusters at all. The results of k-means analysis are also shown. In all cases similarities were based on Euclidean distances between the variables.

6.6. CLUSTERING OF LARGE DATA SETS

The majority of clustering algorithms cannot deal with large data sets[42], as indicated by the time and space complexities of some of these algorithms listed in Table 6.4. For example, for the hierarchical agglomerative clustering methods discussed previously, the time complexity scales with $O(n^2 \log n)$ (Kurita, 1991; Choudhury and Murti, 1990). The space complexity of these algorithms is $O(n^2)$, since a similarity matrix of size n^2 has to be stored. For example, to cluster every pixel in a 100 x 100 image, approximately 200 MB of storage would be required (Jain et al., 1999). Although it is possible to compute the elements of the matrix as needed, instead of storing them, this would increase the time complexity of the algorithm (Anderberg, 1973).

The popularity of the k-means algorithm (equivalent to Kohonen's self-organizing feature map) is also evident from Table 6.4. Its time complexity is $O(nkl)$, where k and l are typically fixed in advance, which means the algorithm has a linear time complexity with regard to the size of the data set (Day, 1992). Its space complexity is $O(k+n)$, as it requires additional space to store the data matrix (storing it in secondary memory, means a huge increase in time complexity). Moreover, the algorithm is order-independent for a given initial seed set of cluster centres. However, the k-means algorithm is senstitive to this initial placement of the cluster centres and can only produce hyperspherical clusters (unlike single link hierarchical methods, for example).

Table 6.4. Computational characteristics of various clustering algorithms, where n is the number of objects, k the number of clusters and l the number of iterations.

Clustering Algorithm	**Time Complexity**	**Space Complexity**
ART/leader	$O(nk)$	$O(k)$
k-means/Kohonen	$O(nkl)$	$O(k+n)$
ISODATA	$O(nkl)$	$O(k)$
Minimum spanning tree	$O(n^2)$	$O(n)$
Single/Complete link	$O(n^2 \log n)$	$O(n^2)$

Incremental algorithms, such as the leader algorithm (Hartigan, 1975) and its neural network equivalent, the adaptive resonance theory neural network (Carpenter and Grossberg, 1990), have space complexities depending on the number of clusters only, as well as time complexities, which are linearly related to the number of data. Although these algorithms can therefore also be used for large data sets, they have the drawback that the clusters they produce are dependent on the order in which the data are presented to the algorithm.

---

[42] Currently, in the order of millions of patterns, as opposed to thousands in the 1960s.

6.7. APPLICATION OF CLUSTER ANALYSIS IN PROCESS ENGINEERING

Although cluster analysis is often used directly by individual analysts to explore process data, it is also used extensively as an integral part of larger systems, such neural networks (radial basis function neural networks, self-organizing feature maps, etc.). One of the areas where it is becoming increasingly important is in the development of expert systems, especially where human expertise is insufficient for the design of the knowledge base. Under these circumstances, process data can be used as an aid to the construction of the data base. The idea is illustrated in Figure 6.11, where relevant data are extracted from a data base and knowledge are extracted from the data, usually in the form of IF-THEN rules. The extraction of rules from the data can be conducted with classification trees, neural networks (see Chapter 7), or some other means. These rules would encode specific plant or process knowledge, as opposed to the human expert's knowledge, which is typically general and specific (deep knowledge and surface knowledge).

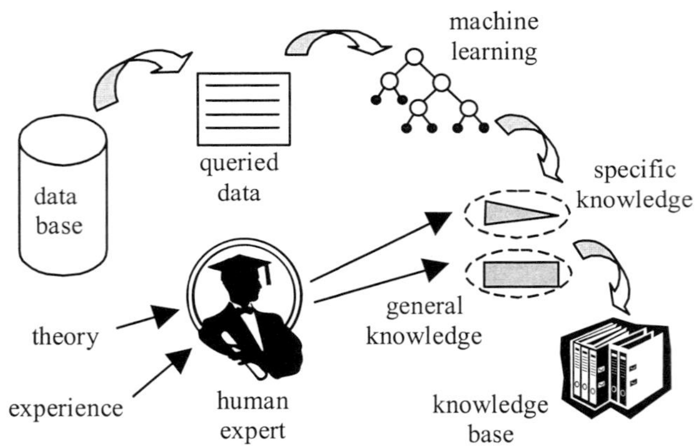

Figure 6.11. Design of knowledge-based systems.

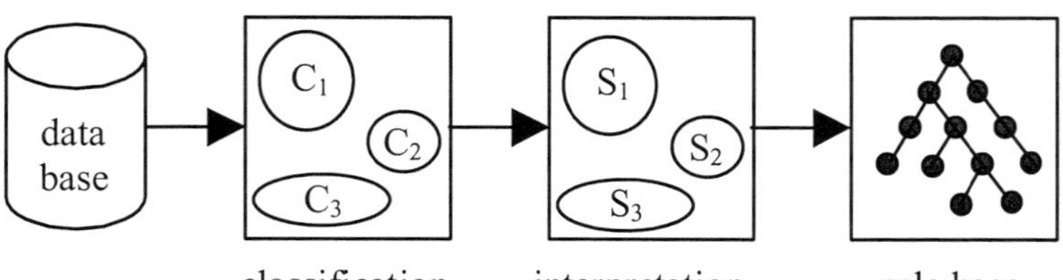

Figure 6.12. Use of cluster analysis to construct knowledge-based systems.

In essence, the interpretation of the data usually implies the finding of clusters in the data, and interpretation of these clusters, as indicated in Figure 6.12. Once the clusters are interpreted or labelled, they can be integrated with the elements of a knowledge base. An example of such a process is given by R-Roda et al. (2001) for dealing with complex biological wastewater treatment plants.

For example, Pulkkinen et al. (1993) applied an integrated approach by use of cluster-based models in conjunction with an expert system, for controlling a grinding and flotation circuit. Likewise, Laine et al. (1995) have used cluster analysis for the on-line classification of different ore types at the Hitura nickel mine in central Finland. The algorithm (Ylinen et al., 1992) formed multidimensional ellipsoids in input space and was used in conjunction with an expert system used to control the plant.

Ginsberg and Whiten (1991) have discussed the use of cluster analysis in mineral processing applications, including the assessment of plant performance, calibration of equipment for on-stream analyses and the analysis of semi-autogenous grinding mills. They have used a leader type partitioning algorithm, using a distance measure similar to the Mahalanobis distance, except that it was based on a local covariance matrix.

Huang et al. (2000) have made use of cluster analysis in conjunction with structured partial principal component analysis for the isolation of faults in nonlinear systems. They have used both k-means and nearest neighbour hierarchical clustering algorithms.

Wang and McGreavy (1998) have used a Bayesian automatic classification method developed by NASA (Cheeseman et al., 1988; Cheeseman and Stutz, 1996) for the analysis of fluid catalytic cracking processes in refineries - a problem previously considered by Whiteley and Davis (1992). The method finds the most probable set of class descriptions given the data and their prior expectations by assuming variables to be independent in each class. Unlike other methods, the variables can be real or discrete, with missing values and the observations are assigned to classes with a degree of certainty, rather than absolutely.

6.8. CLUSTER ANALYSIS WITH NEURAL NETWORKS

Competitive neural networks, such as adaptive resonance theory, learning vector quantization and self-organizing maps, lend themselves naturally to cluster analysis. For example, in self-organizing maps the nodes are distributed in the data space and the weight vector of each node could be interpreted as the mean of a cluster. This idea has served as the basis for a number of networks designed to do cluster analysis. Although neural networks have the advantage that they can be implemented in parallel as high-speed neurocomputers, to accommodate large volumes of data, the different algorithms suffer particular drawback as well. For example, in the case of self-organizing maps and their variants, the number of classes is predefined by the user, similar to that of the k-means algorithm. This presupposes some knowledge of the data distribution in the first place. Although this is not the case where adaptive resonance theory neural networks are used, the specification of a vigilance parameter again requires some knowledge of the distribution of the data. Moreover, both approaches are suitable for detecting hyperspherical clusters only (Hertz et al., 1991).

Examples of the application of neural networks and conventional algorithms for the clustering of data are illustrated in the case studies below.

6.8.1. Cluster analysis with autoassociative neural networks

The methodology for the analysis of clusters in a given data set or matrix $\mathbf{X} \in \Re^{m \times n}$, is based on the use of a two-layer autoassociative neural network with at least one single-node hidden layer and an output layer of size n (Aldrich, 2000). The network is trained on the data to be clustered, and once it has converged, the outer layer of the neural network is removed and the outputs of the single-node hidden layer are recorded. The network therefore maps the n input vectors of dimension m, $\mathbf{x} = (x_1, x_2, \ldots x_m)$ to n (scalar) features z of dimension 1. A feature curve is consequently constructed by ranking the n features extracted by the hidden layer. This amounts to an ordering of the original observations according to the similarities of the observations, i.e. similar observations are located in similar regions on the feature curve. Unlike the case with other clustering techniques, neither the explicit definition of a similarity criterion, nor the a priori specification of the number of clusters is required. Since clusters are delineated by discontinuities or low-density regions on the feature curve, a more intelligent analysis of clustered data is possible, based on post priori information on the distribution of the clusters. The methodology is demonstrated by way of example below.

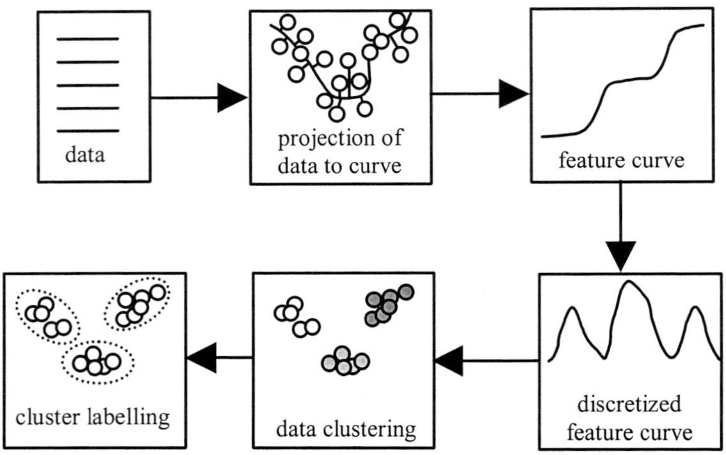

Figure 6.13. Cluster analysis by projection of the data to a curve, ranking of the scores to generate a feature curve, discretization of the feature curve, clustering of the data and interpretation of the clusters.

6.8.2. Example 1: Sn-Ge-Cd-Cu-Fe-bearing samples from Barquilla deposit in Spain

Tetrahedrite samples were taken from a vein, as well as the surrounding host rock, in the Barquilla deposit in Salamanca, Spain (Pascui et al., 1997). The samples consisted of 11 measurements of the various elements (mass percentages), as shown in Table 6.5. These samples were presented as input vectors of the form $\mathbf{x}$ = (%Cu, %Ag, %Zn, %Fe, %Mn, %Cd, %Sb, %As, %Bi, %Ge, %S) to an autoassociative feedforward network. The network consisted of an input layer with 11 nodes, a single-node sigmoidal hidden layer, as well as a sigmoidal output layer with 11 nodes. The network was trained with the generalized delta rule and an epoch size equal to the size of the data set (Hinton, 1989). It converged after approximately

50 000 iterations. After convergence, the output layer of the network was removed, and the features (**s**) were generated by passing all the data through the bottom half of the network, similar to the one shown in Figure 6.13. The values of the ranked features are shown in Figure 6.14.

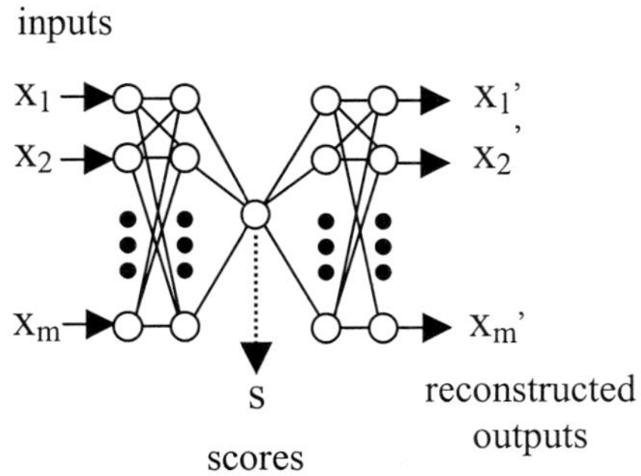

Figure 6.14. Use of autoassociative neural networks for cluster analysis.

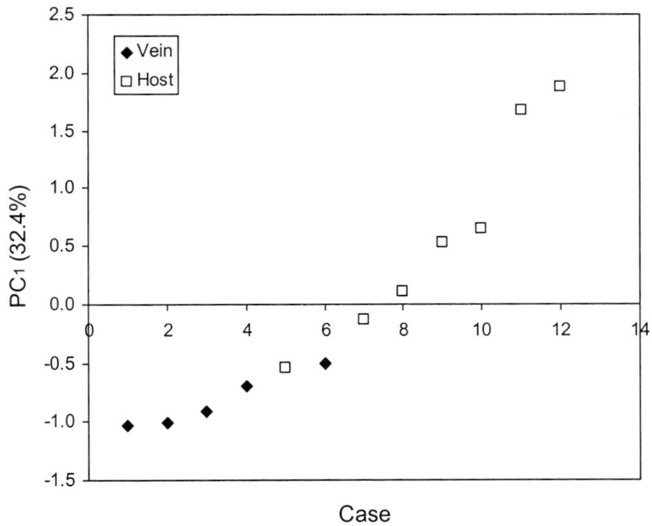

Figure 6.15. Values of ranked features.

The values above the markers correspond to the case or sample numbers indicated in Table 6.6. From these features, two major clusters can be identified. Except for case 9, the first cluster consists of all the measures related to the vein samples, while the second cluster is composed of all the samples from the host rock. As indicated by the feature curve, the second cluster consists of samples (cases) 6, 7, 8, 10, 11, 12. Samples 11 and 12 appear to be relatively distant from samples 6, 7, 8 and 10 in the cluster, and could be seen as a subcluster within the second cluster.

The results summarized in Table 6.5 can be compared with those obtained with other methods, viz. single-link and centroid-based hierarchical algorithms, as well as a k-means algorithm, as described in more detail below. Note that in the latter case two clusters specified a priori, which may not be realistic in practice. From these results it can be seen that the autoassociative neural network can therefore be used as a more accurate means of identifying the source of an ore during exploration.

Table 6.5. Composition of tetrahedrite samples from the Barquilla deposit in Salamanca, Spain.

Case	Source	%Cu	%Ag	%Zn	%Fe	%Mn	%Cd	%Sb	%As	%Bi	%Ge	%S	Total
1	Vein	36.37	1.96	2.39	1.96	0.00	4.71	27.98	0.97	0.16	0.00	23.40	99.90
2	Vein	36.38	1.93	2.56	1.92	0.00	4.65	28.27	0.91	0.20	0.00	22.75	99.57
3	Vein	34.93	3.51	2.00	2.11	0.00	4.83	27.63	1.11	0.18	0.00	23.30	99.60
4	Vein	34.72	3.89	1.75	2.01	0.00	4.97	27.21	1.08	0.35	0.00	23.49	99.47
5	Vein	34.72	2.84	1.77	2.11	0.00	5.12	27.43	1.03	0.29	0.00	23.27	98.58
6	Host	38.18	0.24	1.32	1.44	0.83	6.03	22.07	5.34	0.37	0.00	24.62	100.44
7	Host	38.46	0.94	0.00	5.02	0.00	0.77	28.67	1.03	0.00	0.00	24.65	99.54
8	Host	38.05	2.75	1.47	1.39	0.00	3.51	25.03	1.45	0.80	0.72	24.09	99.26
9	Host	34.98	2.09	0.35	0.79	0.00	10.96	24.85	2.24	0.00	0.27	23.56	100.09
10	Host	36.38	2.33	1.94	2.50	1.87	2.53	25.60	1.39	0.54	0.00	25.14	100.22
11	Host	35.66	1.12	0.77	0.60	0.00	9.60	22.97	2.05	3.69	0.00	22.84	99.30
12	Host	39.10	0.47	2.86	2.09	0.00	1.86	22.61	5.06	1.04	0.00	25.21	100.30

Table 6.6. Allocation of samples to two different classes by different clustering methods.

		Classification							
		ANN		Single Link		Centroid Link		k-Means	
Real	Vein	5	0	5	0	5	0	5	0
	Rock	1	6	5	2	6	1	4	3
Total correct		11		7		6		8	

6.8.3. Example 2: Chromitite ores from the Bushveld Igneous Complex

The chemical compositions of 47 samples of chromitite ores from the Bushveld Igneous Complex in South Africa (De Waal, 1975) are summarized in Table 6.7 below. Each ore is composed of 13 oxides and the structure of the data is shown in Figure 6.15. The multidimensional scaling map of the data suggests the presence of two or more clusters, as indicated in Figure 6.15. This data set was presented to an autoassociative neural network consisting of an input and an output layer of 13 nodes each (for each of the input variables), as well as three sigmoidal hidden layers, with a 3:1:3 node configuration. As before, the network was trained with the generalized delta rule, and converged after approximately 80 000 iterations (the epoch again consisting of the entire training data set). The feature curve generated by the network is shown in Figure 6.15. As can be seen from this figure, the presence of three clusters is strongly suggested.

As discussed in Chapter 5, multi-dimensionally scaled (MDS) maps project the data to lower dimensions, while preserving the structure of the data (configuration of interpoint distances) as best as possible, as opposed to the preservation of maximum variance pursued with principal components.

In order to quantify the solutions obtained with the various algorithms, a clustering index I was computed for each, which provided a measure of similarity with the clusters portrayed in Figure 6.15.

$$I = (\Sigma_i \Sigma_j a_{ij} - n)/(n^2 - n), \text{ with } i = 1,2, \ldots n \text{ and } j = 1,2, \ldots n \qquad (6.9)$$

In the two reference groups shown in Figure 6.15, an n x n relationship matrix was computed, with n (= 47) the number of observations or ore samples. The elements a_{ij} of the matrix assumed binary values of either zero or one, that is $a_{ij} = 1$ if the i'th and the j'th observations belonged to the same cluster, otherwise $a_{ij} = 0$. These matrices were set up for all the algorithms, and the cluster index (equation 6.9) simply consisted of the percentage of matches in the algorithm, corrected for the diagonal elements, which will always have unity values.

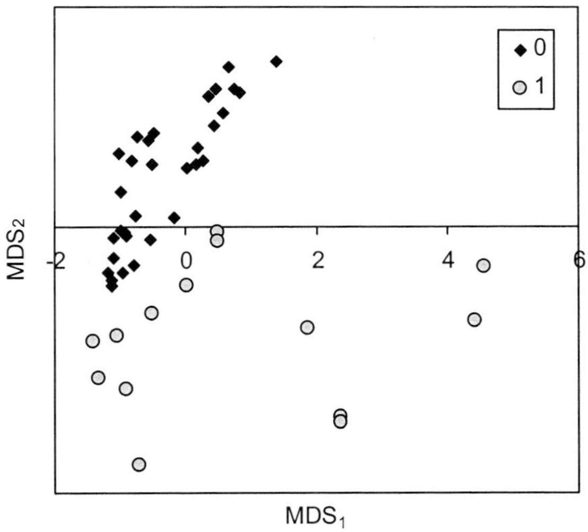

Figure 6.16. Multidimensional scaling map of the oxide compositions (13 components) of the ores from the Bushveld Igneous Complex in South Africa, summarized in Table 6.7.

The classification of the data shown in Table 6.7 (with actual classes denoted by a zero or a one, is summarized in Table 6.8. As can be seen from this table, the clusters (and classification) by use of the feature maps generated with the autoassociative neural network were better than those obtained with the other hierarchical methods, as well as the k-means approach. The clusters generated by the autoassociative neural network provided an average match of 61.7% with the reference clusters, while those of the single linkage, centroid linkage and k-means algorithms yielded indices of 46.8%, 46.8% and 48.9% respectively.

Table 6.7. Chromitite ores from the Bushveld Igneous Complex.

ORE	SiO_2	Al_2O_3	Fe_2O_3	FeO	MgO	CaO	TiO_2	Cr_2O_3	MnO	ZnO	NiO	V_2O_5	CoO	Class
G7BX	27.82	4.92	5.30	6.62	23.59	0.30	0.13	19.97	0.15	0.05	0.58	0.06	0.019	1
G7CX	8.70	9.82	4.61	10.79	14.32	0.23	0.25	47.60	0.22	0.12	0.29	0.14	0.018	1
G7DX	28.26	5.07	5.80	6.73	24.27	0.32	0.13	20.26	0.15	0.06	0.59	0.07	0.015	1
G8AX	14.37	12.23	3.21	15.65	13.74	7.73	0.37	31.41	0.22	0.08	0.13	0.21	0.017	1
G9BX	14.32	12.20	3.65	15.35	13.92	7.53	0.38	31.7	0.21	0.08	0.13	0.21	0.017	1
G9FX	46.40	3.91	0.86	10.47	28.90	2.90	0.18	5.62	0.21	0.03	0.10	0.04	0.013	1
G9GX	2.02	14.23	6.30	19.63	11.18	0.29	0.57	44.70	0.22	0.10	0.13	0.31	0.026	0
G10AX	17.05	18.41	0.22	19.55	10.38	4.36	0.65	27.31	0.16	0.09	0.17	0.22	0.028	0
G10BX	4.13	15.30	5.20	20.36	10.10	0.65	0.75	42.56	0.19	0.12	0.16	0.37	0.026	0
G10CX	18.44	17.33	6.58	14.91	8.66	4.60	0.80	26.56	0.18	0.10	0.18	0.31	0.024	0
G10DX	15.56	17.58	2.56	19.69	9.64	4.30	0.81	28.78	0.18	0.12	0.15	0.35	0.026	0
G11AX	2.65	14.77	6.89	17.53	11.47	0.46	0.62	44.79	0.19	0.08	0.15	0.34	0.015	0
G11BX	3.26	14.51	6.24	18.11	11.03	0.50	0.66	45.00	0.19	0.08	0.16	0.37	0.024	0
G11CX	2.56	13.92	7.28	19.56	10.42	0.78	0.65	43.43	0.22	0.08	0.16	0.40	0.029	0
G11EX	9.01	12.24	5.03	16.27	13.24	0.90	0.50	41.95	0.18	0.05	0.13	0.26	0.019	1
G11GX	5.77	14.62	5.27	17.41	11.23	0.87	0.57	42.45	0.19	0.07	0.14	0.38	0.019	0
G11HX	12.01	12.68	5.17	16.34	15.37	0.68	0.52	36.36	0.20	0.06	0.13	0.22	0.018	1
G189MX	4.14	13.71	6.02	18.96	11.31	0.42	0.60	43.33	0.21	0.08	0.15	0.46	0.024	0
G191BX	47.94	3.76	1.97	11.18	26.02	3.88	0.19	3.49	0.22	0.02	0.11	0.07	0.013	1
G191DX	4.57	15.34	5.70	19.36	9.92	0.71	0.67	42.22	0.19	0.10	0.14	0.39	0.029	0
G191FX	4.88	15.16	5.98	19.12	9.95	0.77	0.69	42.36	0.19	0.09	0.15	0.41	0.024	0
G192PX	7.96	13.85	2.36	21.25	11.95	0.93	0.58	39.34	0.21	0.08	0.16	0.34	0.016	0
G192RX	2.37	15.33	2.87	21.04	10.04	3.93	1.13	40.54	0.25	0.09	0.15	0.47	0.022	0
G192SX	3.97	18.57	2.04	22.22	8.66	0.80	1.01	40.92	0.20	0.11	0.15	0.40	0.031	0
G192TX	7.21	17.63	2.33	20.61	10.50	2.06	0.82	37.19	0.19	0.07	0.18	0.38	0.020	0
G192UX	6.86	14.07	1.02	25.35	9.40	0.91	1.67	38.76	0.20	0.11	0.25	0.59	0.022	0
G192VX	5.16	17.96	0.65	21.80	10.23	1.17	0.90	39.61	0.19	0.08	0.16	0.42	0.020	0
G193AX	1.46	14.17	4.26	17.11	12.23	0.33	0.48	49.10	0.17	0.06	0.17	0.28	0.022	1
G193AX(2)	1.26	14.15	4.65	16.96	12.08	0.34	0.50	49.17	0.17	0.08	0.15	0.30	0.020	1
G193DX	4.82	12.30	4.71	16.76	11.98	0.97	0.47	46.43	0.18	0.07	0.16	0.27	0.016	1
G193EX	3.37	11.89	3.22	17.03	12.47	0.60	0.42	50.02	0.17	0.06	0.14	0.24	0.022	1
G193KX	6.90	11.16	2.52	15.76	13.81	1.23	0.39	47.36	0.18	0.06	0.15	0.21	0.021	1
G193MX	25.14	9.12	2.32	12.79	22.61	2.63	0.33	23.29	0.17	0.03	0.18	0.13	0.016	1
G193QX	3.22	15.20	5.74	19.80	10.07	0.63	0.63	43.63	0.20	0.08	0.17	0.43	0.024	0
G193TX	14.52	12.69	8.78	19.97	10.18	1.65	2.81	27.30	0.33	0.11	0.17	0.55	0.022	0
G193UX	16.59	18.98	5.58	15.16	10.61	4.61	0.67	26.68	0.16	0.06	0.17	0.35	0.020	0
G194AX	13.77	16.58	4.91	17.36	9.21	2.52	0.74	32.02	0.18	0.07	0.17	0.38	0.026	0
G194BX	23.47	15.96	4.94	15.14	9.00	5.54	0.81	22.36	0.17	0.08	0.17	0.22	0.015	0
G194CX	15.43	15.37	6.01	18.72	6.93	3.16	1.01	29.96	0.21	0.09	0.16	0.37	0.026	0
G194DX	13.80	16.23	7.00	16.02	9.84	2.25	0.71	31.51	0.17	0.08	0.15	0.36	0.018	0
G194EX	14.61	18.23	6.05	15.66	9.98	2.74	0.59	29.44	0.16	0.08	0.16	0.30	0.020	0
G194FX	11.44	16.15	8.41	16.47	9.96	1.98	0.76	33.14	0.19	0.08	0.17	0.35	0.018	0
G194GX	12.73	15.53	6.43	17.60	9.82	2.08	0.81	32.06	0.18	0.08	0.16	0.36	0.022	0
G194HX	11.01	13.94	6.76	17.33	10.01	1.60	0.72	36.49	0.20	0.08	0.15	0.31	0.018	0
G201AX	7.12	20.19	3.82	19.18	10.52	1.43	0.58	35.98	0.18	0.09	0.19	0.40	0.020	0
G201AX(2)	6.19	20.09	6.04	17.85	10.00	1.20	0.59	36.68	0.18	0.09	0.19	0.40	0.020	0
G201BX	5.32	15.71	6.66	19.15	10.47	0.66	0.77	40.06	0.20	0.08	0.23	0.37	0.018	0

Table 6.8. Allocation of samples to two different classes by different clustering methods, viz. an autoassociative neural network (ANN), single-link and centroid link hierarchical clustering methods, as well as the k-means.

		Classification											
	Class	**ANN**			**Single Link**			**Centroid Link**			**k-Means**		
Real	**A**	2	2	1	0	1	4	0	1	4	1	3	1
	B	1	12	7	0	1	19	0	1	19	0	1	7
	C	2	5	15	1	0	21	1	0	21	1	1	9
Total		**29**			**22**			**22**			**23**		

6.8.4. Example 3: Data from an industrial flotation plant

The third example is based on data obtained from an industrial platinum flotation plant in South Africa, that formed part of a project to investigate the feasibility of an on-line monitoring system based on estimation of the state of the plant from the features of the froths in the cells.

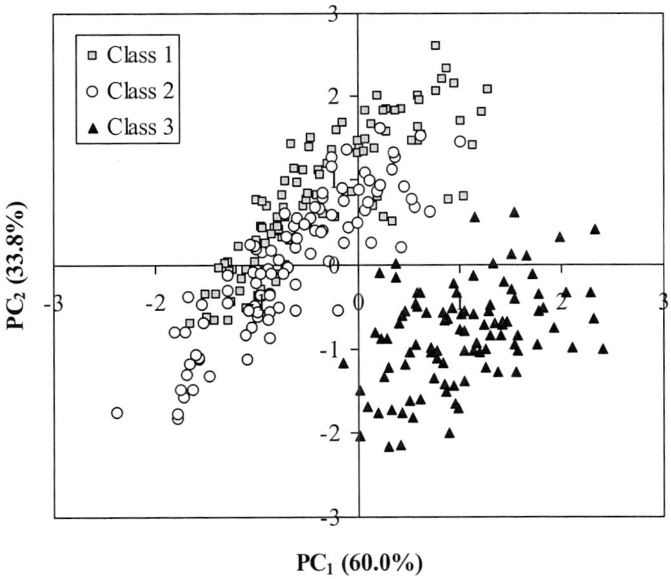

Figure 6.17. Principal component map of three froth classes on a platinum flotation plant. The first two principal components collectively explain 93.8% of the variance of five image feature variables extracted from digital images of the froths.

A set of images of the froths was consequently classified by a plant expert in order to serve as exemplars for the development of a machine vision plant monitoring system that extracts statistical features from digitised images of the froth (Moolman et al., 1995a, 1995b). These sets of neighbouring grey level dependence matrix (Sun and Wee, 1983) features could be related to the performance of the plant, i.e. the recoveries and grades.

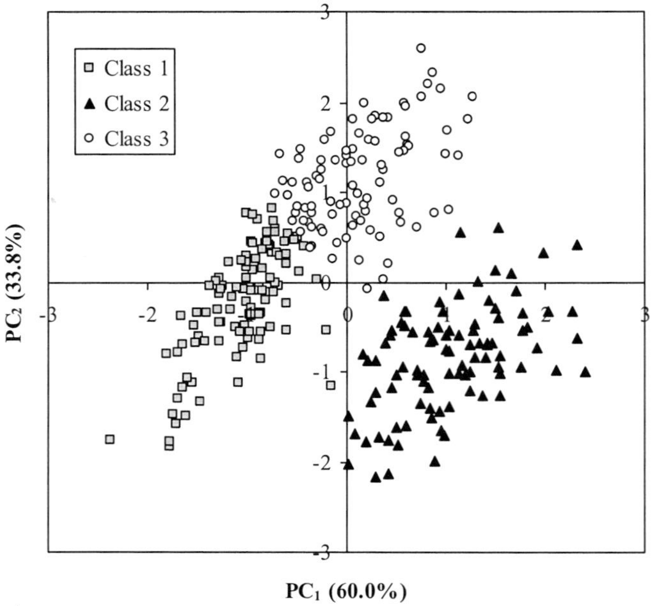

Figure 6.18. Principal component map of three clusters identified by use of k-means cluster analysis. The first two principal components collectively explain 93.8% of the variance of five image feature variables extracted from digital images of the froths.

The data consisted of a set of 297 samples of five features x_1, x_2, x_3, x_4 and x_5. The samples were obtained from three different operational regimes, referred to as classes 1, 2 and 3 respectively. Classes 1 and 2 are relatively similar, the main difference between the two classes being the difference in the concentration of reagents in the flotation cell, resulting in a froth with a higher viscosity. The higher viscosity is among other reflected in the somewhat different shapes of the bubbles. The third regime (class 3) involved froths with numerous small bubbles, depleted in minerals, and generally clearly distinguishable from classes 1 and 2. Correlation among the five froth features allowed accurate representation of the froths by means of two principal components, respectively accounting for 60.0% and 33.8% of the total variance of the features, as shown in Figure 6.17. Clustering of the data with a k-means algorithm is shown in Figures 6.18. Approaches based on hierarchical methods did not perform significantly better than the k-means algorithm (where 3 clusters were specified a priori).

In Figure 6.19, the feature curve generated by an autoassociative neural network with a 2:1:2 configuration is shown (open circles). The location of the curve in the data space is shown by means of a principal component map in Figure 6.20. Inspection of the feature curve clearly shows the presence of three distinct clusters in the data. For comparative purposes, a feature curve based on the 1st principal component scores of the data is also shown in Figure 6.19 (grey '+' markers). No structure is apparent from the latter curve and like the k-means algorithm, the data can only be clustered by specifying the number of clusters to be identified (regardless of whether these clusters are supported by the natural grouping of the data).

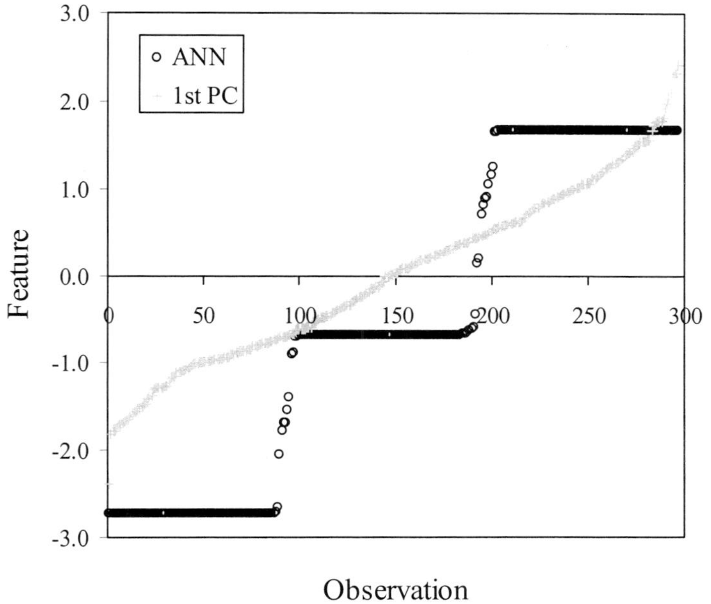

Figure 6.19. Feature curve (open circles) derived with an autoassociative neural network with a (2:1:2) hidden layer configuration. The feature curve derived from the scores of the first principal component is shown with grey '+' markers.

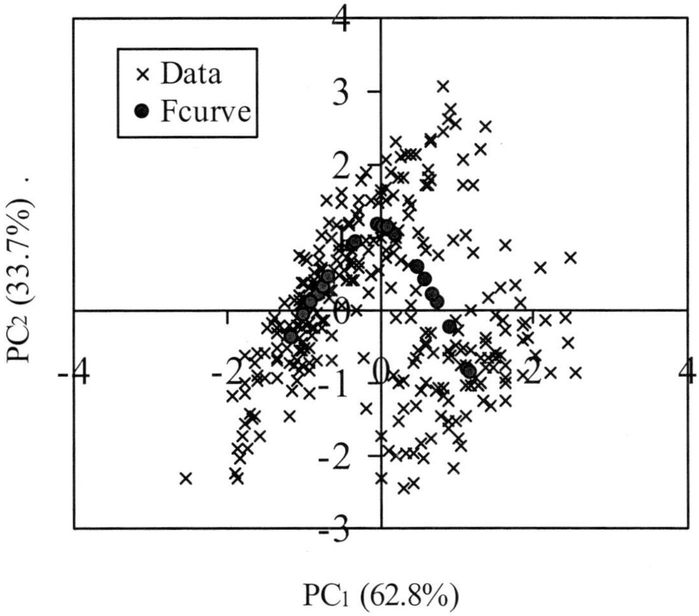

Figure 6.20. Fitting of the feature curve (solid circles) based on the nonlinear principal component scores to the data (cross marks).

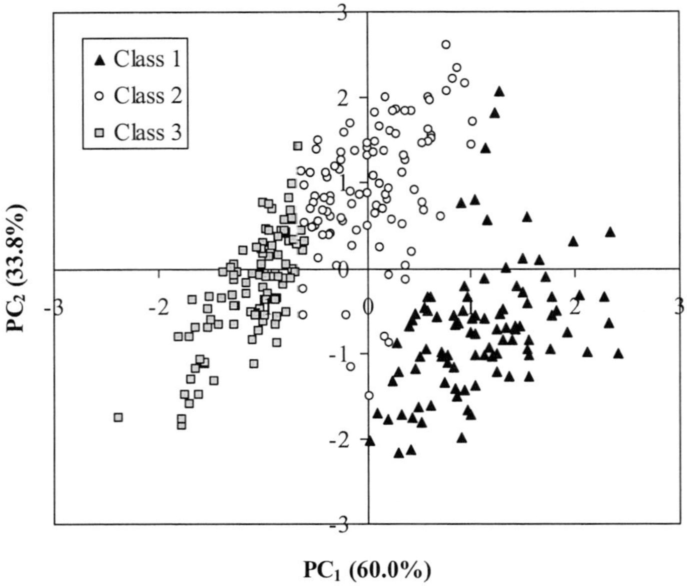

Figure 6.21. Clusters identified by the three percentiles of the scores on the first principal component of the data.

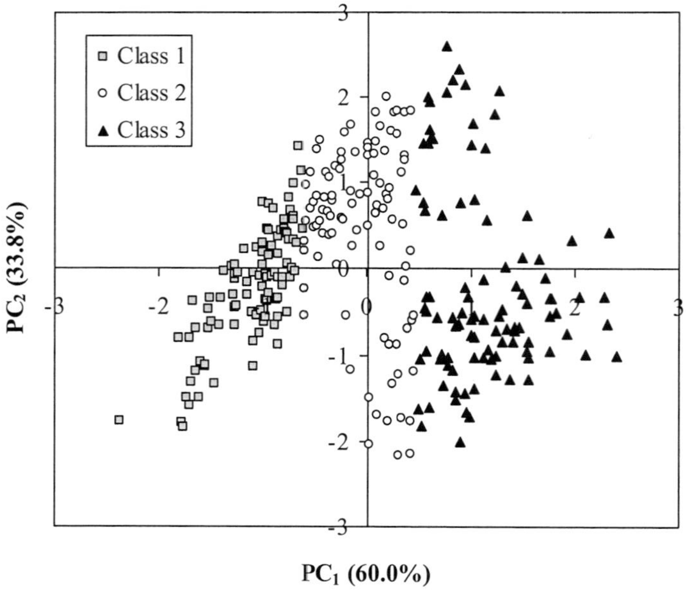

Figure 6.22. Principal component map of three clusters identified based on the feature curve shown in Figure 6.19. The first two principal ccmponents collectively explain 93.8% of the variance of five image feature variables extracted from digital images of the froths.

The clusters forced onto the data with the principal component feature curve are shown in Figure 6.21. This grouping is similar to the one obtained with the k-means algorithm. Figure 6.22 shows the clusters identified with the nonlinear curve generated by the neural network, which had three hidden layers with bipolar sigmoidal nodes. The first and the third hidden layers consisted of two nodes, while the second hidden layer consisted of one node only. The network was trained with the generalized delta rule. The squared correlation between the original data and the reconstructed data was $R^2 = 0.66$.

The use of neural networks to perform cluster analysis of data as outlined above can be classified as a non-parametric technique, similar to other methods used for local density estimation of data (Wishart, 1969, 1987). The results of the case studies considered above suggest that autoassociative neural networks provide a robust method for the identification of multivariate clusters in data. Moreover, no similarity criterion or mode of agglomeration or division needs to be specified beforehand. The interpretation of clusters is based on the analysis of a feature curve, or alternative a differenced form of the feature curve, generated by the bottleneck layer of the neural network.

As mentioned above, natural groupings in the data are viewed as regions of relatively high density surrounded by regions of relatively low density, similar to what has previously been more formally expressed by Hartigan (1975). Although intuitively appealing, this definition may not always be satisfactory in all situations where cluster analysis is used. For example, the number of modes may not be invariant under nonlinear transformation of the data, or the identification of modes can be notoriously difficult, even in the case of univariate data (Silverman, 1981). In these conditions, neural networks may not be the appropriate way of performing cluster analysis.

Other problems involve training of neural networks with more than two hidden layers. In the last case study for example, neural networks with 12:1:12 hidden layer configurations sometimes failed to converge during training. These problems need not be serious, since the mapping and demapping parts of the neural network can possibly be trained separately, as proposed by Tan and Mavrovouniotis (1995). Alternatively, alternative approaches to the mapping of the data can be used, such as principal curves or local principal component analysis, whereupon a neural network can be trained to represent the scores.

6.8.5. Iris data set

The *Iris* data set considered in Chapter 5, consists of four measurements on three iris species, viz. *Iris setosa*, *I. virginica* and *I. versicolor*. The measurements are sepal length (x_1) and width (x_2) and petal length (x_3) and width (x_4). Other researchers have used this data set extensively as a benchmark data set for various clustering and classification algorithms. In this case study, it will be used to compare the cluster analysis method of (FM) Aldrich (2000), a cluster-detection-and-labelling (CDL) network (Eltoft and deFigueiredo, 1998), a winner-takes-all network (WTA), an adaptive resonance theory (ART) neural network, as well as the k-means algorithm (k-M) and two hierarchical clustering algorithms (HCA-s and HCA-c).

In the winner-takes-all network, training was similar to that of a Kohonen network, in that exemplars were competitively assigned to one of three initially specified clusters. However, only the most similar weight vector to a given input was updated, so that the weights tended to converge to the cluster centres.

The cluster-detection-and-labelling network had a two-layer feedforward structure similar to that of a multilayer perceptron (not counting the input layer). The first layer represented the cluster prototypes, while the second layer represented the cluster labels, so that the network itself represented a nonlinear mapping of an m-dimensional input vector to a k-dimensional binary output vector (the cluster labels). Exemplars were considered to be new cluster prototypes if they were sufficiently dissimilar from previously stored prototypes (similar to training of ART networks), as described in more detail by Eltoft and deFigueiredo (1998). The results are summarized in Table 6.4.

Table 6.4. Clustering of the IRIS data set.

Algorithm	*setosa*	*versicolor*	*virginica*	*Total*
k-M	100	92	72	88
HCA-s	98	100	0	66
HCA-c	98	100	0	66
FM	100	94	94	96
WTA	100	92.4	79.6	90.7
CDL	100	90	98	96

6.8.6. Cluster analysis with ART neural networks

The use of an adaptive resonance theory (ART) neural network is illustrated as a final example of the application of neural networks to cluster analysis. Clustering was done on a data set consisting of 497 exemplars of the form $x = (x_1, x_2, x_3, \ldots x_{10})$, which represented the states of a copper froth flotation cell, as derived from images of the froth structures.

A plant expert classified these images into four classes, as indicated in Figure 6.23. In this figure, the 10 variables are portrayed in the score space of the first two principal component derived from the data. These components collectively explained approximately 89.5% of the variance in the data. An adaptive resonance theory neural network was used to cluster the data into four classes, with appropriate forcing of the vigilance parameter and these results are shown in Figure 6.24.

As can be seen from this figure, the clusters identified by the network do not correspond with the actual classes. This can be attributed partly to the network's inability to identify non-spherical clusters, as well as the presence of the two outlier points, which were assigned to a separate cluster by the network. The other reason is that the actual classes A, B and C do not correspond with natural groupings in the data, although the classes themselves form contiguous blocks or regions in the score space of the data.

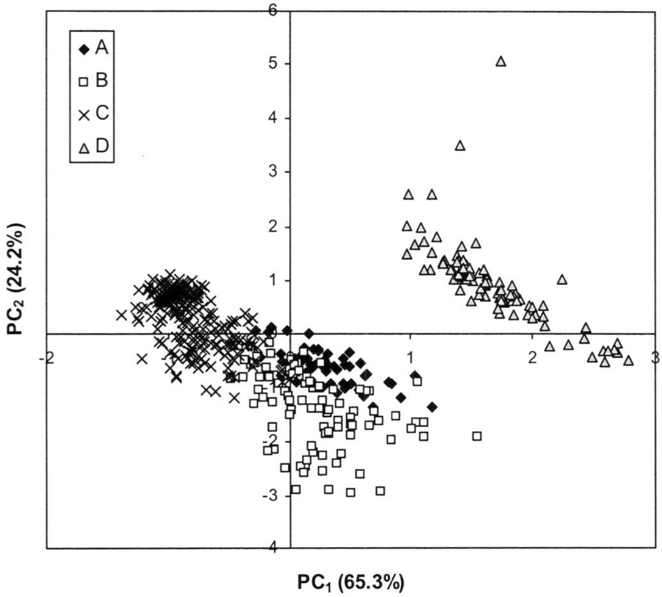

Figure 6.23. Actual classes of froths on a copper flotation plant.

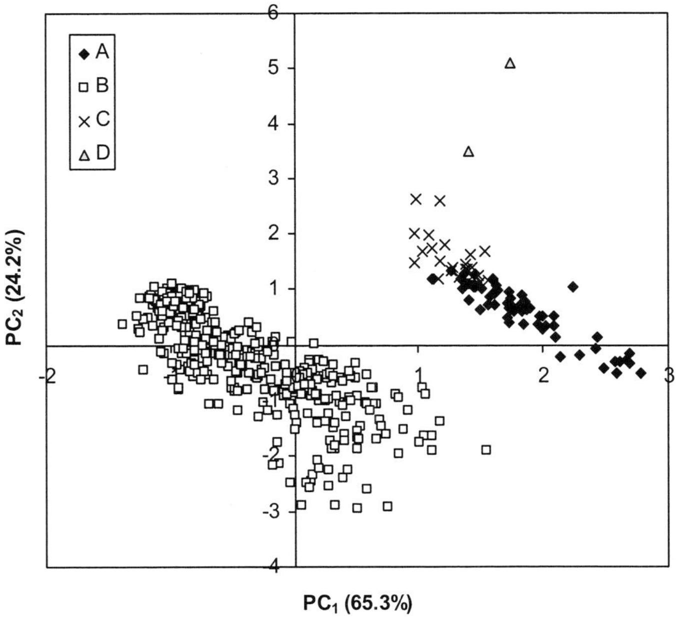

Figure 6.24. Clusters identified with an adaptive resonance theory neural network.

Chapter 7
Extraction of Rules from Data with Neural Networks

7.1. BACKGROUND

Although artificial neural networks have been applied successfully in diverse contexts over the last decade and more, such as pattern recognition, function approximation and process control, they have failed to penetrate areas where a high premium is placed on the transparency of the model. It is for this reason that considerable effort has been devoted to the general problem of interpreting the behaviour of neural network models. In particular, a large part of this effort has been centred on methods for the extraction and representation of neural networks in terms of sets of symbolic rules. These developments have since broadened to include subtly different, but equally important efforts focussed on the use of neural networks to develop rule-based systems, and this is also the focus of this chapter. The extraction of rules from neural networks can be *eclectic*, *decompositional* or *pedagogical* (Craven and Shavlik, 1994; Andrews et al., 1995) depending on the way in which the internal model of the neural network is interpreted in terms of a set of explicit rules.

7.1.1. Decompositional methods

Decompositional techniques such as those used by Towell and Shavlik (1993), Fu (1991) and Gallant (1993) focus on extracting rules separately for each unit of the neural network. With these techniques threshold functions are used to approximate the activation of the individual neurons by assuming that the neurons are either maximally active or inactive. Each node in the neural network (input nodes excluded) can therefore be interpreted as a step function or a Boolean rule, which reduces the rule extraction problem to the identification of conditions under which certain rules are valid. Unfortunately, this means that the neural network model and training procedures may have to be simplified to realize this approximation, which may not be feasible. Moreover, nodes with hard thresholds can be overly sensitive to noise in the data. With soft thresholding, this problem can be alleviated (Sethi, 1995), but again this approach may not be adequate.

7.1.2. Pedagogical methods

With pedagogical techniques, neural networks are treated like black box input-output devices and rules that map inputs to outputs are extracted directly. For example, Thrun used a method called validity-interval analysis (Thrun, 1994, 1995) based on the use of linear programming to determine if a set of constraints placed on a network's activation values is consistent. The activation levels of the hidden nodes are not approximated as threshold functions, but it is assumed that these activations are independent of one another. Since this assumption is not always appropriate, the algorithm does not always find maximally general rules. Other techniques such as those developed by Pop et al. (1994), as well as by Craven and Shavlik (1994) are in their current state limited to discrete inputs. Furthermore, Craven and Shavlik's technique is implemented using either the validity interval analysis (Thrun, 1995) or the KT method described by Fu (1991), whereby it is implicitly assumed that the activations of the hidden layers are independent, or that they can be treated as threshold functions.

7.1.3. Eclectic methods

Eclectic techniques simply combine elements of the two basic approaches discussed above. These methods (Seisto and Dillon, 1993; Tickle et al., 1994) draw inferences from the magnitudes of the weights in a neural network. This has to be done carefully, as certain weights in a neural network can be large, but nevertheless insignificant to the final outcome, owing to cancellations of their contributions in higher layers of the network. When the units are interdependent, the problem is complicated even further (Masters, 1993). The following examples illustrate the ways in which neural networks can be used to extract rules from process data.

7.2. NEUROFUZZY MODELING OF CHEMICAL PROCESS SYSTEMS WITH ELLIPSOIDAL RADIAL BASIS FUNCTION NEURAL NETWORKS AND GENETIC ALGORITHMS

As is well known, neural networks with kernel-based activation functions lend themselves naturally to equivalent descriptions in terms of fuzzy rule systems. The advantage of fuzzy systems lies in their ability to quantify linguistic inputs and to facilitate the rapid development of working approximations of complex or unknown systems in terms of input-output rules. Since neural networks have a powerful ability to learn from data, there is a natural synergy between neural networks and fuzzy systems.

7.2.1. Radial basis function networks and fuzzy systems

As discussed previously, the general radial basis function (RBF) neural network model can be expressed as a function $f:\Re^d \to \Re$, where $f(\boldsymbol{x}) = w_0 + \Sigma_i w_i \Phi_i(\boldsymbol{x},c_i)$ and $c_i \in \Re^d$ is the centre of the i'th basis node. $\Phi_i(\boldsymbol{x},c_i)$ is usually a Gaussian of the form $\exp(-\|\boldsymbol{x}-c\|/2s)$, where $\|\cdot\|$ is the Euclidean norm defined in $\Re^d$ and s is the width of the basis node.

Although clustering of the data exemplifying the functional relationship is computationally efficient and is often used to identify a suitable set of radial basis functions (Moody and Darken, 1989), this method suffers from two drawbacks. First, the relationship between the independent variables **x** and the dependent variables **y** is not taken into account. The clustering algorithms tend to place most centres at those locations where many input vectors can be found, instead of reserving them for those regions that show great variations in the output, for example the class boundaries of a classification problem. Second, an increase in the dimensions of the input space requires an exponential increase in the number of kernels to be fitted to the data. In order to alleviate these problems, Schmitz and Aldrich (1999; Schmitz, 1999) have made use of hyperellipsoidal Gaussians (Saha et al., 1991; Haykin, 1999) that can have a high resolution in certain directions or dimensions and lower resolutions in other dimensions.

By using a weighted sum defuzzifier, a Sugeno fuzzy system can be expressed as $y(\boldsymbol{x}) = \Sigma_i a_i(\boldsymbol{x})\mu_i(\boldsymbol{x})$ or $\Sigma_i a_i \mu_i(\boldsymbol{x})$ for 0th order systems, where the function $a_i(\boldsymbol{x})$ becomes a constant. Moreover, the weight of the i'th rule, $\mu_i(\boldsymbol{x})$, can be expressed as $\Pi_j \mu_{ij}(\boldsymbol{x}_i)$, if the product aggregation method is used and the Sugeno system can be represented by a function of the form $y(\boldsymbol{x}) = \Sigma_i a_i \Pi_j \mu_{ij}(\boldsymbol{x}_i)$. Although the underlying design philosophy differs, this functional form is the same as that of a radial basis function neural network, if the basis function $\Phi(\boldsymbol{x},\boldsymbol{c})$ can be factorised into functions of its components: $\Phi(\boldsymbol{x},\boldsymbol{c}) = \Pi_j \Phi(x_j,c_j)$. This can be done if the ellipsoidal

Gaussians are axis-parallel (Haykin, 1999). Under these circumstances, each node of the radial basis function neural network can this be translated direct into a fuzzy rule, so that collectively, the nodes of the radial basis function model are equivalent to a complete fuzzy rule base. The same result holds for normalized radial basis function networks and 0th order Sugeno systems with weighted averages. In this example, orientated ellipsoidal basis functions are used, which can in principle also be translated into a system of so-called ellipsoidal fuzzy systems (Dickerson and Kosko, 1993). The algorithm used to construct the ellipsoidal radial basis function network is described in more detail below.

7.2.2. Development of hidden layers

The hidden layer in the radial basis function neural network was constructed by use of the CERN (combinatorial evolution of regression nodes) algorithm developed by Schmitz and Aldrich (1999). The algorithm essentially consists of three steps that are repeated until a desired minimum error level is reached. It is initialized by placing a new group of Gaussian nodes into the hidden layer of the neural network. A genetic algorithm is subsequently used to adapt the parameters of the newly added nodes, so that the output error of the network is minimized. Existing nodes in the hidden layer are left unchanged during the search for the best new group of nodes to add to the network. Once the new nodes are optimized by the genetic algorithm, all the outgoing weights are determined by the solution to the least-mean-square problem (Seber, 1977).

Although the CERN algorithm optimizes groups of nodes simultaneously, each individual of the CERN algorithm is a single basis node and not a group of basis nodes as would be the case if an evolutionary algorithm such as the differential evolution algorithm (Storn and Price, 1995) were applied directly.

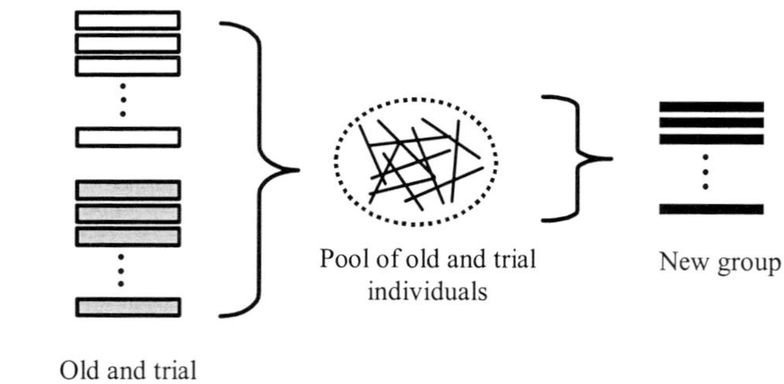

Figure 7.1. The selection of individuals in the CERN algorithm.

Since g such nodes or individuals are selected at a time, the total number of individuals (nodes) in the population is a multiple of g. At each iteration a new trial population of nodes is formed, as in the DE algorithm. As a result, each individual in an original group of individuals has a corresponding partner in the new group of individuals in the next generation. The best possible combination of nodes is therefore propagated to the next generation, instead of selecting either the entire old or the new group of nodes collectively. This is accomplished by pooling both the original the new group of nodes, as indicated in Figure 7.1. The best possible combination of these nodes is subsequently determined by the *Regression by Leaps and Bounds* algorithm of Furnival and Wilson (1974). This is to ensure that if a given Gaussian becomes more useful than its predecessor, it will be retained and a better group of nodes will be formed. This is the case even when all other nodes in the group from the new generation have become less useful than those of the former generation.

7.2.3. Post-processing of membership functions

The membership functions are formed by factorising each of the ellipsoidal Gaussians into a product of one-dimensional Gaussians, that is for each ellipsoidal Gaussian there will be *d* one-dimensional Gaussian projections or one for each input feature. In a fuzzy system these Gaussian projections form the membership functions. In order to simplify the model, a simple algorithm is used to combine membership functions that are sufficiently similar in terms of their Euclidean distances in the data space $\Re^n$, where *n* is the number of data points.

After this, the output weights of the neural network (the constants of the fuzzy system) are recomputed. If the RMS error of the whole network does not increase above a certain threshold, the new membership function replaces the previous two membership functions; else the original membership functions are retained. This process continues for each input, until all its membership functions are found to be sufficiently dissimilar or important in terms of reducing the RMS error. Note that the CERN algorithm allows the widths of the ellipsoidal Gaussians to grow unlimited in every direction. A membership function of an input can therefore be almost constant. This means that changes in this input do not cause significant alterations of the activation of the rule. If required, the proposed post-processing algorithm can remove such membership functions from a rule.

7.2.4. Case study: Induced aeration in liquids in agitated vessels

Gases and liquids are often contacted by means of induced aeration in agitated vessels in among other fermentation processes, bioleaching of minerals and the treatment of waste water (Gianetto and Silverstone 1986; Aldrich and Van Deventer, 1995).

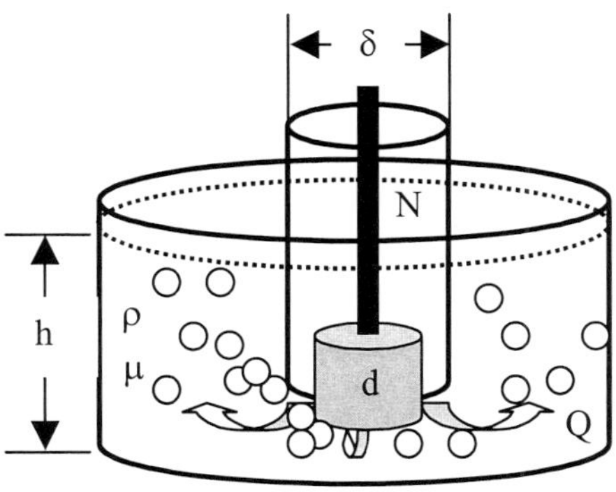

Figure 7.2. Induced aeration in an agitated vessel.

The rate of induced aeration can be expressed in terms of a dimensionless aeration number ($Ae = Q/Nd^3$), which is related to a modified Froude number $Fr\text{-}Fr_c = (N^2\text{-}N_c^2)d^2/gh$, the density ($\rho$) and viscosity ($\mu$) of the liquid, as well as the geometry of the vessel. In the above, Q denotes the volumetric flow rate of air (m^3/s), N the impeller speed (rad/s), N_c the critical impeller speed (rad/s), that is the impeller speed at which aeration starts, d is the impeller diameter, g is gravitational acceleration (m/s^2) and h is the depth (m) of the tip of the impeller underneath the surface of the fluid.

Table 7.1. The percentage of unexplained variance of the various Gaussian network models. The last column indicates the significance of the difference between the orientated ellipsoidal basis function (OEBF) model with 7 hidden nodes and the other radial basis function (RBF) and ellipsoidal (radial) basis function (EBF) models at a 95% significance level.

Model and training method	**Number of hidden nodes**	**% Unex-plained Variance**	**Significant difference with OEBF* at 95%**	**Significant difference with OEBF* at 99%**
RBF with k-means	40	2.2	YES	NO
	30	1.9	YES	NO
	20	3.4	YES	YES
	7	9.0	YES	YES
OEBF with CERN	7	1.0*	-	-
	4	1.6	YES	NO
EBF with CERN	12	1.1	NO	NO
	7	1.2	NO	NO
(simplified)	7 Rules	1.3	NO	NO

A set consisting of 180 data exemplars of the form $\{Fr\text{-}Fr_c, \rho, \mu, \delta \mid Ae\}$ was collected from experiments with two different impeller types with six blades (T6) and twelve blades (T12). The vessel geometry was fixed, except for the diameter (δ) of the draft tube. Since the viscosity spanned several orders of magnitude ($\approx 1 \times 10^{-3}$ to 1 Pa.s), it was scaled by taking the square root of the viscosity in order to expedite training of the neural network model. Apart from the viscosity, all the data were rescaled so that every input variable had a standard deviation between 0.25 and 0.3. In particular, the density ρ was multiplied by 2, the draft tube diameter δ was multiplied by 30 and the viscosity by 10.

Networks containing groups of 4 and 7 orientated ellipsoidal Gaussians, as well as networks with 7 and 12 axis-parallel Gaussians were evolved with the CERN algorithm (Schmitz and Aldrich, 1999). The algorithm used 150 individuals for every Gaussian that had to be optimized. The best group of individuals was selected after 150 generations.

The neural network with the 7 axis-parallel Gaussians was translated to a fuzzy system and the membership functions were simplified by the post-processing algorithm as described above. In order to validate the different models, five-fold cross-validation was used, that is the data were randomised and subdivided into five training (144 observations in each case) and five testing sets (36 observations in each case). Table 7.1 gives the percentage of unexplained variance of the various algorithms on the combined tests sets. The unexplained variance was calculated as $100(1-R^2) = 100\Sigma(y_p-y_t)^2\}/\Sigma(y_t-y_{t,avg})^2$, where y_p is the predicted value of the outcome, y_t is the target value of the outcome, and $y_{t.avg}$ is the average target value of the outcomes. The last two columns in the table indicate the significance of the differences of the correlation coefficients (variances explained) compared to the OEBF model with 7 nodes, at 95% and 99% significance levels. The data in the table suggest that the networks consisting of ellipsoidal basis nodes evolved by CERN tend to be superior to those containing spherical basis nodes obtained by *k*-means clustering, both in terms of accuracy and the number of nodes required to achieve good network performances. The only radial basis function (RBF) network that achieved an R^2 of over 0.98 required 30 basis nodes. As each Gaussian is defined by four centre co-ordinates, a width as well as an outgoing weight the model has a total of $6\times30+1 = 181$ free parameters (including the bias). This is considerably more than the solution of the ellipsoidal basis function (EBF) network with 7 nodes.

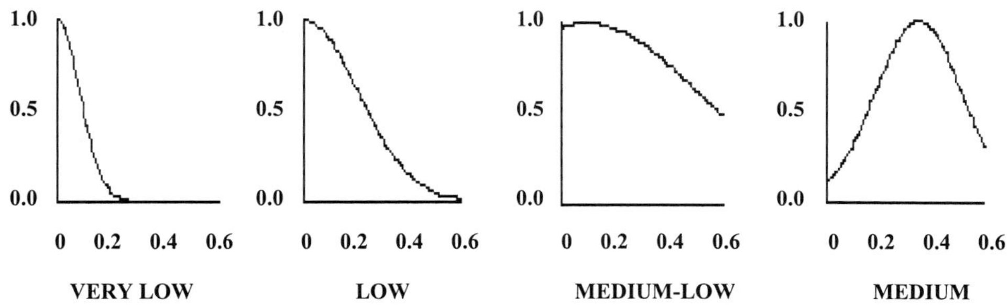

Figure 7.3. Membership functions of the Froude number.

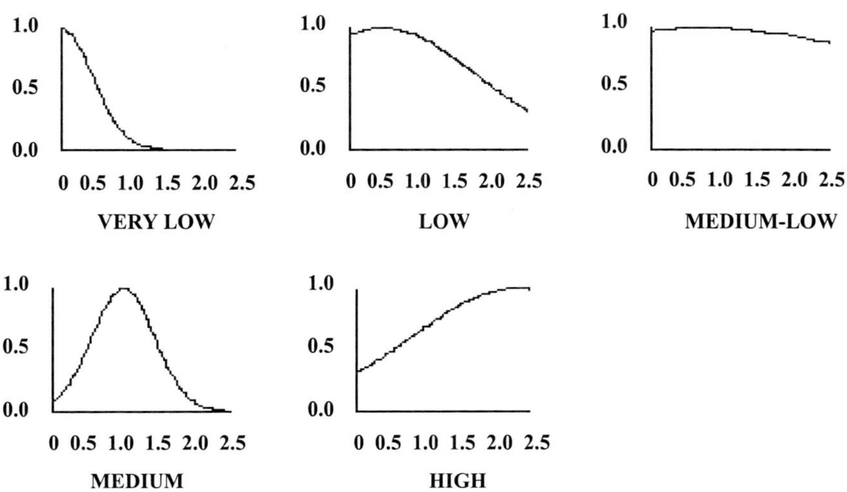

Figure 7.4. Membership functions of the density.

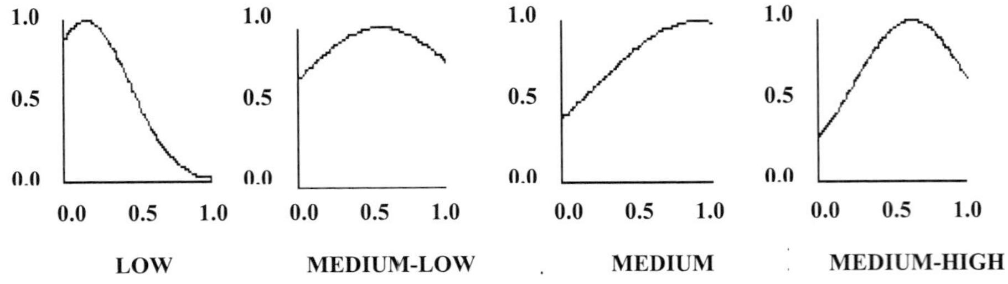

Figure 7.5. Membership functions of the viscosity.

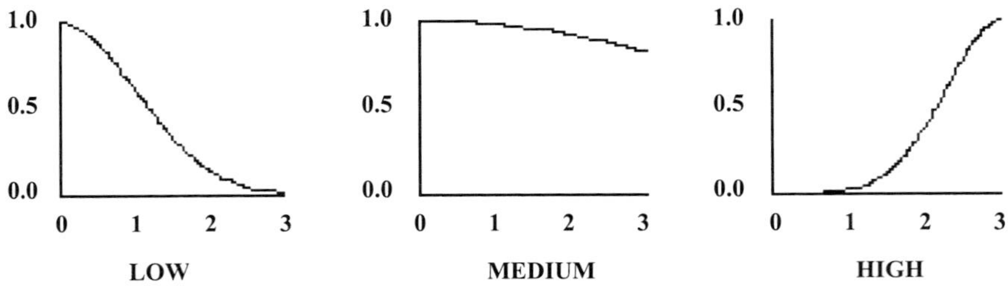

Figure 7.6. Membership functions of the draft tube diameter.

Here each Gaussian has 4 centre co-ordinates, 4 widths and an outgoing weight giving a total of 7*(8+1)+1 = 64 free parameters. After post-processing the membership functions, the number of free parameters is reduced even further. These membership functions of the input variables of the axis-parallel ellipsoidal Gaussian network are shown in figures 7.3-7.6.

The corresponding rules for the six- blade impeller are given in figure 7.7. Unlike using a pure fuzzy logic design philosophy the rules overlap considerably and all the rules have to be considered together. The first two rules are particularly striking as they extend over the whole data space. From figure 7.7 it can be seen that the other Gaussians have much smaller outgoing weights (numbers in brackets after each rule) and explain deviations in the decrease of the aeration rate.

- IF **FR-FR$_c$** medium-low AND **Draft-Tube Diameter** medium AND **Density** medium-low AND **Viscosity** medium
  THEN **Aeration rate** increases (62.7)
- IF **FR-FR$_c$** medium AND **Draft-Tube Diameter** low AND **Density** high AND **Viscosity** low
  THEN **Aeration rate** decreases (9.4)
- IF **FR-FR$_c$** medium AND **Density** very low AND **Viscosity** low
  THEN **Aeration** rate increases (7.9)

- IF **FR-FR$_c$** high AND **Draft Tube Diameter** high AND **Density** medium AND **Viscosity** low THEN **Aeration rate** decreases (9.4)
- IF **FR-FR$_c$** medium AND **Draft Tube Diameter** low AND **Density** increasing and **Viscosity** medium THEN **Aeration rate** decreases (16.5)
- IF **FR-FR$_c$** low and **Viscosity** decreasing **Aeration** increases (7.9)
- IF **Density** low AND **Viscosity** low THEN **Aeration** decreases (9.3)

Figure 7.7. The seven post-processed rules for the six blade impeller from the ellipsoidal radial basis function neural network induced by the CERN algorithm.

7.3. EXTRACTION OF RULES WITH THE ARTIFICIAL NEURAL NETWORK DECISION TREE (ANN-DT) ALGORITHM

The ANN-DT (Artificial Neural Network Decision Tree) algorithm proposed by Schmitz et al. (1999a, 1999b) can be seen as a pedagogical technique, although it does not suffer from the drawbacks of these strategies. Instead of attempting to extract rules from the original data or interpreting the internal structure of neural networks, the ANN-DT algorithm is based on sampling of a (neural network) model of the original data, which is subsequently used as a basis for the extraction of rules. Recently, Craven and Shavlik (1994) have introduced the idea of sampling the neural network. Their TREPAN algorithm exemplifies the idea. It grows nodes in a best first order (1996) and evaluates M-of-N splits based on a greedy gain ratio criterion (Quinlan, 1993). As a result, it is limited to discrete outputs or classification problems. Moreover, the TREPAN algorithm does not make any assumptions with regard to the internal structure or the training of the neural network model, does not assume that the activations of the nodes in the network are independent and does not draw explicit inferences from the distributions of the weights in the neural network.

In contrast, the ANN-DT algorithm can be applied to data sets where both the inputs and outputs can assume discrete or continuous values. An earlier version of the ANN-DT algorithm applicable to discrete output data only has been applied previously to characterize froth structures in a flotation plant (Schmitz et al., 1996). The current version of the algorithm makes use of significance analysis that has a distinct advantage over greedy attribute selection criteria in problems requiring look-ahead criteria.

The ANN-DT algorithm generates a binary decision tree from a trained neural network, as indicated in Figure 7.8. A given decision node in the tree returns 'true' if the node's single specified variable has a higher value than a certain threshold and 'false' otherwise. For each node the algorithm decides on which variable to partition the set of data, after which the threshold of that variable has to be determined. This is accomplished by examining the responses of the neural network in the feature space and conducting a sensitivity or significance analysis of the different attributes or explanatory variables pertaining to these responses in order to construct the decision tree.

Figure 7.9 gives a schematic representation of the ANN-DT algorithm. The first stage of the algorithm consists of training a suitable neural network on a set of training data. The ability of the neural network to generalize the underlying trends in the data is typically assessed via cross-validation on a separate set of test data. In this particular investigation both backpropagation and radial basis function neural networks were used, each of which is briefly described below.

7.3.1. Induction of rules from sampled points in the feature space

A systematic search through the input or attribute space, as can be done by the methods of Thrun (1994, 1995), Fu (1991), Gallant (1993) and Saito and Nakano (1988), can lead to an overwhelming number of rules for many real-world problems.

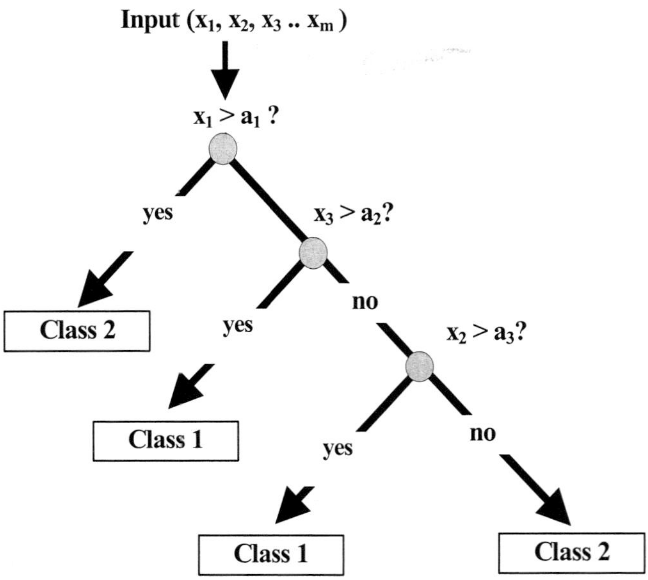

Figure 7.8. A simple binary decision tree.

Even in relatively small bounded attribute spaces, decision boundaries generated by the neural network that are not parallel to the boundaries of the attribute space can lead to an intractably large number of rules. For example, the movements of a robot arm required more than 8000 rules for proper description (Thrun, 1994). Even when these searches are restricted, the number of rules can remain large. For example, Saito and Nakano (1988) found that more than 400 rules were extracted for just one output unit in a neural network with 23 possible outputs. It may therefore be necessary to increase the granularity of the rules in order to improve their intelligibility. This can be done by limiting the depth to which the rules can be induced or by pruning the rules afterwards as will be discussed later.

The ANN-DT algorithm uses constrained sampling, i.e. the sampled points are restricted to those regions of the (often high-dimensional) search space on which the neural network model is based. This is accomplished by taking the distribution densities of points in the feature space into account, not only during sampling of the network, but also when the resulting branches of the trees require pruning.

7.3.2. Interpolation of correlated data

An artificial data set is generated by randomly sampling the input or feature space based on the density of the feature space and computing target or class labels for these sampled points by means of a neural network assumed to be a representative model of the underlying trends or decision boundaries in the data.

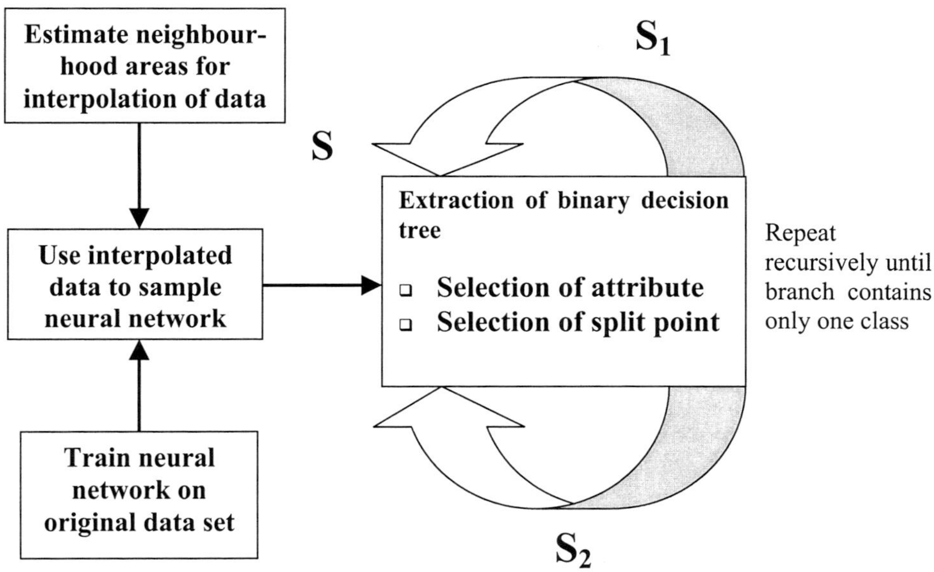

Figure 7.9. A diagrammatic representation of the ANN-DT algorithm used to extract rules from an artificial neural network.

To ensure that the newly generated data have the same distribution as the original training data and do not only reflect the behaviour of the particular neural network model, it is essential that sampling is only allowed in the neighbourhood of points or clusters present in the training data set. This can be accomplished by various means. If the data are normally distributed, a principal component analysis model can be used to evaluate sampled data, as indicated in Figure 7.10. In this case the residuals of the sampled data are used as an indication of the validity of the samples.

This approach may not be valid if the data are non-Gaussian or if the input variables are discrete. In this case, other measures of similarity have to be used (Krzanowski and Marriott, 1995). For mixed data the Euclidean distance or a metric defined by Gower (1971) can be used. In cases where there are few discrete variables, the data sets belonging to the various possible combinations of the discrete variables can be treated separately. The reader is referred to (Krzanowski and Marriott, 1995) for a discussion on this topic.

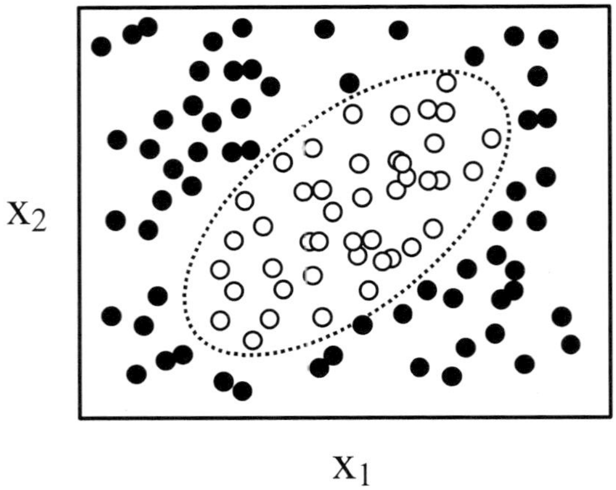

Figure 7.10. Constraining of sample data by use of principal component model of the training data. Invalid sampling data outside the 95% Q-limit (broken line ellipse) are indicated by solid circles, while valid sampling data in this limit are indicated by open circles.

For example, Schmitz et al. (1999a, 1999b) have used a critical distance that was based on the distances between points in the training data set. In relatively small data sets with n data points, all n(n-1)/2 distances could be calculated. In larger data sets this becomes computationally expensive, in which case a random sample of a number of interpoint distances was used. The critical distance was set equal to the average distance between the points and their respective k (= 10) nearest neighbours in the training data set. Each artificially generated exemplar was subsequently presented to the neural network, from which a corresponding output value was computed.

7.3.3. Selection of attribute and threshold for splitting

a) Gain ratio criterion

Consider a set S of n exemplars with m corresponding discrete output classes $\{C_1, C_2, \ldots C_m\}$. Let $P(C_i,S)$ be the proportion of exemplars in S which are in class C_i. The class entropy of any subset S_k of S is defined in by Quinlan (1986) as:

$$I(S_k) = -\sum_{i=1}^{m} P(C_i,S) \times \log_2(P(C_i,S)) \tag{7.1}$$

Assume that a value T of a particular attribute X has been chosen which splits S into two subsets S_1 and S_2. The average class entropy after such a partitioning is:

$$E(X,T,S) = -\sum_{k=1}^{2} \frac{|S_k|}{|S|} \times I(S_k) \tag{7.2}$$

The information gain generated by the split can now be defined as:

$$G(X,T,S) = I(S) - E(X,T,S) \tag{7.3}$$

It was found that when equation (7.3) is maximized by the selection of the attribute and the threshold, it is biased towards features with large number of values (Quinlan, 1988; White and Liu, 1994). In contrast, a normalized measure of information gain performed more satisfactory. Quinlan (1988) defined the normalization factor for an attribute X as

$$I(X,S) = -\sum_{k=1}^{n} \frac{|S_k|}{|S|} \log_2\left(\frac{|S_k|}{|S|}\right) \tag{7.4}$$

where n is the number of possible outcomes of the attribute. The normalized information gain, or gain ratio, is now given by

$$G_R = \frac{G(X,T,S)}{I(X,S)} \tag{7.5}$$

The attribute, together with the numeric threshold T, which will result in the maximum information gain ratio is selected for the Boolean splitting test.

For continuous outputs and a least absolute deviation measure the attribute and threshold are selected in order to minimize the normalized standard deviation (NSD) averaged over the different branches in the same way as the entropy.

$$NSD = \sum_{k=1}^{2} \frac{|Sk|}{S} Stdv(O_k) \tag{7.6}$$

where O_k are the outputs of the data set S_k.

For continuous output and a least squares error measure the attribute and threshold are selected which causes the most decrease in the normalized variance (NS^2) over the two branches.

$$N = \sum_{k=1}^{2} \frac{|Sk|}{S} Stdev^2(O_k) \tag{7.7}$$

By minimizing the normalized variance or standard deviation the resulting branches effectively reduced the RMS error or the sum of absolute errors (Breiman et al., 1984). For discrete outputs equation (7.5) maximizes a normalized measure of information entropy gain. If on of these criteria is used to select the attribute as well as the threshold, the algorithm will be referred to as ANN-DT(e).

b) Analysis of attribute significance

An alternative method to those discussed above can be used for the selection of attributes. This method examines the significance of the various inputs on the behaviour of the neural network. Consider a neural network model or functional relationship f between attributes (inputs) and classes (outputs) that is evaluated at a set of points S lying inside a domain D. If the magnitudes of the partial derivatives of the function with respect to the inputs are to be a measure of the significance, it is implicitly assumed that the variables can change freely and independently from one another. For the analysis of experiments where the influencing factors can be varied independently, this assumption is valid. However, if the measured attributes are correlated this is not appropriate as far as the system represented by the neural network is concerned, as the change in one input feature may be accompanied by a change in another covariant feature.

These interrelationships need to be taken into account by focusing on the variations of f that actually occur inside the domain D. This can be done by looking at the variation of f when moving between the points of S. Define the *absolute variation* $v(f)$ of the function $f(\mathbf{x})$ between the points i and j as the absolute value of the directional derivative of $f(\mathbf{x})$ integrated along a straight line between the two points. Thus

$$v_{ij}(f) = \int_{x_i}^{x_j} |\nabla f(x) \bullet u \,| \, dx \tag{7.8}$$

where $\mathbf{u}$ is the unity vector in direction $\mathbf{x}_i$-$\mathbf{x}_j$.

This variation can be computed between all pairs of points in S. When an attribute is insignificant to the function for the domain D, the variation in the function will be unrelated to the variation in the attribute. Note that for a function where the effect of one attribute is cancelled out by another covariant attribute (e.g. $f(\mathbf{x}) = x_1 - x_2 + \sin\{x_3\}$ and a domain in which $x_1 \approx x_2$) only variations in the other attributes (in this case x_3) will cause notable variations in $f(\mathbf{x})$. Therefore variations in the attributes with more influence (x_3) will correlate with the absolute variations in $f(\mathbf{x})$, while the variation in the attributes such as x_1 and x_2 will be uncorrelated with $v(f)$. Thus a measure of the *significance* $\sigma(f)_a$ of an attribute a for a function f over a data set S would be the correlation between the *absolute variation* of the function and the absolute variation of that attribute taken between all possible pairs of points in S:

$\sigma(f)_a$ = correlation $(\{v_{ij}(f)\}, \{v_{ij}(a)\})$ for all pairs i,j

$$= \frac{\sum_{i}^{N}\sum_{j>i}^{N}\{v_{ij}(f) - \bar{v}(f)\}\{v_{ij}(a) - \bar{v}(a)\}}{\sqrt{\sum_{i}^{N}\sum_{j>i}^{N}\{v_{ij}(f) - \bar{v}(f)\}}\sqrt{\sum_{i}^{N}\sum_{j>i}^{N}\{v_{ij}(a) - \bar{v}(a)\}}} \tag{7.9}$$

At a given node the attribute with the maximum significance for the neural network function over the data set of the particular node was selected. In cases where equation (14) led to excessive computations, the result was approximated using a randomly selected subset of data pairs. This attribute selection approach is henceforth referred to as the ANN-DT(s) variant of the algorithm.

The threshold at which the selected attribute is split is chosen by minimizing the gain ratio, the normalized standard deviation or the normalized variation depending on the type of error that is to be minimized.

c) *Stopping criteria and pruning*

A two-way split is generated, dividing the current set of data into two subsets. This splitting process continues recursively, successively splitting the data into small subsets. For discrete output data recursion halts when a node contains data with only one output class. Where data with continuous outputs are being used, recursion is terminated when the standard deviation or the variance is zero.

Recursion can also be prematurely terminated when a certain stopping criterion is met. Such stopping criteria prevent tree branches from being created where the outcome of one of the sub-branches would not be significantly different from the outcome of the other. These so-called pre-pruning methods are usually implemented for two reasons. The first is to prevent the tree from modelling noise in the data and the second is to improve tree intelligibility. Many neural network architectures and training algorithms are available which can adequately compensate for noise in the training data. Therefore in the context of rule-extraction from neural networks pre-pruning techniques are primarily used to improve the intelligibility of rules.

Statistical tests are applied to the outcomes of the data contained in the two new branches. For discrete outcomes the Pearson's χ^2 (chi-square) test (Hays, 1988) can be used to find out whether the outcome class of a record is not independent from the branch into which the record is to be put. Where continuous outputs are concerned, an F-test (Hays, 1988) can be used to ascertain whether the mean output of the records of each of the two sub-branches are significantly different from each other. Both these tests show whether continued recursion would be meaningful or not. If the selected termination criterion fails at some confidence level α, the current node is converted into a terminal node. Note that the chi-square test becomes less accurate as the number of data points per subset decreases (Hays, 1988). For the ANN-DT algorithm the outcomes stem from a neural network, which usually has continuous outcomes even when trained on data with discrete outputs. Therefore both the χ^2-test and the F-test may be applied simultaneously as stopping criteria. In the ANN-DT algorithm both tests had to fail before a terminal node was formed. These stopping criteria can fail at a certain tree depth, even though future splits further down the tree could become statistically significant again. To prevent premature cessation of tree growth as a result of the above tests failing, these tests were only applied to nodes occurring below a certain minimum depth in the tree.

Three other criteria were applied to prevent unnecessarily large, and therefore incomprehensible trees from being formed. The first was that the tree could only grow to a user-defined maximum depth. The second was a so-called lower branching limit. This criterion prevents a node containing less than s_{min} exemplars from splitting, or nodes containing less than n_{min} exemplars from being formed. The third criterion, applicable to data having continuous outputs, prevents nodes from splitting if the variance of the output of the node falls below some threshold (v_{th}).

7.3.4. Illustrative examples

The performance of both the ANN-DT(e) algorithm and the ANN-DT(s) variant was compared to that of the ID3 and C4.5 algorithm on a number of classification problems and to an algorithm using CART's splitting criteria on continuous output data. Three-fold cross validation was used where the data sets contained few exemplars.

The neural networks used for the problems considered in this chapter were back propagation networks with hyperbolic transfer functions and weights trained with the generalized delta learning rule with momentum, as well as radial basis function neural networks. The hidden layers of the radial basis function neural networks were trained with the k-means clustering algorithm and their output layers with the generalized delta learning rule.

Except for the problems pertaining to artificial data (Examples 1 and 4) the ANN-DT algorithm sampled the trained neural network 20 times for every training point. In other words, for every original training point, 20 sample points were synthesised for querying the trained neural network. It should be noted that during the integration steps, the ANN-DT(s) algorithm made additional queries to the neural network in order to determine the most significant inputs.

Two trees were built by the ANN-DT algorithm for each classification problem. The growth of the first tree was not constrained by any stopping criterion, whereas specific stopping criteria and pruning methods were applied during the formation of the second tree. In particular the maximum depth to which the second tree could grow was limited to 6. Statistical pre-pruning commenced at a depth of 3 with a confidence level α of 95%. The lower branching limit criterion was also implemented. The minimum number of exemplars required for a split, s_{min}, was set at 10 and the minimum number of exemplars that a particular node had to contain, n_{min}, was set at 2.

The same stopping criteria were used for those examples consisting of data with continuous outputs. A further stopping criterion, i.e. the minimum output variance required to continue splitting, v_{th}, was also implemented. The threshold value was 0.0025. The experiments with ID3 and the CART variant used the same settings as ANN-DT, except that s_{min} was set to 4. This is in fact implicitly enforced because n_{min} was set to 2. For all examples other than example 4 (discussed below) the results of the various algorithms on the validation sets were found to be fairly insensitive to changes in the above-mentioned settings.

a) Characterization of gas-liquid flow patterns

Transitions between various fluid flow patterns are of substantial importance for many technical applications that rely on pressure, as well as heat and mass transfer. Reiman et al. (1992) have conducted experiments with multiphase fluid flow systems (air and water) in horizontal pipes yielding 175 experimental observations. Depending on the flow pressure (P), diameter of the pipe (D), superficial liquid velocity (v_L) and superficial gas velocity (v_G), flow regimes characterized by slug flow or non-slug flow could develop (including transitional phases such as slug-annular and slug-wave flow). A back propagation neural network with 4 input nodes, 3 hidden nodes and a single sigmoidal output node was trained with the generalized delta rule to distinguish slug flow from non-slug flow, given exemplars of the form [P, D, v_L, v_G | FLOW], where FLOW denoted either slug or non-slug flow.

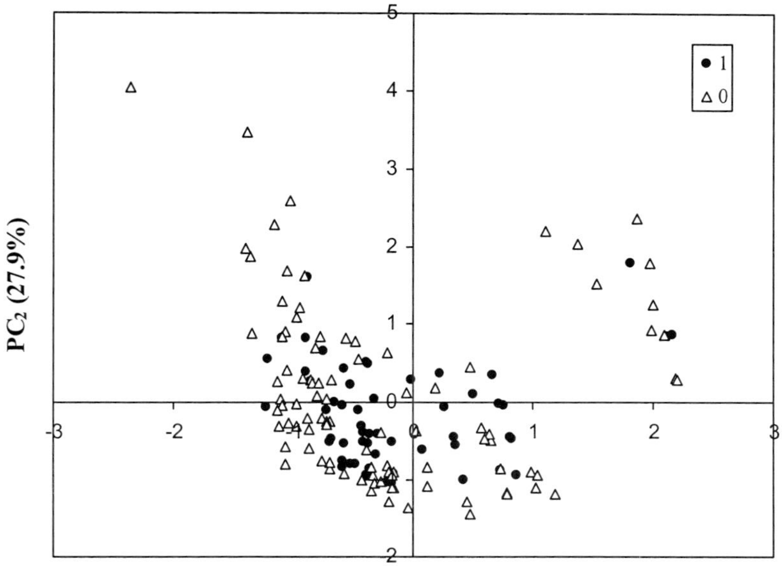

Figure 7.11. The first two principal components of the variables d, p, v_s and v_g. Collectively, the two principal components explained 81.8% of the variance of the data. Classes 0 (solid circles) and 1 (open triangles) denote the absence and presence of slug flow conditions respectively.

The factor loadings for the data indicate that the 2nd PC mainly represented the solids velocity, while the first principal component more or less represents a combination of the remaining variables (see Table 7.2. below).

Table 7.2. Principal component (factor) loadings of the gas-liquid flow pattern data.

	Factor 1	Factor 2
D	.72133	.46395
P	.91912	.10926
VG	-.85550	.24957
VS	-.24380	.90887

b) Solidification of $ZnCl_2$

This data set consisted of 108 exemplars obtained from an experiment where zinc chloride is hydrolysed in watery ammoniacal-ammonium chloride solution (Limpo et al., 1995). Three phases of zinc chloride can occur, viz. $Zn(NH_3)_2Cl_2$, $Zn(OH)_2$ and $Zn(OH)_{1.6}Cl_{0.4}$.

Usually only one phase is present, but under certain conditions more than one phase can occur. The objective of this example is to predict the expected formation of the three phases from the temperature of the solution (°C), the concentration of chloride anions (Cl^-), the concentration of zinc cations (Zn^{2+}) and the ammonia concentration (NH_3). The principal component scores of the data are shown in Figure 7.11 below. The first two principal components could explain 84.3% of the variance of the data. From the loadings in Table 7.3, it can be seen that the factor loadings for the data indicate that the second principal component mainly represented the temperature, while the first principal component more or less represented the remainder of the variables.

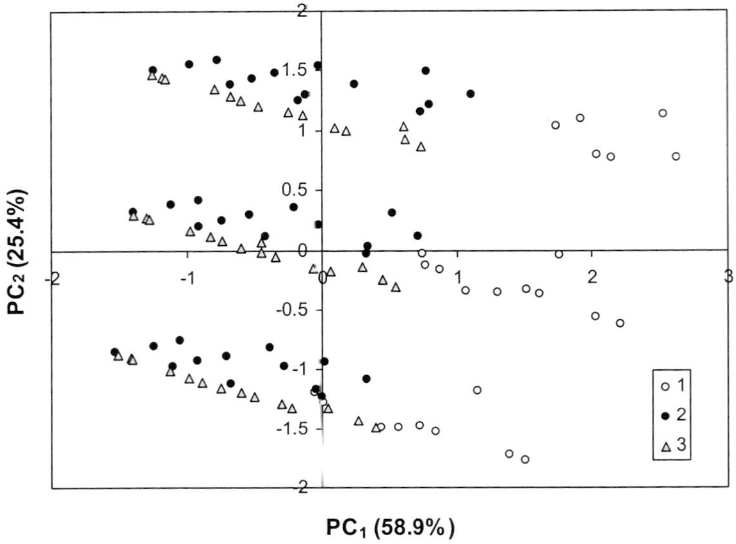

Figure 7.12. The first two principal components of the variables Temp, Cl^-, Zn and NH_3. Collectively, the two principal components explained 84.4% of the variance of the data. Classes 1, 2 and 3 denote the precipitation of $Zn(NH_3)_2Cl_2$, $Zn(OH)_2$ and $Zn(OH)_{1.6}Cl_{0.4}$ respectively.

Table 7.3. Principal component (factor) loadings of the $ZnCl_2$ solidification data.

	Factor 1	Factor 2
TEMP	.20327	.96686
ZN	.88747	-.10904
CL	.94644	-.22398
NH3	.79572	.14104

The outputs are non-exclusive and therefore for each of the three phases a back propagation neural network with 4 input nodes, 5 hidden nodes with hyperbolic tangent transfer functions and a hyperbolic tangent output node was trained on the data. Separate decision trees were likewise induced on each of the three outcomes to indicate whether a specific phase was present or not.

7.3.5. Performance of the ANN-DT algorithm

The results obtained with the various algorithms as specified previously are summarized in Tables 7.4 and 7.5. Table 7.4 contains the classification score out of a 100. The standard deviation of the accuracies over three cross-validation runs are indicated in parenthesis in Table 7.4. Table 7.5 contains the number of leaves in the decision trees, which is an indication of the complexity of the trees. For these binary trees the number of internal nodes is equal to one less than the number of leaves.

Note that for Examples 2 and 3 the rules evolved by the ANN-DT(e) and the ANN-DT(s) algorithms are more accurate than those induced by ID3 and C4.5 algorithm but by a smaller margin than in the previous examples. Note too that the ANN-DT(e) algorithm performed slightly better than the ANN-DT(s) variant. A paired t-test was performed containing each of the three cross-validation results of the real-world classification problems (9 in total). This was to see whether the higher accuracy of the unpruned decision trees evolved by the ANN-DT(e) method is significantly different from the accuracy of the trees evolved by the ANN-DT(s), ID3 or C4.5 algorithm. With the double-sided t-test (Hays, 1988) it was found to be true with a 94% confidence level for ID3, with 91% confidence for ANN-DT(s) but with less than 90% confidence for the C4.5 algorithm. Furthermore, the improvement of the results obtained by ANN-DT(s) over those of the ID3 algorithm was not significant at a 90% confidence level.

The number of rules derived by the ANN-DT algorithm is much greater than the number of rules generated by ID3 and C4.5 respectively. Although this number could be reduced by statistical pruning it was found in Example 3 that after pruning the trees evolved by the ANN-DT algorithm are no longer more accurate than the C4.5 decision trees or the unpruned ID3 decision tree. It seems that for this case study the extra information that is contained in the neural network and not in the ID3 or C4.5 decision trees cannot be represented by both intelligible and accurate axis-parallel rules.

In contrast, the ANN-DT(e) algorithm achieves a higher classification score on Example 3, both with and without statistical pruning. As three independent decision trees and neural networks were induced for each of the three non-exclusive outcomes, the three outcomes will be assumed independent. Under this assumption a paired t-test can be performed on the nine results obtained (three outcomes for every cross-validation run) to see whether the ANN-DT(e) algorithm performs significantly differently from the ID3 or C4.5 algorithms. The classification scores of the pruned ANN-DT rules were significantly higher (at over 95% confidence level for a two-sided t-test) than both those of the unpruned and pruned sets of rules derived by ID3 or C4.5. The same result was also true for the ANN-DT(e) algorithm without pruning. A typical set of statistically pruned rules derived by the ANN-DT(e) algorithm indicating the formation of $Zn(OH)_2$ is given in Figure 7.13, which can be compared with rules for the same data generated by the CART algorithm in Figure 7.14.

ANN-DT successfully extracted a faithful rule representation from the trained neural network on all the case studies. In particular the ANN-DT(e) variant seemed to be more capable with regard to classification problems, while the ANN-DT(s) method was shown to be more robust for continuous output data.

An interesting result is that the rules induced by the ANN-DT algorithm are more accurate than those induced by the ID3 algorithm or variant of the CART algorithm using the same pruning techniques. A similar result was also found by Craven and Shavlik (1996), who compared the TREPAN algorithm to classification trees induced by C4.5 (Quinlan, 1993) and ID2-3 (Quinlan, 1986). The results indicate that for many problems the inductive techniques like C4.5 and CART do not use all the information that is contained in the original data. A possible source of this loss of information is that the technique splits the data recursively into branches in such a way that the data to be processed in the underlying branches are isolated from another. This means that any trend that might exist between the inputs and the output of the data, which are distributed over points belonging to different branches, will not be discovered by the algorithms. It also means that points not complying with this trend as a result of noise cannot be identified and a rule can arise out of these exceptions that does not generalize very well.

If it is assumed that the neural network detects these trends and does not overtrain on outliers in the data, the ANN-DT(e) and ANN-DT(s) algorithms are evolved on data where these exceptions are already removed. Moreover, the more densely sampled points help in identifying the exact threshold value at which the data should be split. The C4.5 and CART algorithm can only estimate this value to lie somewhere between two points of that subset of the original data points that belongs to the branch in which the next split is to be made. This subset is much denser in the case of the ANN-DT(e) and ANN-DT(s) algorithms.

Moreover contours, such as class boundaries that extend over different branches of the tree, can be continued over the regions where there is very little training data or no data to support them. This is because the neural network does not split the data and can extend such a decision boundary between the training points via interpolation. The ANN-DT algorithm can sample in these regions and produce additional rules to cover these regions, but unfortunately the higher accuracy in such cases is also associated with more rules.

From the classification results (Case Studies 1-4) it can be seen that the unpruned ANN-DT(s) algorithm produced significantly larger trees than the ANN-DT(e) algorithm, while also yielding a slightly lower classification. This can possibly be attributed to the fact that the significance analysis for classification problems could find high correlations between the output and one of the variables, even when the variations in the output that correlate with the variations in the attribute do not cause a change in the class. Consider the case in which all outputs of the sampled neural network with a value of higher than 0.5 will belong to class 1 and those with a lower output value than 0.5 to class 0. Then a change from 0.1 to 0.4 in the output is weighted by the significance analysis as much as a change from 0.4 to 0.7, despite the fact that in the latter case a shift in class occurred and in the former not. This causes splits in the data that are non-optimal in terms of separating points of different classes from one another. The gain ratio criterion on the other hand is computed only from the class values (either 0 or 1).

Both ANN-DT(s) and ANN-DT(e) can be applied to any non-parametric model, other than feed-forward neural networks, without making any assumptions about the model's internal states or the nature of the data. The computational time of the ANN-DT algorithm scales linearly with the neural network size and is only dependent on the time it takes the neural network to assign a label to a data point. However, the algorithm's computational time does suffer from the curse of dimensionality. In order to achieve a higher density of points than that of the training points, progressively more sample points are required as the dimensionality of the data increases. This problem can be reduced somewhat by initially using fewer sample points and growing the tree from a node in a *best-first* manner. This is performed in the TREPAN algorithm Craven and Shavlik, 1996) by presenting the node that is most likely to increase fidelity with sufficient samples. Naturally the number of these sample points also needs to grow exponentially with the dimensionality of the data in order to achieve the same accuracy, as once a split is made in the tree it cannot be adjusted later.

Table 7.4. The percentage of examples classified correctly on the test data set by the various algorithms. The standard deviations of this percentage over the 3 cross-validation sets (where applicable) are indicated in brackets. Where indicated, statistical pruning was done with an α value of 0.05. C4.5 used error reduction pruning with a 75% confidence limit.

Case Study	**ANN**	**ANN-DT(e)**	**ANN-DT(e) Pruned**	**ANN-DT(s)**	**ANN-DT(s) Pruned**	**ID3**	**ID3 Pruned**	**C4.5**
Slug-Flow	73.2 (3.4)	72.8 (4.2)	68.3 (5.5)	72.4 (2.55)	69.9 (3.1)	69.8 (2.2)	63.9 (4.5)	72.1 (4.1)
Solidification of $ZnCl_2$	92.0 (1.9)	92.9 (2.3)	91.4 (2.0)	91.4 (2.2)	80.6 (3.7)	87.7 (3.5)	88.0 (2.8)	89.2 (2.4)

Table 2(a): The fidelity of various algorithms. The fidelity is given as the percentage of examples classified with the same accuracy as the neural network (ANN).

In case study 5 it was seen that using the significance analysis in attribute selection could hold significant advantages over the greedy variance criterion. For this particular problem specification we know that although single splits on attributes ϕ and θ do not cause a significant decrease in the normalized variation of the data, changes in these attributes are nevertheless correlated with changes in the output of the neural network and will therefore have a high σ_f value. This is again based on the assumption that the neural network has learnt the unknown function represented by the data. The significance analysis therefore learns from the trained neural network function which attributes have the most influence over the data set covered by a particular node.

Table 7.5. The average number of leaves of the decision trees for the case studies with discrete outcomes of each decision tree algorithm.

Case Study	**ANN-DT(e)**	**ANN-DT(e) Pruned**	**ANN-DT(s)**	**ANN-DT(s) Pruned**	**ID3**	**ID3 pruned**	**C4.5**
Slug-Flow	65	33	179	33	33	8	8
Solidification of $ZnCl_2$	38	20	41	15	6	4	4

RULE 1

if Zn ≥ 0.54
then 0 Prob.= 0.98, Size = 43.1%

RULE 2

if $Cl^- < 3.11$
and Zn < 0.54
and $NH_3 \geq 0.24$
then 1 Prob.= 0.94, Size = 30.9%

RULE 3

If Zn < 0.54
and $NH_3 < 0.24$
then 0 Prob.= 0.89, Size = 18.7%

RULE 4

if Temp. ≥ 35
and $Cl^- \geq 3.11$
and $Cl^- < 3.61$
and Zn < 0.54
and $NH_3 \geq 0.57$
then 1 Prob.= 0.82, Size = 3.4%

RULE 5

if $Cl^- \geq 3.61$
and Zn < 0.54
and $NH_3 \geq 0.24$
then 0 Prob.= 1.00, Size = 1.8%

RULE 6

if $Cl^- \geq 3.11$
and $Cl^- < 3.61$
and Zn < 0.54
and $NH_3 \geq 0.24$
and $NH_3 < 0.57$
then 0 Prob.= 1.00, Size = 1.0%

RULE 7

if Temp. < 35
and $Cl^- \geq 3.11$
and $Cl^- < 3.61$
and Zn ≥ 0.32
and Zn < 0.54
and $NH_3 \geq 0.57$
then 0 Prob.= 0.95, Size = 1.0%

RULE 8

if Temp. < 35
and $Cl^- \geq 3.11$
and $Cl^- < 3.61$
and Zn < 0.32
and $NH_3 \geq 0.57$
then 1 Prob.= 1.00, Size = 0.2%

Figure 7.13. The rules evolved by the ANN-DT(e) algorithm example 4 for the presence (1) or absence (0) of the $Zn(OH)_2$ phase in the solidification of $ZnCl_2$. The 'Size' following each rule indicates the percentage of the exemplars covered by the rule. The probability is the ratio of the number of exemplars in the dominant class to the total number of exemplars. The rules could predict 89% of the classes correctly. The first 3 rules covered 92.7% of the cases with an accuracy of 87.9%.

The greedy splitting criteria of CART and ANN-DT(e) on the other hand, do not compensate for the periodicity of the function with respect to the attribute θ. The ANN-DT(e) algorithm could use additional sample points to obtain a respectable performance, whereas the CART algorithm no longer has sufficient data points to identify the periodic functional behaviour, after performing the initial greedy splits. Although case study 5 concerned a synthetic data set, such periodic response of the output to one of the attributes with relatively sparse noisy data can also occur in real world data sets. The fact that both ANN-DT(e) and especially ANN-DT(s) could overcome such a pitfall is at least in theory a significant benefit of the ANN-DT algorithm compared to the CART algorithm.

RULE 1

If $Zn \leq 0.512$
and $NH_3 \leq 0.233$
and $Cl^- \leq 3.095$
then 1 Prob.= 0.85, Size = 37.0%

RULE 2

If $Zn > 0.512$
then 0 Prob.= 1.00, Size = 27.8%

RULE 3

If $Zn \leq 0.512$
and $NH_3 \leq 0.233$
and $Zn > 0.062$
then 0 Prob.= 0.97, Size = 26.9%

RULE 4

If $Zn \leq 0.512$
and $NH_3 \leq 0.233$
and $Cl^- > 3.095$
then 0 Prob.= 1.00, Size = 5.6%

RULE 5

If $Zn \leq 0.062$
and $NH_3 \leq 0.23$
then 1 Prob.= 1.00, Size = 2.7%

Figure 7.14. The rules evolved by the CART algorithm in example 4 for the presence (1) or absence (0) of the $Zn(OH)_2$ phase in the solidification of $ZnCl_2$. The 'Size' following each rule indicates the percentage of the exemplars covered by the rule. The probability is the ratio of the number of exemplars in the dominant class to the total number of exemplars. The classification tree could predict 87.0% of the classes correctly. The first 3 rules covered 81.7% of the cases with an accuracy of 85.3%.

7.4. THE COMBINATORIAL RULE ASSEMBLER (CORA) ALGORITHM

The combinatorial rule assembler (CORA) algorithm (Gouws, 1999) constructs models using the rules generated by an existing fuzzy rule construction technique in order to obtain a new set of rules with maximum predictive performance and improved comprehensibility and intelligibility. The algorithm is predominantly designed to model regression problems, but can also be used for classification.

The CORA algorithm is based on Fritzke's growing neural gas (GNG) version of the radial basis function neural network training algorithm (Fritzke, 1994a) and the reactive tabu search (RTS) algorithm (Battiti and Tecchiolli, 1994). A set of fuzzy rules is first constructed by using the growing neural gas algorithm. The antecedent parts of these fuzzy rules are subsequently combined using the reactive tabu search combinatorial optimization technique to form a smaller set of fuzzy rules. Finally, this modified set of rules is further simplified to reduce rule complexity without significantly affecting accuracy.

7.4.1. Construction of fuzzy rules with the growing neural gas algorithm

The growing neural gas (GNG) algorithm (Fritzke, 1994a, 1995) combines the topology generation of the neural gas (NG) algorithm (Martinetz and Schulten, 1991) with the growth mechanism of the growing cell structures (GCS) algorithm (Fritzke, 1992, 1994b). The growing neural gas algorithm (Fritzke, 1994a) has two variants for training neural networks. Unlike other self-organizing learning methods, such as the self-organizing map of Kohonen (1982), the first variant automatically determines both the structure and size of the neural network model.

Both variants of the growing neural gas algorithm construct lateral connections between the nodes of the neural network model based on the use of Hebb's competitive learning rule (Hebb, 1949; Martinetz, 1993). Unlike Fritzke's growing cell structures algorithm (Fritzke, 1992, 1994b), the growing neural gas algorithm can adapt the network architecture locally to the intrinsic dimensionality of the attribute manifold. Growing neural gas network learning is thus able to generate a topology preserving attribute map, regardless of the actual dimensionality of the data.

For the construction of fuzzy rules, a variant of the growing neural gas algorithm is first used to train a radial basis function neural network. This particular network structure contains one hidden layer and a single summation node in the output layer. Cell nodes with hyperspherical Gaussians are used in the hidden layer and no normalization is performed. This has the advantage that outliers in the data do not activate the Gaussian hidden nodes and therefore have little, if any, adverse effect on the overall performance of the network (Fritzke, 1994b).

After training, the radial basis function neural network is translated into a set of purely conjunctive 0th-order Sugeno rules. The Gaussian of a given cell node forms the antecedent part of a rule and the weight of the connection between the cell node and the output node forms the consequent of the rule. The connection structure of the hidden layer generated by the growing neural gas algorithm is stored for later use by the reactive tabu search algorithm. The connection structure indicates neighbouring (overlapping) cell nodes in the hidden layer of the network, while the extent of overlapping in the attribute space is also recorded for later use.

7.4.2. Assembly of rule antecedents with the reactive tabu search algorithm

Next, the antecedent parts of each fuzzy rule are broken up and distributed amongst a new fixed set of user-defined fuzzy rules, i.e. the one-dimensional membership functions of an original fuzzy rule are distributed individually among the fuzzy rules in the new set. The antecedent part of a new fuzzy rule therefore contains membership functions taken from a number of original, created by the growing neural gas-created fuzzy rules. This implies that the antecedent part of each new fuzzy rule can contain internal disjunction. Moreover, there are fewer of these new fuzzy rules than originally obtained from the radial basis function neural network.

The reactive tabu search algorithm is then used to repeatedly exchange these membership functions in the new set of fuzzy rules. The problem of finding the best partitioning of membership functions in the new set of fuzzy rules is modelled as a set-covering problem. Specifically, one membership function from each of two rules is swapped with one another. Note that the total number of membership functions of each fuzzy rule remains constant. After each swap of a pair of membership functions between two rules, the consequent of each fuzzy rule is determined by use of linear least squares methods. The aim of this iterative process is to find membership function exchanges that improve the accuracy of the set of rules. The reactive tabu search algorithm selectively implements swaps that improve the performance of the set of rules, i.e. minimize the fitness of the rule model, with the aim of finding the best combination of membership functions amongst the fixed set of fuzzy rules. Repeated swapping continues until an acceptable level of predictive performance is attained or a user-defined maximum number of search iterations have been completed.

Of particular importance is that only certain membership function swaps are permitted. The choice of permissible swaps is based on the adjacency matrix information obtained from the radial basis function neural network trained with the growing neural gas algorithm. Swaps are permitted only when they place a membership function in a fuzzy rule where at least one of the existing membership functions of the rule is in the direct topological neighbourhood of the candidate swap membership function. In other words, a membership function is only swapped to a rule if it is in the direct topological neighbourhood of (or overlaps with) any of the membership functions currently assigned to the antecedent part of the fuzzy rule. This overlapping requirement must be valid for both fuzzy rules under consideration. The dual purpose of this swap restriction is first to bias the swapping process towards creating fuzzy rules with highly overlapping membership functions in each attribute dimension. Second, restricting swaps significantly reduces the part of the search space that the reactive tabu search algorithm has to cope with.

7.4.3. Membership function merging and rule reduction

The next stage of the CORA algorithm considers each fuzzy rule in turn. In each attribute dimension of the rule antecedent part, those sets of membership functions that overlap to a large extent are merged into single membership functions. The purpose of merging is first to simplify the antecedent part of each fuzzy rule. This increases the comprehensibility of the set of fuzzy rules. Second, the number of membership functions is reduced. This means that if the new set of membership functions is presented to the reactive tabu search algorithm for further membership function swapping, the size of the corresponding combinatorial problem is reduced. Merging of membership functions and searching for an optimal partitioning of these membership functions continues until an acceptably comprehensible and accurate set of rules is generated.

The final stage of the CORA algorithm performs subset selection to determine the optimal subset of the trained and merged fuzzy rules. Rules that do not significantly contribute to the training accuracy of the rule model are discarded. The number of rules that are discarded is determined either by the user or according to a user-defined limit on the degree of accuracy that may be lost during training.

7.4.4. Calculation of a fuzzy rule consequent and solution fitness

Cholesky decomposition is used to solve overdetermined, linear systems of the form $\mathbf{Ax} = \mathbf{b}$. In both cases described above, $\mathbf{A}$ represents the $m \times n$ (with $m > n$) firing strength matrix of the given set of n fuzzy rules. $\mathbf{x}$ represents the $n \times 1$ solution vector, i.e. the set of fuzzy rule consequents. The $m \times 1$ vector $\mathbf{b}$ represents the true output of the m training exemplars. Let $\mathbf{C}$ be the augmented matrix $[\mathbf{A}\ \mathbf{b}]$. The linear system is solved by first calculating the $n \times n$ product matrix $\mathbf{P} = \mathbf{C}^T\mathbf{C}$. Thereafter LU decomposition based on Gaussian elimination with partial pivoting is used to factorise $\mathbf{P}$ to determine the solution vector $\mathbf{x}$ as well as the estimated output $\mathbf{b}'$, as discussed in more detail by Golub and van Loan (1993), as well as Datta (1995).

In the first case mentioned above where the outgoing weights of the network need to be calculated, the entire set of matrix operations are performed. In addition to the above calculations, the reactive tabu search technique also calculates the residual sum of squares $||\mathbf{Ax}\text{-}\mathbf{b}||^2{}_2$, with the residuals being $(\mathbf{Ax})_j - \mathbf{b}$ for the j'th data exemplar. This residual sum of squares is used as a component of the fitness (cost function value) of the current solution.

In the calculation of the network weights or the fitness of a solution generated by the reactive tabu search technique, the set of fuzzy rules is forced to fit the complete input-output relationship. In other words, the bias term in the radial basis function neural network is permanently set to zero and no constant term is calculated by the Cholesky decomposition method. The reason for this is to improve the comprehensibility of the given set of fuzzy rules and to speed up the reactive tabu search technique.

7.4.5. Optimal-size models

Simplification of the antecedent parts of the rules is performed once the reactive tabu search algorithm has found a solution with satisfactory accuracy or a maximum number of reactive tabu search search iterations has been completed. The criteria used to determine the best level of simplification are identical to the ones used to determine the best radial basis function neural network size.

A number of methods have been proposed for simplifying a set of fuzzy rules (e.g. Setnes, 1995; Setnes et al., 1998). The purpose of such methods is three-fold. The first is to simplify the antecedent part of a fuzzy rule by merging membership functions that are similar to each other into a single membership function of the same type. The second is to reduce the number of membership functions used in a set of fuzzy rules, and if possible delete copies of identical rules. This is done by redefining similar membership functions of different fuzzy rules in terms of a single, representative membership function. The third purpose is to delete membership functions that make an insignificant contribution to the fuzzy rule output.

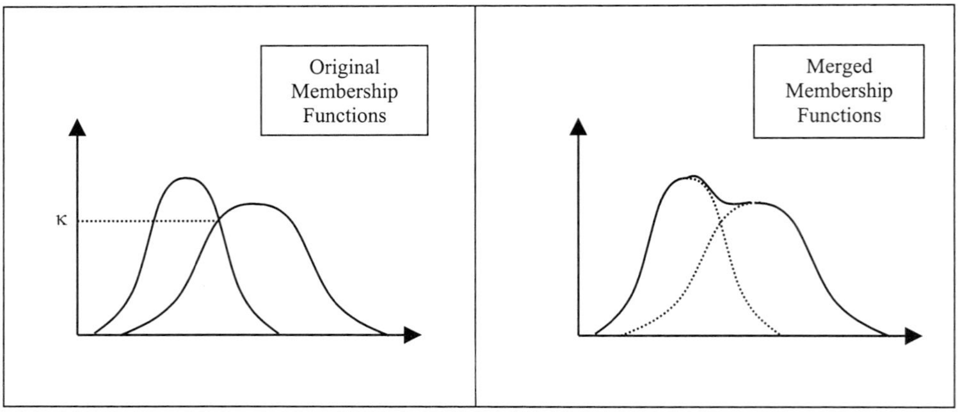

Figure 7.15. Merging of similar membership functions.

The purpose of fuzzy rule simplification, as it is used by the CORA algorithm, is most like, but not identical to, the first purpose described above. The difference is that merging by the CORA algorithm repeatedly attempts to replace two membership functions that are similar, with one that best fits the profile of the original two. The dual purpose of such merging is to minimally alter the firing strength properties of the fuzzy rule antecedent part while attempting maximum simplification. The merged membership function can be of a different type to the original two. This is in contrast to methods that merge two membership functions into a single one of the same type. The procedure is illustrated in Figure 7.15.

In some cases, such as the one illustrated in Figure 7.16, the determination of a proper width value is more complicated. In Figure 7.16 the lefthand shoulders of the two original Gaussians intersect. In this case the width value is chosen as the one that generates a lefthand shoulder of the two-sided Gaussian that best approximates (in the least squares sense) the maximum lefthand profile formed by the original two Gaussians. The same method is used to find the best righthand shoulder width value.

Such merging is sequentially attempted for all pairs of membership functions (original and previously merged) whose intersection point κ exceeds a certain threshold.

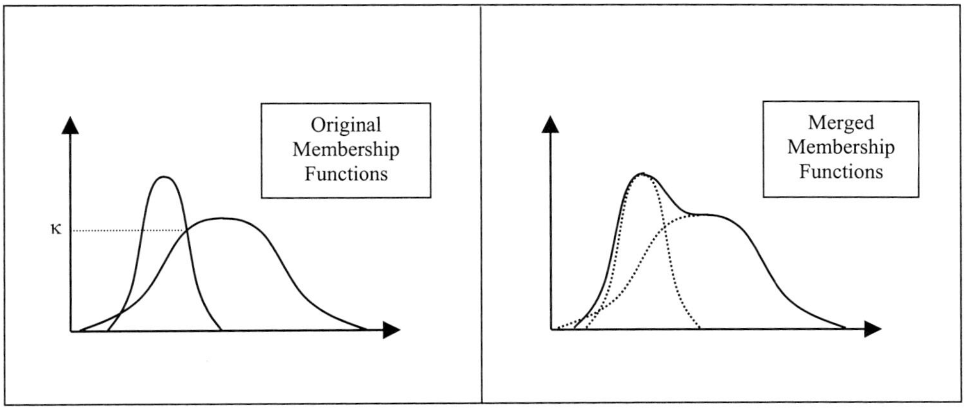

Figure 7.16. Merging of two membership functions that intersect twice.

Owing to the nonlinearity of the antecedent part of each fuzzy rule, as well as the interaction between the antecedent parts of different fuzzy rules, it is difficult to determine the optimum value of this threshold. Therefore, the following method is followed to determine the optimal size of the model. In this sense optimality implies the model that best approximates the true, underlying system that generated the data.

The threshold at which merging is allowed to take place is sequentially lowered by small decrements from a maximum value of one to a user-defined lower limit. For a threshold value of one no merging occurs. As the threshold decreases, so the degree of merging increases. For each threshold value the antecedent parts of the *unmerged* fuzzy rules obtained from the reactive tabu search technique are merged using the method described above. Two criteria are then calculated and subsequently used to find the best level of merging and in turn the optimal model[43]. These are Akaike's AIC information criterion (Akaike, 1973, 1974), given by

$$AIC = \log(RSS) + 2\left(\frac{p}{m}\right) \tag{7.10}$$

where *RSS* is the residual sum of squares calculated for the particular set of merged fuzzy rules on the training data. p is the total number of parameters required by the merged set of rules (two for each symmetrical Gaussian, four for each two-sided Gaussian, one for the consequent of each fuzzy rule). m is the total number of training exemplars. Akaike's BIC information criterion (Akaike, 1977) is given by the following equation with p and m defined as before

[43] Other criteria for determining the optimal size of the model were also investigated. These included criteria based on minimum description length (Oliver and Hand, 1994; Judd and Mees, 1995), minimum message length (Oliver and Hand, 1994), generalized prediction error (Moody, 1991) and Akaike's final prediction error (Akaike, 1970). Preliminary experiments showed that the AIC and BIC criteria produced satisfactory results. These two easily implemented information criteria were therefore used to find the optimal size of the model.

$$BIC = \log(RSS) + \frac{p\log(p)}{m}. \tag{7.11}$$

The level of merging that produced the minimum AIC and BIC values (the two criteria produced similar results) is chosen as the final level of fuzzy rule merging.

The reactive tabu search algorithm is run until the training accuracy of the best solution found thus far is satisfactory or until a maximum number of search iterations have been completed. Membership function merging then takes place. Thereafter the merged set of fuzzy rules can again be presented to the reactive tabu search technique for further membership function swapping.

The AIC and BIC criteria are also used to determine the optimal size of the radial basis function neural network. In particular, these criteria are used to find the minimum number of cell nodes to place in the network's hidden layer. A series of networks are generated, each with an increasing number of cell nodes. The network associated with the minimum AIC and BIC values, and that is of reasonable size to produce good generalization characteristics, is then used for further computation by the reactive tabu search technique.

7.4.6. Fuzzy rule set reduction

Rule set reduction is performed after rule antecedent assembly and merging have taken place. The Regression by Leaps and Bounds technique of Furnival and Wilson (1974) (see also Seber (1977)) is used for this purpose. The particular technique is used because it is guaranteed to find the best subset r of n fuzzy rules, given the integer r. In addition, the technique has been shown by Furnival and Wilson (1974) to be amongst the fastest available.

This combinatorial search technique is used to select the best 1, 2, ... $n - 1$ subsets of fuzzy rules from the assembled and merged set of n fuzzy rules. The smallest subset of fuzzy rules that retains a user-defined fraction of the training accuracy of the original n rules is then selected as the final trained, merged and reduced set of fuzzy rules.

7.4.7. Rule model output prediction surface smoothing

Preliminary experiments with the CORA algorithm showed that the reactive tabu search technique tended to overfit models and consequently magnitude penalty factor, also called a weight penalty factor, was used to circumvent the problem. The magnitude of this factor was calculated as the sum of the squared consequents of each fuzzy rule in the current rule model. A user-defined fraction of this factor was added to the fitness of the current rule model. The initial fitness was simply the residual sum of squares obtained by the current rule model.

The phenomenon of disproportionately large fuzzy rule consequents, or in different terms the generation of linear regression coefficients that exhibit large variances, is often the result of multicollinearity as discussed in Chapter 4. Although other methods, such as ridge regression could also have been used to ameliorate the effects of multicollinearity, the above-mentioned weight penalty factor method in the reactive tabu search algorithm was used, as it was easy to implement, was relatively fast and gave an explicit indication of how the reactive tabu search technique handled multicollinearity.

7.4.8. Overlapping of fuzzy rules in the attribute space

Preliminary experiments also indicated that fuzzy rules generated by the algorithm tended to overlap in the attribute space to a considerable degree. This could be attributed to the inherent bias in membership function swapping, i.e. to create fuzzy rules with highly overlapping membership functions. In addition, the method used by the supervised growing neural gas algorithm to determine the width of the hyperspherical Gaussians did not take attribute space overlapping explicitly into account. Furthermore, the CORA algorithm in effect used the weighted sum operator (Jang et al., 1997) to determine the overall fuzzy rule set output. The result of these three factors was that the set of fuzzy rules turns into a distributed processing system. In other words, more than one rule needed to be inspected to determine the overall fuzzy output.

One way of dealing with the problem was to determine the overall output of the set of fuzzy rules via the weighted average defuzzification operator (Jang et al., 1997; Lee and Park, 1997). If this operator is used, there is no loss of linguistic meaning associated with the membership functions. Unfortunately this operator is significantly more computationally expensive than the weighted sum operator. In addition, normal linear regression methods such as Cholesky decomposition cannot be used to determine the fuzzy rule output and solution fitness without considerable modification.

A different method of reducing attribute space overlapping was therefore used by the CORA algorithm, in particular by the reactive tabu search technique (no changes have been made to the growing neural gas algorithm). For each training exemplar, the sum of the firing strengths of all the rules with respect to the specific exemplar was calculated. The least loss of membership function linguistic meaning was attained if the sum of firing strengths were at most unity (Jang et al., 1997). Consequently if the sum of the firing strengths for a particular training exemplar were larger than unity, the squared difference (between one and the sum of the firing strengths) was added to an initially zero penalty term. The final overlap penalty factor was calculated over all the training exemplars.

Subsequently, the magnitude of the overlap penalty factor was compared to the value obtained by the growing neural gas algorithm using exactly the same calculation. If the value for the current rule model were greater than that of the GNG-trained rule model, a user-defined fraction of this overlap penalty factor was added to the existing fitness value of the current rule model. The effect of this form of overlap penalisation was to discourage rule models that exhibited higher attribute space overlap in comparison to that of the GNG-trained rule model.

The reasons why this particular form of overlap penalisation was used are as follows. First, preliminary experiments showed that any form of overlap penalisation reduced attribute space overlap, but also started to reduce the training accuracy of generated rule models. In the extreme, too much overlap penalisation made it impossible for the reactive tabu search technique to build rule models with good training accuracy. Second, the above-mentioned overlap penalisation methodology was followed, because it made overlap penalisation relatively insensitive to the factor by which it is multiplied before being added to the current model fitness. In other words, the user did not have to tune the multiplication factor each time a rule model was constructed for a different data set.

In summary, the fitness value was therefore a linear combination of the residual sum of squares obtained by the fuzzy rule model, the weight penalty factor and the overlap penalty factor. During fuzzy rule antecedent assembly, the reactive tabu search component of the CORA algorithm attempted to minimize this composite fitness value. The purpose of such fitness minimization was to find a fuzzy rule model that exhibited good training accuracy, had small rule consequents and at worst, had the same attribute space overlapping as the overlapping obtained by the growing neural gas algorithm.

7.4.9. Performance of the CORA algorithm

The characteristics of the CORA algorithm can be demonstrated by means of two examples. The first example (*Sin-Cos* data set) concerns an artificial data set with continuous outputs and the second (*Slug Flow* data set) involves the classification of multiphase flow systems. These data sets are briefly described below.

a) Sin-Cos data

The *Sin-Cos* data set previously used by Schmitz et al. (1999a), consists of four variables, three (θ, x, s) of which are continuous and the fourth (ϕ) that can take on one of two numeric values. The single, continuous output is given by

$$y = \sin(4\pi\theta - \phi) + ax + bs + c\varepsilon \tag{7.12}$$

where θ, x and s can range between zero and one. ϕ can be either 0 or $\frac{1}{2}\pi$. ε is a Gaussian random variable with a zero mean and standard deviation of one.

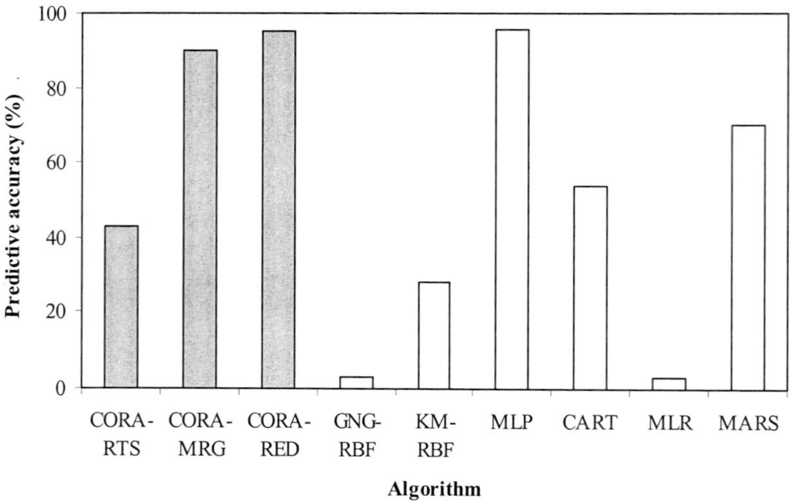

Figure 7.17. Comparative accuracy of the CORA algorithm on the Sin-Cos problem.

The data used in this investigation were generated using values of 0.3, 0 and 0.2 for a, b and c respectively. Note that even though b is 0 and subsequently s plays no role in determining y, the algorithm still has to deal with this nuisance variable. The validation set of data was generated with c set to 0.

b) The Slug Flow Data Set

The *Slug Flow* data set was reported by Reimann et al. (1992) who studied the transitions between flow patterns of multiphase fluid flow systems in horizontal pipes. Two primary flow regimes are possible, viz. slug flow or non-slug flow. The latter regime includes transitional phases such as slug-annular and slug-wave flow. The flow regime is determined by four variables, viz. flow pressure, the diameter of the pipe, the superficial liquid velocity and the superficial gas velocity.

Each data set was randomised and split into a training and validation set and then normalized to have a zero mean and a standard deviation of one. This normalization was done individually for all the attributes as well as the output of each problem. This was done, as both the growing neural gas and the CORA algorithms assume that no bias terms are used. In other words, these algorithms force the regression function that is constructed (the set of fuzzy rules can also be interpreted as a function consisting of a linear combination of nonlinear functions) to have an output of zero when no rule fires. Moreover, the data are scaled because the growing neural gas algorithm uses the Euclidean metric to determine the distance between a hidden layer, multidimensional Gaussian and the input of data exemplars. If the data were not scaled an attribute with a high standard deviation would be emphasised disproportionately compared with an attribute with a smaller standard deviation.

The results of the CORA algorithm on each of the two problems are compared with those of other algorithms in Figures 7.17-7.18. For the continuous output (regression) problem, CORA was compared with a growing neural gas (GNG) algorithm, a radial basis function neural network (RBF), a multilayer perceptron with linear output nodes (BPLinear), a regression tree model (CART), multiple linear regression (MLR) and a multi-adaptive regression spline (MARS) model. Three versions of CORA are indicated in Figure 7.17, viz. CORA-RTS, CORA-MRG and CORA-RED. These versions simply indicated the performance of the CORA algorithm with reactive tabu search only (RTS), performance after merging of the fuzzy rules (MRG) and performance after the final stage of rule reduction (RED), as described previously in sections 7.4.2 and 7.4.3.

For the classification problem (Slug Flow data), CORA was compared with two radial basis function neural networks, multilayer perceptrons with linear (BPlinear) and nonlinear output nodes (BPtanh), a regression tree model (CART) and a rule induction algorithm (BEXA). The radial basis function neural networks were used to construct fuzzy rules from the data, as they were equivalent to fuzzy rule-based models. The conditions under which radial basis function neural networks are equivalent to fuzzy inference systems are discussed inter alia by Hartman et al. (1990), Park and Sandberg (1991) and Jang and Sun (1993). These networks and the BEXA algorithm are briefly described below.

c) Radial basis function neural networks (GNG-RBF, KM-RBF)

The growing neural gas algorithm was used to develop a radial basis function neural network with a hidden layer consisting of spherical Gaussian kernels and a single-node linear output layer. In one case, the hidden layer was trained with the growing neural gas algorithm (Fritzke, 1994a) and in the second case k-means clustering (Moody and Darken, 1989) was used to train the hidden layer. The weights to the output layer of the growing neural gas network were determined by using a combination of the delta rule (Widrow and Hoff, 1960; Stone, 1986) and multiple linear regression. The weights of the output layer of the k-means, radial basis function network were trained using linear regression only.

The k-means radial basis function network had a bias and the input nodes were also directly joined to the output layer by weighted connections. This means that the network function contained both linear terms and a constant in addition to the nonlinear terms represented by the network's hidden layer. The growing neural gas network had neither a bias, nor any linear terms. The growing neural gas network was therefore directly equivalent to a set of 0th-order Sugeno fuzzy rules. The k-means, radial basis function network also generated a set of 0th-order Sugeno fuzzy rules, comparable to those generated by the CORA algorithm, but these were combined with a constant fuzzy rule (the bias term) and a linear function (the set of linear terms).

d) Rule-induction algorithm (BEXA)

The BEXA algorithm (Theron, 1994; Theron and Cloete, 1996) is a state-of-the-art crisp rule induction algorithm that induces classification rules from data by using a covering approach. Each condition involves only one attribute. For categorical attributes the condition is an equality test. For numeric attributes the condition is inclusion within a one-sided, e.g. "$X \leq x_i$" or "$X \geq x_i$", or two-sided interval. In contrast with other algorithms, such as RISE (Domingos, 1994), both internal disjunction and negation of the antecedents for categorical and numerical attributes are allowed. In addition, BEXA employs a postpruning scheme (Quinlan, 1987) that prunes rules in two steps. The first step simplifies individual rules by deleting conditions that are insignificant according to Fisher's exact test. The second step discards those rules that do not reduce the accuracy of the entire rule set on the complete set of training data.

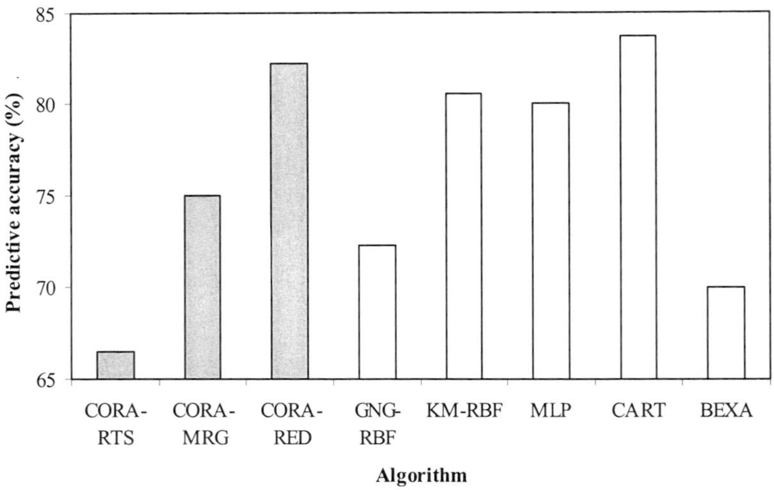

Figure 7.18. Predictive accuracy of the CORA algorithm on the Slug Flow problem.

The CORA algorithm was capable of constructing rule models with comparable or better accuracy than all the other techniques considered. The growing neural gas algorithm (upon which the CORA algorithm is partly based) in particular, appears to have difficulty with problems where the attribute space is primarily discrete (Gouws, 1999). The growing neural gas algorithm attempts to construct a topology preserving map of the manifold formed by the training data in attribute space. The lack of numeric attributes and the consequent polarisation of exemplars in the attribute space make it difficult for the growing neural gas algorithm to generate such a mapping.

The CORA modelling approach that encompasses internally disjunctive fuzzy rules is generally more powerful than the modelling approach used by the growing neural gas algorithm (which does not allow internally disjunctive rules). This is aptly illustrated by the experiments performed on the Sin-Cos problem. The best results that were obtained using the growing neural gas algorithm were based on a hidden layer size of 120 cell nodes. For such a growing neural gas network the training R^2 is 0.83 with a corresponding validation R^2 of 0.52. In contrast, the CORA algorithm was able to construct fuzzy rule models with a median validation R^2 of 0.90 (best validation R^2 is 0.96) based on a growing neural gas model containing only 30 cell nodes. (All CORA experiments were performed based upon a growing neural gas model containing only 30 nodes.) However, the best model that the growing neural gas algorithm could build if only 30 rules were used, exhibited a training R^2 of 0.24 and a validation R^2 of 0.024, i.e. the CORA algorithm was able to construct more accurate models than the growing neural gas algorithm, even if the growing neural gas algorithm was allowed to use four times as many cell nodes as was allowed the CORA algorithm.

More detailed inspection of the models derived by the growing neural gas and CORA algorithms for the Sin-Cos problem showed that the restrictions on the type of fuzzy rules that the growing neural gas algorithm can construct (no internal disjunction allowed) make it difficult for the algorithm to effectively model the primarily sinusoidal nature of the Sin-Cos output. In contrast, the ability of the CORA algorithm to build internally disjunctive rules allows the algorithm to selectively use membership functions in the appropriate attribute dimensions to approximate the sinusoidal output. Membership functions from the same original growing neural gas radial basis functions from which the above membership functions were obtained, but that represent different attributes, can be used to model the secondary linear nature of the Sin-Cos output surface.

In addition, it was found that the CORA algorithm was forced to use relatively large fuzzy rule consequents in order to build accurate rules from the inaccurate 30 fuzzy rules obtained from the growing neural gas algorithm. In particular, rule consequent values ranged between +27.7 and -33.0. In normal circumstances, given the preprocessing performed on the data, such large and small consequents imply that the predicted output surface is irregular.

7.5. SUMMARY

In summary, in this chapter the equivalence of neural networks to rule-based models or the use of neural networks to extract rules from data has been discussed. Under certain conditions radial basis function neural networks are equivalent to fuzzy rule-based systems and by way of case studies it was shown that

- Radial basis function neural networks evolved with the CERN algorithm (Schmitz and Aldrich, 1999, Schmitz, 1999) constituted significantly more accurate and compact representations than those obtained by means of networks containing spherical Gaussians trained with the k-means clustering algorithm.

- Moreover, ellipsoidal radial basis function (ERBF) networks could be translated into more parsimonious sets of fuzzy rules, which could be condensed even further through appropriate post-processing.

- The fuzzy rules generated by the ERBF neural networks exhibited considerable overlap, which could be eliminated by an algorithm such as the combinatorial rule assembler (CORA) discussed in the chapter.

The extraction of rules are not restricted to radial basis function neural networks, as was shown by way of example with the ANN-DT algorithm (Schmitz et al., 1999a, 1999b), which had the following features.

- The ANN-DT algorithm facilitates the extraction of explicit IF-THEN rules from trained feed-forward neural networks of arbitrary structure. These rules appear to be significantly more representative of the behaviour of the neural networks than rules extracted from training data only.

- Alternatively, the algorithm can be used as a method to extract rules from data sets. These rules appear to be of comparable accuracy to those obtained with C4.5 and an algorithm using procedures similar to CART's splitting and pruning algorithm. However in some cases a significant improvement could be obtained with the ANN-DT algorithm. Pruning could in some cases simplify the trees, while retaining relatively high accuracy.

- Also, in order to focus the extracted rules on the relevant domain, it is important to sample the neural network on data similar to the original training data set.

- The ANN-DT algorithm tended to produce somewhat more complex trees for classification problems than ID3 and C4.5

- For classification problems the ANN-DT(e) algorithm, which selects attributes and split points using a gain ratio criterion, appeared to perform better than the ANN-DT(s) algorithm, especially in terms of the number of rules extracted. The latter algorithm bases the selection of attributes and split points on a significance analysis of the input attributes. Unlike a sensitivity analysis that only considers the partial derivatives, the significance analysis takes the correlational structure of the data into account.

- For problems with continuous outputs, it was demonstrated that the significance analysis of the ANN-DT(s) could correctly identify the most important attributes, whereas a greedy error driven procedure, such as used in ANN-DT(e) and standard decision tree techniques (e.g. CART), failed to identify these attributes. The significance analysis would therefore seem to be a suitable splitting criterion near the root of the decision tree, whereas a greedy splitting criterion would better split the lower branches of the tree.

Finally, the combinatorial rule assembler (CORA) algorithm, proposed by Gouws (1999) could be used to build compact fuzzy rule-based systems based on the use of different versions of radial basis function neural networks. In short

- The CORA algorithm is capable of constructing rule models with comparable or nominally better predictive accuracy than other advanced modelling techniques, but is not suitable for problems characterized few observations on discrete variables.

- The CORA models were more accurate and less complex than the growing neural gas models upon which they were based. They also contained fewer rules with similar accuracy, when compared to models like CART and BEXA.

- Unlike fuzzy systems derived from radial basis function neural networks, the CORA algorithm facilitates the building of conjunctive and disjunctive rules, enabling it to deal with more complex relationships than exclusively conjunctive rule-based systems.

Chapter 8
Introduction to the Modelling of Dynamic Systems

8.1. BACKGROUND

Many real world systems, such as chemical and metallurgical processes, exhibit sequentially changing behaviour. These systems tend to be large, nonlinear and ill-defined in that little knowledge of plant dynamics and chemical reaction mechanisms is available. As a consequence, engineers attempt to develop dynamic process models for these systems direct from input-output data, rather than pursuing the development of time consuming and expensive analytical models. In practice, many different approaches to time series analysis are used, including Fourier spectral analysis (Wornell, 1996), Bayesian modelling (West and Harrison, 1989), neural networks (Elman, 1990; Cheng et al., 1995), various pattern search methods (Hamilton, 1989; Rabiner, 1989), etc. In most of these processes, the analyst exploits prior knowledge of the underlying system to develop the model, or assumes that the process is linear and/or the variables Gaussian.

Even so, in many instances, scientists have discovered features in their data that are not coherent with traditional models, e.g. bursting phenomena that are more frequent and enhanced than expected with usual models, river flows[44], financial data[45], avalanching of sand piles and powders (Kaye et al., 1995; Kaye, 1997), biophysical applications, atmospheric sciences (Dzubay, 1982) and various other applied fields. As early as 1965, Mandelbrot has already discovered that classical models were not in accordance with observations when long-range dependence was introduced. These are processes whose autocorrelation at lag k decays slower than ρ^k, for $0 < \rho < 1$. Such systems, that are weakly understood, are represented well by classical or ad hoc approaches, and are best modelled by so-called parametric methods.

With parametric approaches, the time series is parameterized by embedding in a state space, as will be explained in more detail in the rest of this chapter. Parametric methods constitute a formal and systematic means for the analysis and modelling of dynamic systems and are superior to non-parametric methods in many ways. For example, with proper embedding, an estimate of the order of the underlying dynamic system can be made and it is also possible to construct free-run models that are far more reliable than similar models fitted direct to the time series.

In theory, a single record (time series) from a dynamic system is the result of all interacting variables associated with the process. In principle, it should therefore contain information about the dynamics of all the key variables involved in the evolution of the system. Mathematically, this can be justified by assuming that the variables of a dynamic process system satisfy p 1st order differential equations, i.e.

$$dx_1/dt = f_1(x_1, x_2, \ldots x_p)$$

[44] Scientists have for some time been aware of the so-called Hurst effect (Hurst, 1951), where the range of the observed n times long series is of order n^α, for $0.5 < \alpha < 1$, while usual models yield $n^{1/2}$. The growth is faster than expected, something that is very important if the aim is to perform flood prediction without underestimating risks.

[45] Time series in the economical area are sometimes very long (tick-by-tick data) and many researchers have considered them as prototypes for new modelling and inferential techniques (random walks, chaos, fractals, semimartingales).

$$dx_2/dt = f_2(x_1, x_2, \ldots x_p)$$

$$\ldots$$

$$dx_p/dt = f_p(x_1, x_2, \ldots x_p) \qquad (8.1)$$

This system of differential equations (8.1) can be reduced to a single p^{th} order differential equation via successive differentiation of any one of the 1st order differential equations, that is

$$d^p x_1/dt^p = f(x_1, dx_1/dt, d^2x_1/dt^2, \ldots d^{p-1}x_1/dt^{p-1}) \qquad (8.2)$$

Equation (8.2) represents the entire set of 1st order equations without any loss of information regarding the dynamics of the system. In other words, the single variable system (set of 1st order differential equations) is equivalent to the multivariable system (p'th order differential equation).

Moreover, instead of a continuous variable and its derivatives, a discrete time series and its successive shifts by a lag parameter τ should be sufficient to capture the dynamics of the system (Ruelle, 1980). The lagging is equivalent to first order differencing of the time series, which is analogous to differentiation. In an observed time series, the lagged copies of the series can thus be seen as additional system variables.

One of the difficulties in dealing with empirical systems is that one does not know a priori whether deterministic dynamics underlie the data, that is whether dynamic attractors exist. Nonlinear identification algorithms that calculate the system dimension from a time series (for example, by use of correlation dimension, as discussed in more detail later) has initially been thought to return an infinite value for stochastic processes (that have infinite dimension), but instead converge to a finite, non-integer value for deterministic processes (Osborne and Provenzale, 1989). If so, this could be used as a simple means to classify stochastic and deterministic processes.

However, Osborne and Provenzale (1989) have showed that stochastic data with power-law spectra also yield correlation dimensions with finite values. Some stochastic processes known as coloured noise have fractal curves in phase space, but no dynamic attractors. In fact, many random functions or signals have power spectra obeying definite power law relationships, where the power spectra are plots of the squares of the magnitudes of the Fourier transform of the series, versus the frequency (Addison, 1997). The time trace of a random variable with a Gaussian distribution is known as white noise and its power spectral density P(f) remains constant over the frequency range, i.e. it is $\propto 1/f^0$. The integral of white noise gives regular Brownian motion (brown noise) proportional to $1/f^2$. Pink noise is related to phenomena with power spectral densities proportional to $1/f$, while black noise is related to phenomena with power spectral densities proportional to $1/f^n$, where $n \geq 2$ (Wornell, 1996; Addison, 1997). Most nonlinear identification algorithms cannot distinguish between fractal curves and fractal attractors. Clearly, reliable classification of the data as a first step in system identification is important, otherwise the resulting model is unlikely to generalize beyond the training data set.

A promising method of dealing with systems such as these is by use of global models based on the reconstruction of the attractor(s) of the systems. In practice the state space of a dynamic system is often unknown and all that is available is a set of observations (time series) on one or more variables of the system. Therefore, the problem is to derive the characteristic attractor(s) of the system from the available time series. Packard et al. (1980) were the first to propose the application of state space reconstruction methods to scalar time series. At about the same time Takens (1981) provided a mathematical rationale for this approach by showing that from a time series only, it is possible to (re)construct a new state space diffeomorphically[46] equivalent to the original state space of the observed system. The reconstructed states allow the time series to be analyzed from the viewpoint of deterministic nonlinear[47] dynamics, i.e. the underlying dynamics of the system can be modelled and characterized, e.g. by Lyapunov exponents or the correlation dimension.

Reconstruction of the state space of the system facilitates successful identification of the system and can be applied to all deterministic systems. A state space model ensures the complete description of the dynamics of a system and allows prediction of the system outputs, even without complete knowledge of the fundamental system dynamics.

Broadly speaking, three methods can be used for state space reconstruction from a scalar time series, viz. derivative coordinates, delay coordinates and what is known as singular spectrum analysis (Elsner and Tsonis, 1996). Derivative coordinates (Packard et al., 1980) are based on higher order derivatives of the time series, but are not particularly useful for experimental data, owing to their susceptibility to noise. In contrast, delay coordinate methods and singular spectrum analysis are used extensively on experimental systems and consequently only these approaches will be outlined below.

8.2. DELAY COORDINATES

Consider a smooth (C^2) m-dimensional manifold ($A \in \Re^m$) representing the original state space of the system being investigated and let $\phi^t: A \rightarrow A$ be the corresponding flow. The measured scalar quantity $s(t) = h[x(t)]$, given by the measurement function $h: A \rightarrow \Re$, where $\mathbf{x}(t) = \phi^t[x(0)]$, can be used to construct a delay coordinate map.

$$F: A \rightarrow \Re^m, \quad x \rightarrow y = F(x) = [s(t), s(t+L), \ldots s(t+(m-1)L)] \tag{8.3}$$

that maps a state x from the original state space A to a point y in a reconstructed state space $A' \in \Re^d$. The embedding dimension and the time lag (delay) are denoted by d and L respectively. The idea is portrayed diagrammatically in Figure 8.1, which shows a smooth manifold constituting the original state space of the dynamic system, with corresponding flow[48] ϕ^t.

If some scalar quantity $s(t) = h[\mathbf{x}(t)]$ can be measured, given the measuring function $h: M \rightarrow \Re$, where $\mathbf{x}(t) = \phi^t[\mathbf{x}(0)]$, then a delay coordinate system can be constructed, so that

$$F: M \rightarrow \Re^m$$

$$\mathbf{x} \rightarrow \mathbf{y} = f(\mathbf{x}) = [s(t), s(t-t_L), \ldots s(t-(m-1)t_L)] \tag{8.4}$$

---

[46] Diffeomorphism of a functional mapping $f: u \rightarrow v$, means that the mapping $f(u) = v$ is one-to-one (i.e. invertible) and that the inverse of the mapping $f^{-1}: v \rightarrow u$ is continuously differentiable with regard to its argument.

[47] That is functions that do not have the property $f(x+y) = f(x)+f(y)$ and $f(ax) = af(x)$.

[48] A dynamic system can be associated with discrete or continuous time. The discrete case is defined by a map, $z_1 = f(z_0)$, that gives the state z_1 resulting from the initial state z_0 at the next time value. The continuous case is defined by a *flow*, $z(t) = \phi^t(z_0)$, which gives the state at time *t*, given that the state was z_0 at time 0. A smooth flow can be differentiated w.r.t. time to give a differential equation, $dz/dt = F(z)$. In this case we call $F(z)$ a *vector field*, as it gives a vector pointing in the direction of the velocity at every point in phase space.

maps a state **x** from the original state space M to a point y in a reconstructed state space $\Re^m$, where m is the embedding dimension and t_L is the delay time or lag.

In effect, this means that with lagged copies of a single time series, the coordinates of a phase space can be defined that will approximate the dynamics of the system that has generated the time series. A univariate time series is thus transformed into a multivariate set of observations. The number of lags is the embedding dimension of the system.

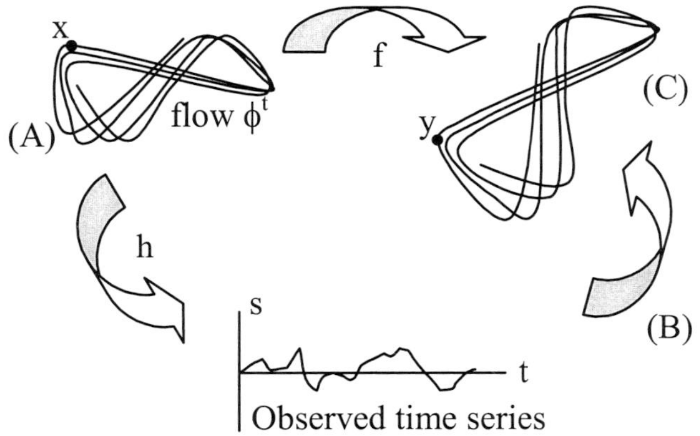

Figure 8.1. Delay reconstruction of states from a scalar time series, (A) attractor in unknown state space M, (B) delay coordinates, $y(t) = [s(t), s(t-t_L), \ldots s(t-(m-1)t_L)]$ and (C) measurement $h:M \rightarrow \Re$, observable $s = h(x)$.

8.3. LAG OR DELAY TIME

Different lags (L) lead to reconstructions that are diffeomorphically equivalent, but geometrically different. Lags that are too small lead to coordinates that do not differ sufficiently from each other, resulting in the state points scattered along the diagonal, as indicated in Figure 8.2. This makes the estimation of the dimensionality difficult, because of the very small scales that have to be investigated, the presence of noise, etc. Too large a lag can lead to excessive folds in the data, bringing states close together in the reconstructed state space that are not close together in the actual state space.

The presence of noise tends to aggravate this situation. Even in the absence of noise, spurious intersection of the attractor can lead to areas where the reconstruction is not one-to-one anymore. This problem can be remedied by increasing the dimensionality of the embedding, but it is more desirable to decrease the lag. Optimal values for the lag are among other determined by directly considering the geometry of the reconstructed attractor, especially its unfolding from the diagonal (Kember and Fowler, 1993; Rosenstein et al., 1994).

In the presence of noise the time lag becomes important. Too small a time lag or window results in too little information extracted from the data and an attractor contracted into the main diagonal line in state space. This phenomenon is called *redundance* (Sauer et al., 1991) and is based upon the relative effect of measurement error to the difference between data in a delay coordinate vector. The effect levels out, because the attractor is finite in size, hence likewise for the differences between data.

On the other hand, too large a time lag results in data without correlation to be included, which causes an attractor with a false complex shape. This phenomenon is called *irrelevance* (Sauer et al., 1991). Underlying this phenomenon is the local divergence in time between adjacent trajectories of the state vector, which causes a dynamic error with increasing time lag. This error levels out owing to the finite shape of the attractor.

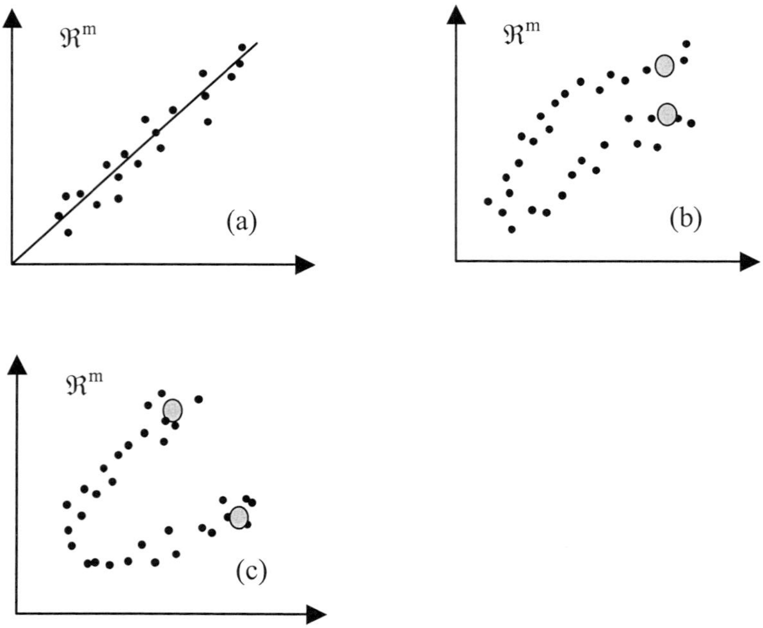

Figure 8.2.Reconstructions with different lags, (a) delay too small, (b) delay too large, and (c) delay appropriate.

8.3.1. Average Mutual Information (AMI)

The most popular methods for calculating the lag are based on the concept of mutual information. Given the delay vector $\mathbf{x}_i = [y_{i+(m-1)\tau}, y_{i+(m-2)\tau}, y_{i+(m-3)\tau}, \ldots y_i]$ for the time series $\mathbf{y} = [y_1, y_2, y_3, \ldots y_n]$ where m is the embedding dimension and n the size (length) of the time series, an accurate reconstruction of system invariants like the attractor and Lyapunov coefficients are ensured by an embedding dimension of $m > 2d_A$, where d_A is the topological or fractal dimension of the attractor (Takens, 1981). Time lag, τ, is not specified in Taken's theorem, because it is based upon the assumption of unlimited noise-free data.

The dominant method for calculating the time lag is the method of mutual information (Frazer and Swinney, 1986). The time lag is fixed heuristically at the point of the first minimum mutual information for the time series. This method estimates the probability to find a measurement again given that the same measurement has been already been made. This statistic is calculated over all the observations in the time series.

From an information theoretical point of view, measurements can be regarded as signals. Therefore, let S be a system of possible messages, $s_1, s_2, s_3, \ldots, s_n$ associated with the probabilities $P_S(s_1), P_S(s_2), P_S(s_3), \ldots P_S(s_n)$, of the occurrence of the messages or events, with subscript S denoting the particular system being observed. The information entropy of the system indicates the average information gained from a measurement,

$$H(S) = -\Sigma_i P_S(s_i)\log\{P_S(s_i)\}, \text{ for all } i = 1, 2, \ldots n \tag{8.5}$$

The dependence of $x(t+\tau)$ on $x(t)$ can be expressed as the uncertainty of finding $x(t+\tau)$, given $x(t)$, i.e.

$$[s,q] = [x(t),x(t+\tau)]$$

$$H(Q|s_i) = -\Sigma_i P_{q|s}(q|s_i)\log\{P_{q|s}(q|s_i)\} \tag{8.6}$$

The average uncertainty of finding $x(t+\tau)$ given $x(t)$ is calculated by averaging $H(Q|s_i)$ over s_i.

$$\begin{aligned} H(Q|S) &= -\Sigma_i P_S(s_i)H(Q|s_i) \\ &= -\Sigma_i P_{sq}(s_i,q_i)\log\{P_{sq}(s_i,q_i)/P_s(s_i)\} \\ &= H(S,Q) - H(S) \end{aligned} \tag{8.7}$$

The reduction in uncertainty about $x(t+\tau)$ gained from the measurement of $x(t)$ is the mutual information, which is

$$\begin{aligned} I(Q,S) &= H(Q) - H(Q|S) \\ &= I(S,Q) \\ &= -\Sigma_i P_{sq}(s_i,q_i)\log\{P_{sq}(s_i,q_i)/P_s(s_i)P_s(q_i)\} \end{aligned} \tag{8.8}$$

The joint probability distribution, P_{sq}, is estimated by the joint histogram of s and q. In this chapter the symbol k denotes embedding lag as a unit of the sampling period of the system observer.

8.3.2. Average Cross Mutual Information (AXMI)

Average cross-mutual information is related to average mutual information discussed above. A formal algorithm to calculate this statistic was first proposed by Frazer and Swinney (1986) for determining embedding lag, used in time series embedding (Frazer 1989a, Abarbanel 1996). The AXMI between the observation $x(t)$, at time t, and the observation $y(t-k)$, at time $(t-k)$, is

$$I_{xy}(k) = \sum P[x(t), y(t-k)]\log_2\left\{\frac{P[x(t), y(t-k)]}{P[x(t)]P[y(t-k)]}\right\} \tag{8.9}$$

where $P[\cdot]$ is the probability function in terms of x or y, $P[\cdot,\cdot]$, the joint probability function in x and y, t the time, and k some lag (multiples of sampling period, τ). Let R_{xy} be AXMI normalized with the average auto mutual information of the selected dependent state, $I_{xx}(0)$ as reference:

$$R_{xy} = I_{xy} / I_{xx} \tag{8.10}$$

For $k = 0$, $R_{xy}(0)$ is a nonlinear equivalent to linear cross-correlation.

8.4. EMBEDDING DIMENSION

The optimal choice of the embedding dimension d is the smallest value that provides a proper reconstruction. Theoretically, the geometric structure of the multivariate dynamics of the system can be unfolded from the observable variables s(t), in the absence of noise, in a d-dimensional space, according to equation (8.3). This can be done provided that $d \geq 2m+1$, where m is the state space dimension of the system. The condition $d \geq 2m+1$ is a sufficient, but not a necessary condition for the reconstruction of the system dynamics (Takens, 1981).

In reality, the reconstruction problem is an ill-posed inverse problem. For an inverse problem to be well-posed, it has to satisfy the following conditions (Tikhonov and Arsenin, 1977; Morozov, 1993; Kirsch, 1996).

- *Existence*: An output vector $y = f(\mathbf{x})$, where $y \in Y$, has to exist for every input vector $\mathbf{x} \in X$.
- *Uniqueness*: For any pair of input vectors $\mathbf{x}_1, \mathbf{x}_2 \in X$, $f(\mathbf{x}_1) = f(\mathbf{x}_2)$, if and only if $\mathbf{x}_1 = \mathbf{x}_2$.
- *Continuity or Stability*: The mapping is continuous (stable), if for any $\varepsilon > 0$, there exists a $\delta = \delta(\varepsilon)$, such that the condition $C_X(\mathbf{x}_1,\mathbf{x}_2) < \delta$ implies that $C_Y[f(\mathbf{x}_1),f(\mathbf{x}_2)] < \varepsilon$. The concept is illustrated diagrammatically in Figure 8.3.

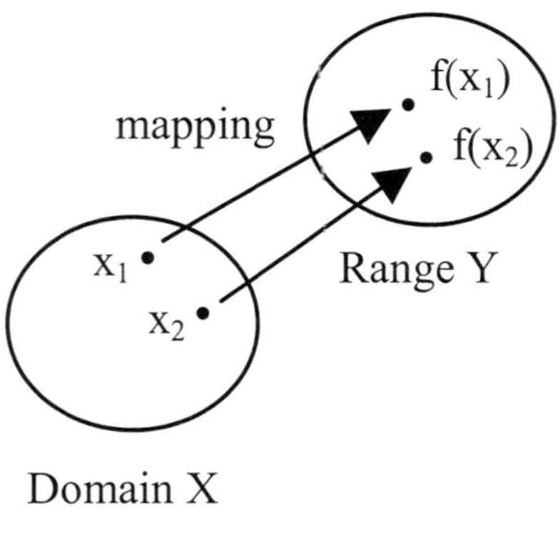

Figure 8.3: Continuous mapping of two observations from domain X to range Y.

In ill-posed inverse problems, one or more of the above conditions are violated. For example, the existence condition may not be satisfied, or there may not be sufficient information in the observations to enable a unique reconstruction of the system dynamics. In addition, the inevitable presence of noise in the time series causes uncertainty in the reconstruction and may also lead to violation of the continuity criterion.

Ill-conditioned problems can be improved by regularization, i.e. imposing conditions on the predictive model, such as smoothness of the input-output mapping, etc. As universal approximators, neural networks such as multilayer perceptrons and radial basis function neural networks lend themselves naturally to this kind of problem. Radial basis function neural networks are especially attractive, since regularization theory is applied as an integral part of their design.

Too large an embedding dimension (m) results in very thin distributions of data in the reconstructed space, while too small an embedding dimension will lead to improperly reconstructed state spaces. Most of the methods for determining the appropriate embedding dimension are based on continuity tests for the induced flow in the reconstructed state space (Alecsic, 1991; Buzug and Pfister, 1992; Gao and Zheng, 1993, 1994; Huerta et al., 1995) or for the embedding itself (Liebert et al., 1991; Kennel et al., 1992; Fredkin and Rice, 1995; Cao, 1997). These approaches ensure that the topology of the map is maintained, i.e. that closely neighbouring points are mapped to neighbouring points and vice versa. The most popular approach is aimed at minimizing the redundancy of the coordinates by use of information theory (mutual information) or linear autocorrelation functions (Frazer and Swinney, 1986; Frazer, 1989a; 1989b).

The diversity of methods for determining the optimal embedding parameters has its likely origin in the fact that no pair of parameters (d and L) is optimal for all systems. In fact, there is some evidence that the window length Ld is more important than the individual values of L and d. Also, many of the methods are computationally expensive and in many instances it may even be less time consuming to compute the parameters direct, i.e. compute several values of L and d, thereby also gaining some idea of the robustness of the embedding (Parlitz, 1999).

8.4.1. False nearest neighbours

An attractor[49] will be completely unfolded in an *m*-dimensional space given that $m > 2d_A$ where d_A is the topological or fractal dimension of the attractor (Takens, 1981). Since this is not a sufficient requirement, a better technique is needed to determine the minimum embedding dimension. With the false nearest neighbours algorithm (Kennel et al., 1992), neighbouring observations or points are identified by unfolding the attractor in space of increasing dimension, and eliminating spurious neighbouring points discovered in the process. False neighbours appear only because one views the attractor in a space of too small a dimension, thereby mistaking two points for being neighbours. Proximity is expressed in terms of the Euclidean distance between two points and neighbours are identified as points of which the Euclidean distance is within a preset limit.

8.4.2. False nearest strands

The false nearest neighbours technique tends to fail on excessively noisy data. Under these conditions, the method of false nearest strands may be considered (Kennel, 1997), as it appears to be better for time series showing a high degree of autocorrelation or sparsely

[49] A set in the phase space that has a neighbourhood in which every point stays nearby and approaches the attractor as time goes to infinity. Informally an attractor is simply a state into which a system settles (thus dissipation is needed). Thus in the long term, a dissipative dynamic system may settle into an attractor.

populated regions of attractors. Oversampling can be heuristically identified by a mutual information time lag in excess of $\tau = 10\tau_s$. Pairs of strands are defined when temporally corresponding points or temporal iterates lie on two adjacent trajectories as nearest neighbours.

8.5. CHARACTERIZATION OF ATTRACTORS

8.5.1. Correlation dimension and correlation entropy

A natural approach towards the characterization of the dynamics of a system is based on the characteristics that are invariant under reconstruction of the attractor of the system. The correlation dimension and correlation entropy in particular are popular measures and can be estimated from correlation integrals. The correlation integral $C_m(\varepsilon)$ of a reconstruction measure μ_m is defined as the probability that the distance between two points drawn independently according to μ_m is less than ε. That is

$$C_m(\varepsilon) = \iint \theta(\varepsilon - \|\mathbf{x}_1 - \mathbf{x}_2\|) d\mu_m(\mathbf{x}_1) d\mu_m(\mathbf{x}_2) \tag{8.11}$$

where $\theta(\cdot)$ denotes the Heaviside function, i.e. $\theta(w) = 0$, for $w < 1$ and $\theta(w) = 1$, for $w \geq 1$ and $\|\cdot\|$ is some norm[50]. For deterministic time series with large m and small ε, the correlation integrals scale according to

$$\log\{C_m(\varepsilon)\} \approx D_2 \log(\varepsilon) - m\tau K_2 \tag{8.12}$$

where τ is the delay used in the embedding, D_2 is the correlation dimension and K_2 is the correlation entropy. The dimensionless correlation dimension can be interpreted as the dimensionality of the attractor. It is bounded by the number of state variables of the system (i.e. a rough indication of the number of variables involved in the generation of the time series) and does not have to be an integer. The correlation entropy K_2 has the dimensions of inverse time and is a measure of the rate at which pairs of nearby orbits diverge or the unpredictability of the time series. It is often measured in bits/second or nats/second, depending on the base of the logarithm. The quantity $1/K_2$ can be seen as the time scale on which errors increase by a factor of e or 2, owing to the system dynamics.

Grassberger-Procaccia (1983) has proposed a method whereby the values of D_2 and K_2 can be estimated for an observed time series. Calculation of correlation dimension requires embedding of the time series so as to reconstruct the dynamic attractor of the original system in state space. For an observed time series x_t ($t = 1, 2, \ldots n$), sets of $N = n-(m-1)\tau$ reconstruction vectors $\mathbf{v}_t = (x_t, x_{t+\tau}, \ldots x_{t-(m-1)\tau})$ are constructed for consecutive embedding dimensions m. The set of delay vectors constitutes a set of points in $\Re^m$, which can be regarded as a sample of the underlying probability measure μ_m (provided that the time series is stationary).

The correlation integrals (defined in equation 8.11) are estimated as

$$\hat{C}_m(\varepsilon) = \binom{n}{2}^{-1} \sum_{i=1}^{n-1} \sum_{j=i+1}^{n} \theta\left(\varepsilon - \|v_i - v_j\|\right) \tag{8.13}$$

[50] The supremum norm $\|(x_1, x_2, \ldots x_m)^T\| = \sup_i|x_i| = \max(|x_1|, |x_2|, \ldots |x_m|)$ or the Euclidean norm is often used.

n is the number of points in the data set. The maximum and minimum value the correlation function can attain are C(ε) = 1 and C(ε) = 2/[n(n-1)] respectively. The idea is illustrated in Figure 8.4, whichs shows a hypersphere of radius ε traversing all the points on the attractor, counting all neighbouring points covered by the volume of the hypersphere.

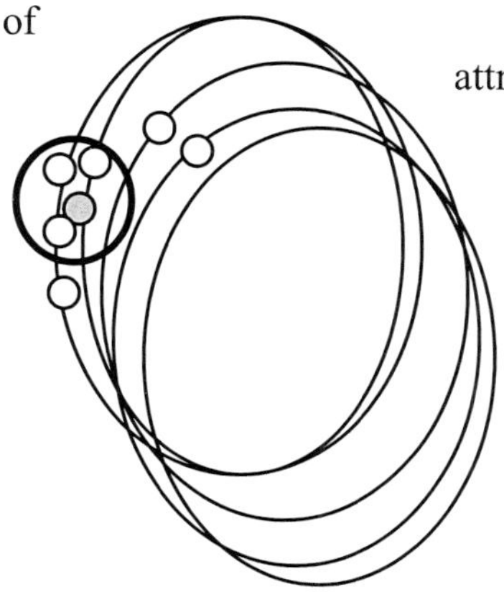

Figure 8.4. Calculation of correlation dimension by use of a probing hypersphere of radius ε moving along the attractor.

The next step is to estimate the correlation dimension and correlation entropy from the sample correlation integral (equation 8.13). Strictly speaking, the scaling law only holds in the limit as $\varepsilon \to 0$, and $n \to \infty$, that is

$$D_2 = \lim_{\varepsilon \to 0} \lim_{n \to \infty} \frac{\log C_m}{\log \varepsilon} \qquad (8.14)$$

This is illustrated in Figure 8.5. Reliable calculation of the correlation dimension is not as straightforward as was first thought when the Grassberger-Procaccia algorithm appeared (Grassberger and Procaccia, 1983; Grassberger et al., 1991). Using this algorithm requires a linear scaling region to reliably calculate the correlation dimension. In practice, estimates of the correlation integral are not very reliable, owing to noise in the data. As a consequence, one has to find a region (ε-interval) where the scaling law holds approximately, that is where curves $C_m(\varepsilon)$ vs ε on a double logarithmic scale are approximately straight and parallel for consecutive values of m. The values of D_2 and K_2 are found by fitting ε to the sample correlation integral within the scaling region. Lai and Lerner (1998) showed that the scaling region is sensitive to the choice of embedding lag. Linear correlation in the data set misleads the algorithm to wrongly show convergence to some low dimension, which could then be misinterpreted for inherently low-dimensional dynamics (Judd, 1994).

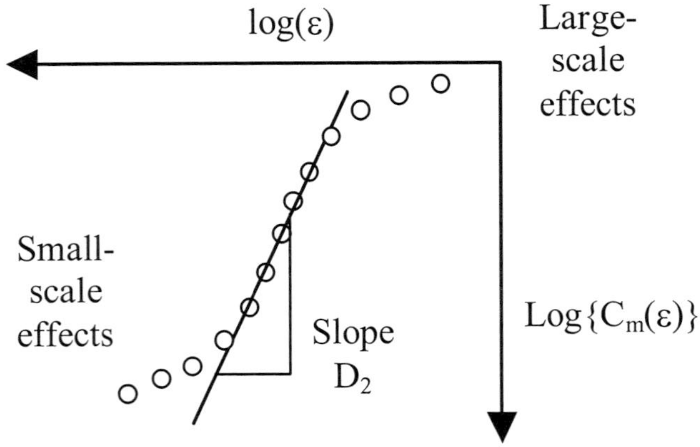

Figure 8.5. Calculation of the correlation dimension (D_2), as the slope of the functional relationship between the correlation sum ($C_m(\varepsilon)$) and the scale (ε).

Earlier, Judd (1992) pointed out the deficiencies of the Grassberger-Procaccia algorithm and proposed a different algorithm to calculate correlation dimension. This algorithm replaces the requirement for a linear scaling region by fitting a polynomial of the order of the topological dimension in that region. It finally expresses correlation dimension for inter-point distances below a specific scale ε_0. Instead of comparing single values of correlation dimension estimates, one rather compares the clustering of correlation dimension estimation curves calculated by the Judd algorithm. This allows correlation dimension to be used for examining the macro- and microscale of the reconstructed dynamic attractor. For large data sets it asymptotically approaches the value of the true correlation dimension as ε_0 goes to zero. In addition, the algorithm is not easily confused by linear correlation present in the data (Judd, 1994).

Judd proposed that the correlation dimension be estimated as a function of scale ε_0 using the equation (8.15), which is valid for $\varepsilon < \varepsilon_0$:

$$C_m(\varepsilon) \approx \varepsilon^{d_c} q(\varepsilon) \tag{8.15}$$

where $q(\cdot)$ is a polynomial of order of the topological dimension.

More comprehensive reviews of the theory and estimation of fractal dimensions can be found in the papers of Cutler (1993), Takens (1985), Theiler (1990), Isliker (1992), among others.

8.5.2 Other invariants

a) Generalized dimensions and entropies

The correlation integrals considered previously can be generalized into q'th order integrals, as follows

$$[C_m(\varepsilon)]^{q-1} = \int[\int\theta(\varepsilon - ||\mathbf{x}_1-\mathbf{x}_2||)d\mu_m(\mathbf{x}_2)]^{q-1}d\mu_m(\mathbf{x}_2) \quad (8.16)$$

with $q \in \Re$. The usual correlation integral is retrieved with $q = 2$. For $q \geq 2$, the integral can be interpreted as the probability that all points in a μ_m random (q-1)-tuple have a distance smaller than ε to a μ_m-random point. The generalized correlation integral can be estimated similar to $C_{2,m}$ outlined above, by using q-tuples of points, rather than pairs of points (Ghez and Vaienti, 1992; Venziano et al., 1995). The spectrum of Renyi dimensions D_q and K_q (Renyi, 1971) is defined by the scaling region[51]

$$C_{q,m}(\varepsilon) \approx \varepsilon^{Dq} e^{-m\tau Kq} \quad (8.17)$$

This spectrum is preserved under smooth coordinate transformations (Ott et al., 1984b). The invariants D_1 and K_1 are referred to as the information dimension and information entropy[52] respectively. D_0 (the fractal dimension of the support of the attractor) is known as the box counting or capacity dimension (Lasota and MacKay, 1994; Ott et al., 1994a).

b) Lyapunov exponents

Once embedded, the features of the dynamic system can be evaluated by various means, among other via calculation of the Lyapunov exponents of the system. For example, in chaotic (nonlinear) systems two states that are nearly identical diverge from each other at an exponential rate, which causes a sensitive dependence on initial conditions. The idea is illustrated in Figure 8.6, which is a schematic of two separation boundaries. ε_0 is the observed separation of two very close points on separate trajectories on an attractor. After time t, the divergent trajectories are separated by distance ε_t. Since the divergence is exponential, it can be expressed as

$$\varepsilon_t = \varepsilon_0 e^{\lambda t} \quad (8.18)$$

where λ is the Lyapunov exponent.

This can be rearranged to give[53]

$$\lambda = (1/t)\ln[\varepsilon_0/\varepsilon_t] \quad (8.19)$$

In order to measure the average divergent properties of an attractor, a reference trajectory is followed through space and time and the divergence of neighbouring trajectories are subsequently considered, as indicated in Figure 8.7 below. An overall Lyapunov exponent (Λ) can be defined as:

Complete characterization of the divergence or convergence properties of the attractor actually requires a set[54] of Lyapunov exponents (one for each orthogonal direction in the phase space). Stable, periodic attractors are characterized by $\lambda \leq 0$, chaotic attractors[55] have at

---

[51] Measures for which D_q is invariant with q are known as monofractals, as opposed to measures where D_q varies with q, which are known as multifractals.

[52] A dynamic system is chaotic if $K_1 > 0$. As both D_q and K_q are monotonous non-increasing functions of q, $K_2 > 0$, if $K_1 > 0$, hence a positive value of K_2 also marks the presence of chaos.

[53] Note that any base logarithm can be used. In practice base 10 and 2 are frequently used. In the latter case the Lyapunov exponent is expressed in terms of information loss per unit time.

[54] This is referred to as the *Lyapunov spectrum* of the attractor.

[55] Attractors with more than one positive Lyapunov exponent are sometimes referred as hyperchaotic.

least one $\lambda > 0$, while random attractors have at least one $\lambda \to \infty$ (instant divergence of trajectories).

$$\Lambda = \frac{1}{n}\sum_{k=1}^{n}\log_2\left[\varepsilon_{t(k)}/\varepsilon_{0(k)}\right] \tag{8.20}$$

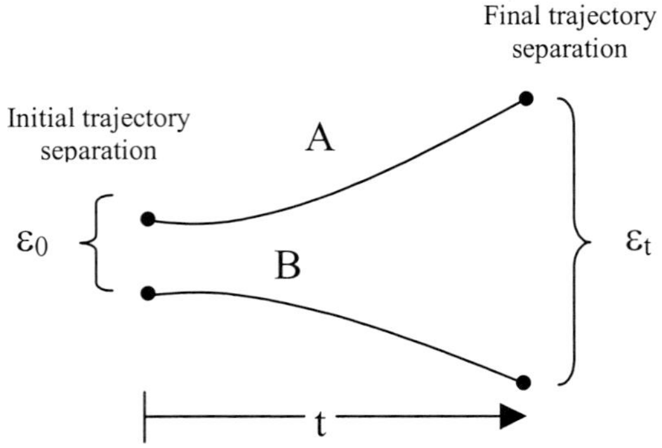

Figure 8.6. Separation of trajectories and calculation of Lyapunov exponents.

The calculation of Lyapunov exponents direct from data can be treacherous, owing to the adverse effect of noise in the data and should be approached with circumspection (Brown et al., 1991; Parlitz, 1992; Abarbanel, 1996). Under these circumstances it is necessary to first smooth the data, before calculation of the exponents. A fairly reliable method to calculate Lyapunov exponents has been proposed by Brown et al. (1991).

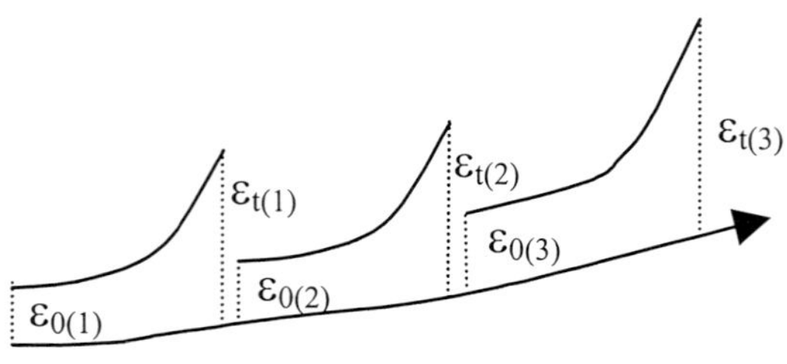

Figure 8.7. Average Lyapunov exponents of attractors via tracking of neighbouring trajectories.

To date, the different methods proposed for the calculation of Lyapunov exponents can be divided into two classes, viz. direct methods and Jacobian-based methods. Direct methods, such as proposed by Wolf et al. (1985) estimate the divergence of the reconstructed states without fitting a model to the data. With Jacobian methods, a model is first fitted to the data and the Jacobian matrices of the model are then used to calculate the Lyapunov exponents (Shimada and Nagasima, 1979; Benettin et al., 1980; Eckmann and Ruelle, 1985; Geist et al., 1990). Local linear approximations are used often (Sano and Sawada, 1985; Eckmann et al., 1986; Stoop and Meier, 1988; Holzfuss and Lauterborn, 1989; Stoop and Parisi, 1991; Zeng et al., 1991, 1992; Parlitz, 1993; Kruel et al., 1993). The basic algorithm has been modified by using

- Nonlinear approximations of the flow by use of polynomials (Bryant et al., 1990; Brown et al., 1991; Abarbanel et al., 1991), neural networks (Gencay and Dechert, 1992) or radial basis functions (Holzfuss and Parlitz, 1991; Parlitz, 1992; Kadtke et al., 1993).

- Neighbouring states from a shell ($r_{min} < r < r_{max}$) instead of a sphere ($r < r_{max}$) to alleviate the effect of noise (Eckmann et al., 1986; Zeng et al., 1992; Parlitz, 1993; Kruel et al., 1993)

- The embedding dimension d for the selection of neighbours and a small local dimension d_L for fitting and computation of the resultant d_L Lyapunov exponents (Eckmann et al., 1986; Bryant et al., 1990; Brown et al., 1991)

- Projections unto linear subspaces, i.e. local Broomhead-King coordinates (Stoop and Parisi, 1991)

Although especially the latter two approaches have been designed to avoid so-called spurious Lyapunov exponents that occur when the dimension d of the reconstructed state space exceeds the dimension m of the original state space, these approaches are not completely reliable, as discussed for example by Eckmann and Ruelle (1992) and Ellner et al. (1991).

8.6. DETECTION OF NONLINEARITIES

If the time series originates from an unknown process, it is important to determine the nature of the data prior to building models. It is especially important to distinguish between (nonlinear) deterministic data and linear (stochastic) data. Several approaches have been proposed to this effect (Savit and Green, 1991; Provenzale et al., 1992; Smith, 1992; Takens, 1993; Palus et al., 1993; Kaplan, 1994; Salvino and Cawley, 1994; Rapp et al., 1994; Theiler, 1995; Schreiber and Schmitz, 1996; Schreiber, 1998). Methods based on the use of surrogate data provide a general and flexible framework for investigating unknown data, as discussed in more detail below (Barnard et al., 2001).

8.6.1. Surrogate data methods

The method of surrogate data (Theiler et al., 1992a, 1992b; Takens, 1993; Theiler and Pritchard, 1996; Theiler and Rapp, 1996; Wayland et al., 1993) involves a null hypothesis against which the data are tested, as well as a discriminating statistic. The data are first assumed to belong to a specific class of dynamic processes. Surrogate data are subsequently generated from this process and various statistics are calculated from both the surrogate and the original data (Theiler et al., 1992a). If the calculated statistics of the surrogate and the original data are different, then the null hypothesis that the process that has generated the original data is of the same class as the system that has generated the surrogate data is rejected.

More specifically, let $\mathbf{x} \in \Re^n$ be a time series consisting of n observations, ψ a specific hypothesis, $\Im_\psi$ the set of process systems consistent with the hypothesis, and $T:\Re^N \rightarrow U$ be a statistic that will be used to evaluate the hypothesis ψ that x was generated by some process $\Im \in \Im_\psi$. Generally the statistic $U \in \Re$, and it will be possible to discriminate between the original data $\mathbf{x}$ and the surrogate data $\mathbf{x}_s$ consistent with the hypothesis given by the probability density of T, given $\Im$, i.e. $p_{T,\Im}(t)$.

a) Pivotal test statistics

Theiler (1995) has suggested that a distinction can be made between so-called pivotal and non-pivotal statistics. A test statistic T is considered to be pivotal, if the probability distribution $p_{T,\Im}$ is the same for all processes $\Im$ consistent with the hypothesis ψ. Moreover, a distinction can be made between simple and composite hypotheses. If the set of all processes consistent with the hypothesis ($\Im_\psi$) is a singleton, then the hypothesis is simple. Otherwise, the hypothesis is composite and can be used not only to generate surrogate data consistent with a particular process $\Im$, but also to estimate $\Im \in \Im_\psi$. In fact, $\Im$ has to be specified when the hypothesis is composite, unless T is a pivotal statistic (Theiler, 1995), meaning that $p_{T,\Im}$ is the same for all $\Im \in \Im_\psi$.

Constrained realization (Schreiber and Schmitz, 1996) schemes can be employed when non-pivotal statistics are applied to composite hypotheses. That is, apart from generating surrogate data that represent typical realizations of a model of the system, the surrogate data should also be representative of a process yielding identical estimates of the parameters of the process compared to the estimates of the process parameters obtained from the original data. Put in a different way, if $\Im_{est} \in \Im_\psi$ is the process estimated from the original data $\mathbf{x}$, and $\mathbf{x}_s$ is a surrogate data set generated by $\Im' \in \Im_\psi$, then $\mathbf{x}_s$ is a constrained realization of $\Im_{est} \in \Im'$.

As an example, if ψ is the hypothesis that x is generated by linearly filtered independent identically distributed noise, then *non-constrained* surrogate data $\mathbf{x}_s$' can be generated from a Monte Carlo simulation based on the best linear model estimated from x. The data $\mathbf{x}_s$' can be constrained by shuffling the phases of the Fourier transform of the data, producing a set of random data $\mathbf{x}_s$" with the same power spectra (and autocorrelation) as the original data $\mathbf{x}$. The autocorrelation, rank order statistics, nonlinear prediction error, etc., would all be non-pivotal test statistics characterizing dynamic manifold structures, since the distributions of these statistics would all depend on the form of the noise source and the type of linear filter. In contrast, the Lyapunov exponents and the correlation dimension (fractal dimension) would be pivotal test statistics, since the probability distributions of theses quantities would be the same for all processes, regardless of the source of the noise or the estimated model. Since recent investigations have shown that Lyapunov exponents can be misleading in the presence of noise, the correlation dimension has gained favour as the pivotal statistic of choice.

b) Classes of hypotheses

Three classes of hypotheses are used widely, viz. those equivalent to the assumption that the data are identically, independently distributed noise (0), linearly filtered noise (1) and a monotonic nonlinear transformation of linearly filtered noise (2), generated by randomizing the data (0), randomizing or shuffling the phases of the Fourier transform of the data (1) and randomizing the phases of amplitude adjusted Gaussian noise (2).

In the latter case (hypotheses of class 2), the procedure for generating surrogate data consists of the following steps

i) Generation of an identical independently distributed Gaussian data set $\mathbf{y}$, and reordering of $\mathbf{y}$ to have the same rank distribution as $\mathbf{x}$.

ii) Generation of a surrogate data set $\mathbf{y}_s$ of $\mathbf{y}$ (by shuffling of the phases of the Fourier transform of the data set $\mathbf{y}$)

iii) Finally, reordering of the original data set **x**, to generate a surrogate data set $\mathbf{x}_S$, which has the same rank distribution as $\mathbf{y}_S$.

Surrogates generated by these procedures are referred to as algorithm 0, algorithm 1 and algorithm 2 surrogates, while algorithm 2 surrogates are also known as amplitude adjusted Fourier transform (AAFT) surrogates (Small and Judd, 1998b).

Surrogate data methods are not infallible and can yield more false rejections of the null hypothesis when the data are strongly correlated (Kantz and Schreiber, 1997). Moreover, while it is true that Gaussian linear processes are fully specified by either the autocorrelation function or the power spectrum, only an *estimate* of the spectra is used with the analysis. This can give rise to additional high frequency elements in the spectrum. Another problem is that nonstationarity of the time series can also cause false positives. For example, spikes or changes in the process over time means the null hypothesis will be rejected. This can lead to a large class of formally correct rejections, which are interpreted in the wrong way (Kantz and Schreiber, 1997).

In order to assess the nature of the data, the surrogates and the original time series are displayed in bivariate plots of the correlation dimension of the data sets against the scale. No rigorous statistical tests are as yet available to discriminate between the original and the surrogate data sets and evaluation takes place by means of inspection. These calculations are outlined below by means of a simple example.

8.6.2. Example: Generation of surrogate data

In this example the generation of surrogate data of different types for time series is illustrated. In all cases, the time series y = 1.9, 3.2, 1.2, 2.2, 0.9, 2.1, 2.9, 2.3 is considered.

a) Generating index-shuffled surrogates (Type 0)

i. Generate normally distributed data

$$\mathbf{y}_r = \text{randn}(\text{size}(\mathbf{y}))$$

as before $\mathbf{y}_r$ = -0.30, -1.28, 0.24, 1.28, 1.20, 1.73, -2.18, -0.23

ii. Sort random data in ascending order and store sequence of indices

$\mathbf{y}_{r2} = \text{sort}(y_r)$, giving $\mathbf{y}_{r2}$ = -2.18, -1.28, -0.30, -0.23, 0.24, 1.20, 1.28, 1.73

indices($\mathbf{y}_{r2}$) = 7, 2, 1, 8, 3, 5, 4, 6

iii. Sort observed data to rank order of random data

$$\mathbf{y}_{s0} = y(7), y(2), y(1), \ldots y(6) = 2.9, 3.2, 1.9, 2.3, 1.2, 0.9, 2.2, 2.1$$

where $\mathbf{y}_{s0}$ is the surrogate data of type 0. Different surrogate data sets will be generated by different random samplings ($\mathbf{y}_r$).

b) Generating phase-shuffled surrogates (Type 1)

To generate phase-shuffled surrogate data, the following procedure can be used:

a. Calculate the Fourier transform of the time series

$\mathbf{y}_{fft} = FFT(\mathbf{y})$

$\mathbf{y}_{fft} =$ 16.70, 1.8485 + 0.9929i, -1.3000 - 0.8000i, 0.1515 - 2.4071i, -2.9000,

1.8485 - 0.9929i -1.3000 + 0.8000i, 0.1515 + 2.4071i

ii. Generate uniformly distributed random complex numbers with modulus 1.

$\mathbf{b} = 2\pi$ rand(size(y))

e.g. $\mathbf{b}$ = 5.9698, 1.4523, 3.8129, 3.0535, 5.6002, 4.7884, 2.8681, 0.1163

$\mathbf{z} = \exp(i\,\mathbf{b})$

$\mathbf{z} =$ 0.9513 - 0.3082i, 0.1182 + 0.9930i, -0.7830 - 0.6220i, -0.9961 + 0.0880i,

0.7757 - 0.6311i, 0.0759 - 0.9971i, -0.9628 + 0.2701i, 0.9932 + 0.1160i

iii. Determine the indices for the first half of the FFT conjugate pairs, i.e. not counting first and symmetry point, indices = 2, 3, 4 and indices of the conjugates = 6, 7, 8

iv. Multiply original Fourier Transform with random complex number vector, as follows:

$\mathbf{ys}_{fft}^{+} = \mathbf{y}_{fft}$(indices) $\mathbf{z}$(indices)

$\mathbf{ys}_{fft}^{-}$ = conj[$\mathbf{y}_{fft}$(indices)][conj($\mathbf{z}$(indices)], where indices are ordered in reverse ranking

$ys_{fft}(1) = y_{fft}(1) = 16.70$

$ys_{fft}(2) = y_{fft}(2)*z(2) = -0.7674 - 1.9529i$

$ys_{fft}(3) = y_{fft}(3)*z(3) = 0.5203 - 1.4350i$

$ys_{fft}(4) = y_{fft}(4)*z(4) = 0.0609 - 2.4111i$

$ys_{fft}(5) = y_{fft}(5) = -2.9000$

$ys_{fft}(6) = conj[y_{fft}(4)]*conj[z(4)] = 0.0609 + 2.4111i$

$ys_{fft}(7) = conj[y_{fft}(3)]*conj[z(3)] = 0.5203 + 1.4350i$

$ys_{fft}(8) = conj[y_{fft}(1)]*conj[z(1)] = -0.7674 + 1.9529i$

v. Finally, calculate the type 1 surrogate data set as the real part of the inverse Fourier transform of $\mathbf{ys}_{fft}$.

$\mathbf{y}_{s1} = real[FFT^{-1}(\mathbf{ys}_{fft})]$

$\mathbf{y}_{s1}$ = 1.6784, 1.1734, 1.7095, 2.1837, 2.0317, 3.0091, 1.4804, 3.4338 where $\mathbf{y}_{s1}$ is a surrogate data set of type 1.

c) *Generating amplitude adjusted Fourier transform surrogates (Type 2)*

i. Generate random data (as with Type 0 surrogates)

$\mathbf{y}_r = \text{randn}[\text{size}(\mathbf{y})]$

$\mathbf{y}_r$ = -0.30, -1.28, 0.24, 1.28, 1.20, 1.73, -2.18, -0.23 (same set used as above)

ii. Randomly shuffle the original data and store the new sequence of indices

$\mathbf{y}_2 = \text{sort}(\mathbf{y})$

$\mathbf{y}_2$ = 0.9, 1.2, 1.9, 2.1, 2.2, 2.3, 2.9, 3.2

$\mathbf{i}_2 = \text{Indices}(\mathbf{y}_2)$ = 5, 3, 1, 6, 4, 8, 7, 2

$\mathbf{y}_{r2} = \text{sort}(\mathbf{y}_r)$

$\mathbf{y}_{r2}$ = -2.18, -1.28, -0.30, -0.23, 0.24, 1.20, 1.28, 1.73

$\mathbf{i}_{r2} = \text{Indices}(\mathbf{y}_{r2})$ = 7, 2, 1, 8, 3, 5, 4, 6

Let $\mathbf{y}_{t1}(\mathbf{i}_{r2}) = \mathbf{y}_{r2}$

i.e. $y_{t1} = y_{r2}(7), y_{r2}(2), \ldots y_{r2}(6)$ = 2.9, 3.2, 1.9, 2.3, 1.2, 0.9, 2.2, 2.1

iii. Shuffle phase of random vector y_{s1} as explained above for generation of Type 1 surrogate data.

$\mathbf{y}_{t2} = \text{shuffle\_phase}(\mathbf{y}_{t1})$,

which gives y_{t2} = -0.0671, 1.1552, -0.9412, 2.2162, -0.5305, -0.7462, -0.9412, 0.3148

iv. Sort phase-shuffled random data in ascending order

$\mathbf{y}_{t3} = \text{sort}(\mathbf{y}_{t2})$;

$\mathbf{y}_{t3}$ = -0.9412, -0.9412, -0.7462, -0.5305, -0.0671, 0.3148, 1.1552, 2.2162

$\mathbf{i}_{yt3} = \text{Indices}(\mathbf{y}_{t3})$ = 3, 7, 6, 5, 1, 8, 2, 4

v. Sort data to rank order of random data

$\mathbf{y}_{s2}(\mathbf{i}_{yt3}) = \mathbf{y}_2 = y_2(3), y_2(7), \ldots y_2(4)$

$\mathbf{y}_{s2}$ = 2.2, 2.9, 0.9, 3.2, 2.1, 1.9, 1.2, 2.3

where $\mathbf{y}_{s2}$ is a surrogate data set of type 2.

The minimum size of the data set is important on at least two levels, the first being to calculate the test statistic (correlation dimension in this case) reliably. Stefanovska et al. (1997)

have shown that too few points in a neighbourhood can lead to overestimation of the correlation dimension when using the Grassberger-Procaccia algorithm. The Judd algorithm is less sensitive than the Grassberger-Procaccia algorithm for the number of samples by an order of magnitude. Typically a data set of approximately 1000 points is sufficient for the Judd algorithm.

The second level of importance concerns fitting a model to the data. When embedding data one attempts to fill the reconstructed equivalent dynamic attractor with as many embedded points as possible. The higher the embedding dimension for a given size of data set, the sparser the distribution is going to be. Too few independent points in any local neighborhood of the attractor lead to weaker generalization for global models.

The point of stationarity of a time series is closely related to the previous remarks on data set size. Nonlinear (chaotic) systems are never stationary in the same sense as linear systems. One could regard a given time series as stationary when a non-adaptive or stationary model fits the dynamics of the time series properly, that is it generalizes well when validated on other, unseen subsets of the time series. One can intuitively expect some minimum data set size to be a lower bound in terms of this definition, while the upper bound would normally be set by practical considerations of cost of data acquisition as well as available computational infrastructure.

8.7. SINGULAR SPECTRUM ANALYSIS

More recently, considerable attention has been given to the use of singular spectrum analysis as a means to embed time series, instead of the use of the delay coordinate approach discussed above. Singular spectrum analysis (Vautard and Ghil, 1989; Vautard et al., 1992) has the advantage that it is simpler and much less computationally intensive than the use of delay coordinates derived by means of false nearest neighbours or false nearest strands methods, for example. Most importantly, it is not as vulnerable to noise as the method of delay coordinates, where especially the estimation of the time lag can be influenced considerably by noisy data.

The term singular spectrum comes from the spectral (eigenvalue) decomposition of a matrix **A** into its spectrum (set) of eigenvalues. The eigenvalues λ are the values that make the matrix $\mathbf{A}-\lambda\mathbf{I}$ singular. Actually, the term singular spectrum is somewhat unfortunate in the context of the analysis of time series, since the traditional eigenvalue decomposition of matrices representing multivariate data is also an analysis of the singular spectrum. The spectral decomposition of matrices has only recently been applied to time series analysis and has had its roots mostly in the application of chaos theory (Fraedrich, 1986; Broomhead and King, 1986; Landa and Rosenblum, 1991).

In essence, the data are embedded in a very high-dimensional reconstruction, followed by the introduction of a new coordinate system, where the origin is moved to the centroid of the reconstructed system states and the axes are represented by the dominant principal components of the states (points).

The idea is explained by way of a simple example. Consider a uniformly sampled univariate time series with $n = 5$ observations $\mathbf{x} = [x_1, x_2, x_3, x_4, x_5]^T$. With an embedding dimension of $m = 2$, there will be $n-(m-1) = 4$ snapshots of the time series, i.e. $y_1 = [x_1, x_2]^T$, $y_2 = [x_2, x_3]^T$, $y_3 = [x_3, x_4]^T$ and $y_4 = [x_4, x_5]^T$. These snapshots can be arranged as row vectors, as shown below.

$$\mathbf{X} = \begin{bmatrix} y_1 \\ y_2 \\ y_3 \\ y_4 \end{bmatrix} = \begin{bmatrix} x_1 & x_2 \\ x_2 & x_3 \\ x_3 & x_4 \\ x_4 & x_5 \end{bmatrix}$$

It is not necessary to use successive values of the observations. For example, with every second observation only two new variables would have been created, $z_1 = [x_1, x_3]^T$, $z_2 = [x_3, x_5]^T$. The construction of the matrix **X** is based on an (m, J) window, i.e. m indicates the number of time series observations in each new snapshot and J indicates the sample times between observed elements. In the first case, a (3,1)-window was used, while a (3,2)-window was used in the second case. If J = 1 is used, then reference is made to an m-window only. Matrix **X** is referred to as the *augmented* or *trajectory matrix* and contains a complete record of patterns that have occurred in a window of size m. That is

$$X_{ij} \in \Re^{n-m+1xm}$$

Since $X_{(i-1)j} = X_{i(j-1)}$ for all i > 1 and j >1, the columns of the matrix are highly correlated. The embedding space $\aleph \in \Re^m$ is the space of all m-element patterns.

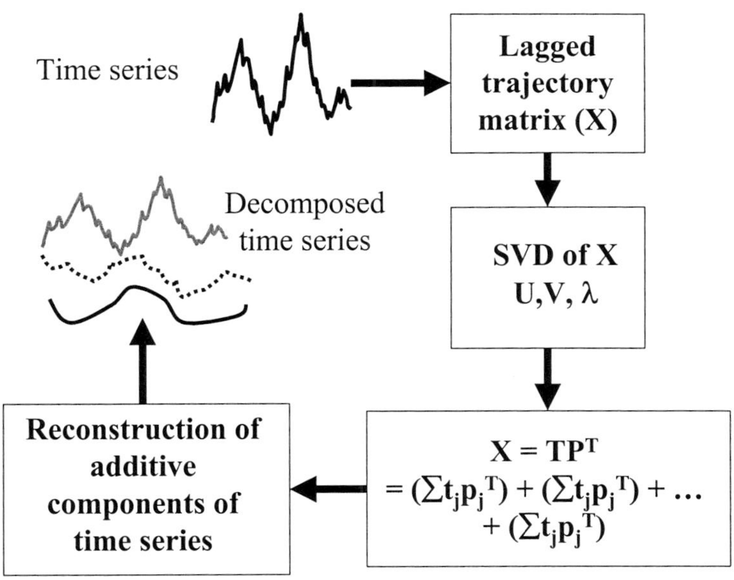

Figure 8.8. Decomposition and reconstruction of a time series by use of singular spectrum analysis.

Rather than investigating the trajectory matrix **X** for repetitive patterns in the original time series, it is more effective to consider the *lagged covariance matrix* computed from the trajectory matrix **X** and its transpose, i.e. $\mathbf{S} = \mathbf{X}^T\mathbf{X}$.

For time series with very peculiar patterns, such as series consisting of short sharp spikes, delay embedding may lead to very inhomogeneous sets of points in the reconstructed state space, that may be difficult to analyze. Under these circumstances, embedding can be based on the pattern features of the time series, e.g. the time intervals between spikes.

More formally, basic singular spectrum analysis can be seen as a four-step process, as indicated in Figure 8.8 (Golyandina et al., 2001). In the first step (embedding), the one-dimensional time series is recast as an L-dimensional time series (trajectory matrix). In the second step (singular value decomposition), the trajectory matrix is decomposed into a sum of bi-orthogonal matrices of rank one. These two steps constitute the decomposition stage of singular spectrum analysis.

The third step concerns splitting of the matrices computed in step 2 into several groups. The trajectory matrix is thus represented as the sum of several resultant matrices. In the final step, a time series is reconstructed for each resultant matrix by means of diagonal averaging (Golynadina et al., 2001). The original time series is thus decomposed into a series of additive components.

8.8. RECURSIVE PREDICTION

Once embedding of the observed data has been completed, predictive models can be built, based on the multiple input single output (MISO) system $f:\Re^m \rightarrow \Re^1$, defined by

$$y_{est}(t+1) = f[\mathbf{y}(t)] \tag{8.21}$$

Equation (8.21) is the crux of the dynamic model. Once it is established, the evolution $y(t) \rightarrow y(t+1)$ is known and this in turn determines the unknown evolution of $\mathbf{x}(t) \rightarrow \mathbf{x}(t+1)$. Unlike linear models, where minimization of the mean square of the prediction error leads to accurate models, minimization of the mean square of the prediction error is a necessary, but not a sufficient criterion for the reliability of the models of nonlinear systems, since trajectories in the same attractor can differ vastly from one sample to another.

Many different models have been used to map the evolution of the time series, including polynomials and locally linear approximations (Farmer and Sidorowich, 1987), radial basis functions (Casdagli, 1989) and neural networks (Small and Judd, 1998a; Barnard et al., 2001).

In addition, the reliability of dynamic models is characterized by dynamic variants that measure the global properties of the attractor, such as Lyapunov exponents and the correlation dimension. Indeed, in practice it can be difficult to tell dynamic systems from stochastic systems via inspection or superficial analysis. A general methodology for the modelling of empirical systems should therefore incorporate some preliminary means of system identification in the first place. Such a methodology is outlined schematically in Figure 8.9. According to this approach, the time series is embedded first (parameterized) by selecting an appropriate lag and embedding dimension. Once the time series is embedded, the attractor is mapped to the time series data by means of a suitable model (a neural network is suggested in the diagram). The validity of the model is subsequently assessed by means of a surrogate data analysis and finally the model is used for free-run prediction of the data.

As can be expected, noisy data (measurement noise, as well as dynamic noise) can have a major influence on the quality of predictive models. In the presence of measurement noise ε_t, the output equation can be expressed as

$$y_t = g(\mathbf{x}_t,\mathbf{u}_t) + \varepsilon_t \tag{8.22}$$

Measurement noise is caused by natural random process fluctuations and instrument inaccuracy, which reduce the signal-to-noise ratio of observations and compromise the quality of the dynamic information provided by the model. In contrast, dynamic noise, δ_t can be mathematically expressed as

$$x_{t+1} = f(x_t,u_t) + \delta_t \tag{8.23}$$

Dynamic noise is caused by inherent disturbances of the process that generates the dynamic orbit of the state vector in state space and tends to increase the model dimension or complexity. It is often necessary to reduce the noise and for linear systems, filtering is a common solution to this problem (Grassberger et al., 1993). However, since nonlinear data appear random under linear analysis, linear filtering can seriously impair model validity by removing interesting high order dynamic information with the noise. Among other, adaptive moving average filters, Volterra filters, bi-linear and multi-linear filters have been applied with various degrees of success in the filtering of nonlinear systems. For example, Rauf and Ahmed (1997) have presented a class of nonlinear adaptive filters based on successive linearization (Rauf, 1997) for the predictive modelling of chaotic systems.

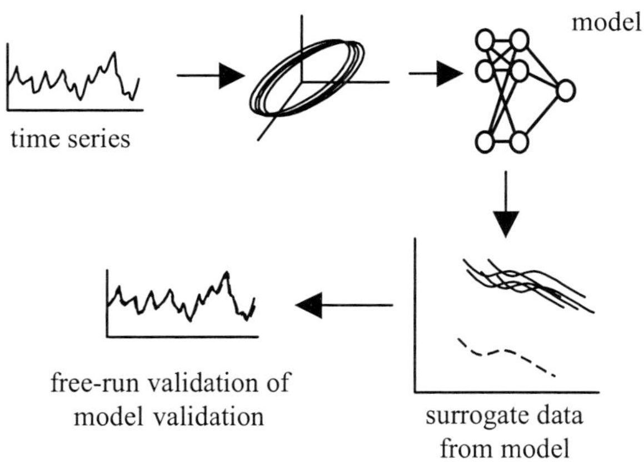

Figure 8.9. Methodology for system identification and modelling by use of surrogate data and neural networks.

Geometric filtering, another form of nonlinear filtering, first embeds the observed time series in a high-dimensional phase space and then fits local linear models on the resultant geometric structure. For example, Kostelich and Yorke (1988) have demonstrated a method whereby points in a local neighbourhood on an attractor, reconstructed from the time series, can be used to find a local approximation of the dynamics. Subsequently, the approximations can be used collectively to produce a new time series, which is dynamically more consistent with the reconstructed attractor. However, geometric filtering can easily be misused through misunderstanding of the techniques (Mees and Judd, 1993). Since geometric filtering imposes locally linear models on data, these processes can force the data to fit a linear model inappropriately, if iterated excessively.

Chapter 9
Case Studies: Dynamic Systems Analysis and Modelling

C. Aldrich and J. P. Barnard*

9.1. EFFECT OF NOISE ON PERIODIC TIME SERIES

In the first case study, the use of surrogate data analysis is demonstrated on a set of artificial curves, generated by the simple function $y = \sin(\theta) + \sin(\theta/2) + \varepsilon$, with $\theta \in [0, 20\pi]$ and ε representing Gaussian noise added to these curves. Three identical curves (y_1, y_2 and y_3) were generated. These curves were consequently corrupted by $\varepsilon_1 = N(0, 0.001)$, $\varepsilon_2 = N(0, 0.05)$ and $\varepsilon_3 = N(0, 0.25)$, i.e. Gaussian noise with zero mean and standard deviations as indicated.

The three time series curves and the associated correlation dimensions of the curves and their surrogate data sets are shown Figure 9.1. The curves were embedded by use of singular spectrum analysis, as the use of the average mutual information or autocovariance function criteria could not cope with the noise in the data. The progressive effect of noise is clearly visible from Figures 9.1(a)-9.1(c). The correlation dimension curves associated with the original observations and those associated with the surrogate data derived from the original observations tend to move closer with progressive noise levels, indicating the growing stochastic nature of the observations. Moreover, the noisier the data, the higher the correlation dimension values become at small scales. In Figure 9.1(a), the original (solid) curve has a correlation dimension of less than 1.2 at a log-scale of -2, but this increases to approximately 1.8 and 6 (Figures 9.1(b) and 9.1(c) respectively) as the noise levels increase from 0.05 to 0.25.

The net effects of the noise are ultimately reflected in the quality of the predictive models, shown in Figure 9.2. These models were derived with multilayer perceptron neural networks that mapped the coordinates on the attractors shown in Figure 9.1, to future observations. The neural network models were generated automatically by use of a constructive algorithm based on the Schwartz information criterion (Schwartz, 1978) and generally consisted of single-hidden layer structures with sigmoidal hidden and output nodes. The models were used to generate the time series by feeding back their own predictions as successive inputs in a moving window. This is a very sensitive approach to the assessment of the quality of the models, as small errors are propagated and compounded and gives an indication of the degree to which the global dynamics of the process are captured by the model. For example, the one-step ahead predictions of the model in Figure 9.2(a) and the corresponding observations were correlated with a Pearson correlation coefficient of $r = 0.9993$ on the validation data set, which makes it very difficult to assess the quality of the model. In the case of models 2 and 3, shown in Figures 9.2(b) and 9.2(c) respectively, the models could predict the trends up to approximately 98 and 93 steps ahead respectively, after which the models broke down completely. With the addition of higher noise levels, the data start to assume a degree of stochasticity that precludes the fitting of dynamic models.

* Department of Chemical Engineering, University of Stellenbosch, Private Bag X1, Matieland, 7602, Stellenbosch, South Africa.

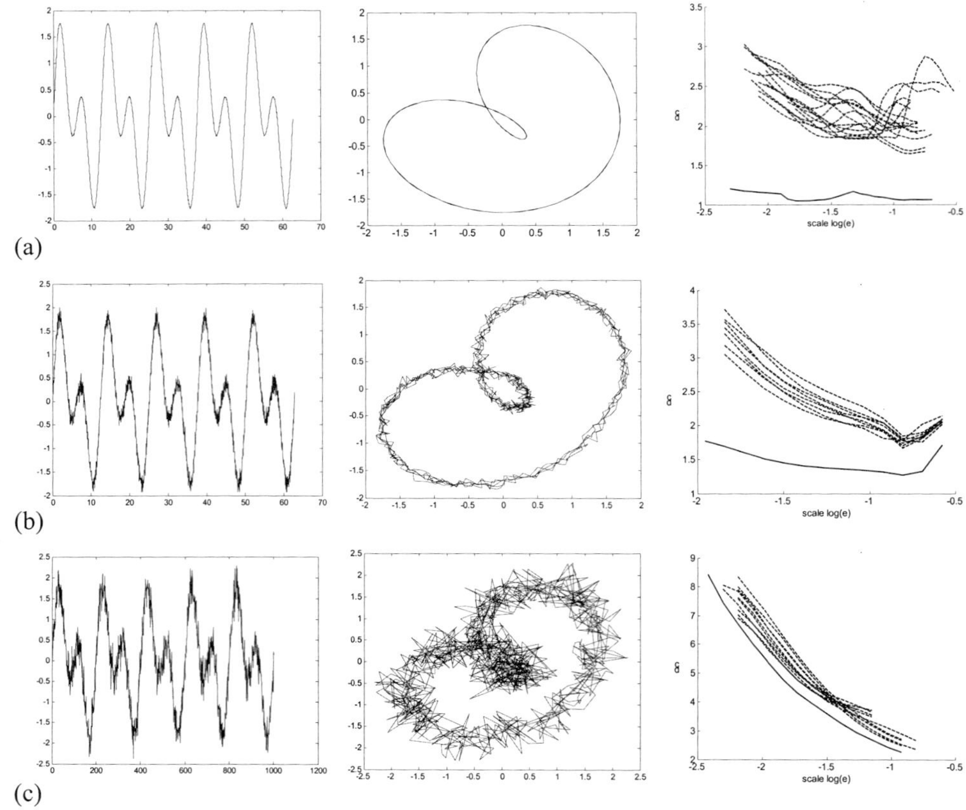

Figure 9.1. Three time series of the form y = sin(θ) + sin(θ/2) + ε, with θ ∈ [0, 20π], with (a) ε = N(0,0.001), (b) ε = N(0,0.05) and (c) ε = N(0,0.25)

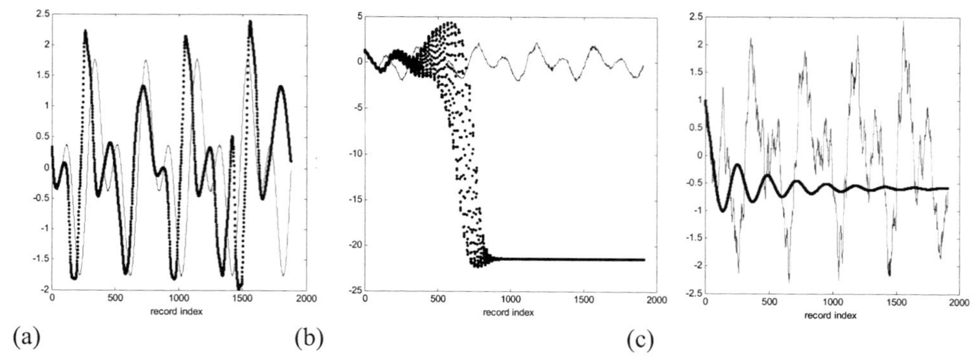

Figure 9.2. Free-run predictions (a-c) of observations in Figure 9.1(a)-9.1(c) respectively.

The use of surrogate data analysis prior to the development of predictive models gives the analyst an idea of the predictability of the time series in the first place, something which he can otherwise only assess by the costly trial-and error development of actual models and even then he may not be sure whether failure to develop predictive models has to be attributed to inadequate models or to indeterministic data.

9.2. AUTOCATALYSIS IN A CONTINUOUS STIRRED TANK REACTOR

The second case study concerns an autocatalytic process in a continuous stirred tank reactor originally considered by Gray and Scott (1983, 1984) and subsequently investigated by Lynch (1992). The system is capable of producing self-sustained oscillations based on cubic autocatalysis with catalyst decay and proceeds mechanistically as follows.

$$\begin{aligned} \mathrm{A} + 2\mathrm{B} &\rightarrow 3\mathrm{B}, \quad -r_\mathrm{A} = k_1 c_\mathrm{A} c_\mathrm{B}^2 \\ \mathrm{B} &\rightarrow \mathrm{C}, \quad r_\mathrm{C} = k_2 c_\mathrm{B} \\ \mathrm{D} + 2\mathrm{B} &\rightarrow 3\mathrm{B}, \quad -r_\mathrm{D} = k_3 c_\mathrm{D} c_\mathrm{B}^2 \end{aligned} \tag{9.1}$$

where A, B, C and D are the participating chemical species and k_1, k_2 and k_3 the rate constants for the chemical reactions. This process is represented by the following set of ordinary differential equations.

$$\begin{aligned} \frac{dX}{dt} &= 1 - X - aXZ^2 \\ \frac{dY}{dt} &= 1 - Y - bYZ^2 \\ \frac{dZ}{dt} &= 1 - (1+c)Z + daXZ^2 + ebYZ^2 \end{aligned} \tag{9.2}$$

where X, Y, and Z denote the dimensionless concentrations of species A, B and D, while a, b and c denote the Damköhler numbers for A, B and D respectively. The ratio of feed concentration of A to that of B is denoted by d and the same ratio of D to B by e. The process is chaotic, with a well-defined attractor for specific ranges of the two parameters, d and e. For the settings $a = 18000$; $b = 400$; $c = 80$; $d = 1.5$; $e = 4.2$, and initial conditions $[0, 0, 0]^T$, the set of equations was solved by using a 5th order Runge Kutta numerical method over 100 simulated seconds. This gave approximately 10 000 observations, which were resampled with a constant sampling period of 0.01 s. The Y state was taken as the output variable. Figure 9.3 shows the attractor of the process reconstructed from the process states X, Y and Z.

Two different data sets were considered in order to assess the effect of the size of the data set on the identification method. The smaller of the two sets consisted of the first 2000 observations of the original data set of 10 000 observations, while the larger of the two sets consisted of the first 8000 observations. In each case the remainder of the data was used to validate subsequent models.

Classification of the data with type 2 surrogates was performed first, as discussed in Chapter 8. With surrogate data methods, model validation is based on criteria related to the topology of the system's dynamic attractor. Instead of comparing models based on single-valued statistical criteria, they can be compared on multiple scales of attractor topology by means of surrogate data methods. This allows better discrimination between models and can in principle also aid in the development of better models at different dynamic scales, where one model is not consistently better than the other over the entire range of scales.

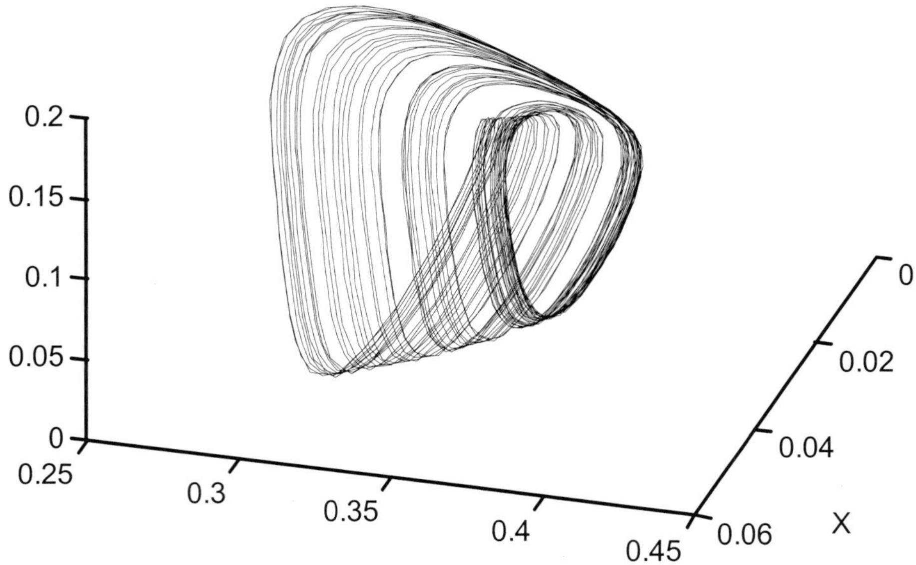

Figure 9.3. Attractor of autocatalytic process constructed from process states *X*, *Y*, *Z*.

The surrogate data analysis entailed calculation of the correlation dimension for each of the two data sets, as well as for 15 surrogate data sets generated from each data set. The results for both data sets are shown in Figure 9.4(a) and 9.4(b). The deterministic character of the data is evident from these figures. In both cases the curve representing the data set is well separated from the cluster of curves representing the corresponding surrogate data, which indicates that the data are not random. It is also interesting to note the sharp increase in dimension at small scales compared to noiseless data, which can probably be attributed to measurement noise.

The next step involved the embedding of each of the training and validation data sets in an appropriate state space. By making use of the method of false nearest neighbours, both the smaller and the larger data set could be optimally embedded in a three-dimensional space ($m = 3$). The average mutual information criterion indicated a time lag between the embedding variables of $k = 7$, and two nonlinear models were subsequently fitted to the data.

9.2.1. Multi-layer perceptron network model

The first set of model structures consisted of multi-layer perceptron neural networks, each with an input layer of three nodes, a hidden layer with six bipolar sigmoidal nodes (activation functions of the form $g(\cdot) = [1-\exp(\cdot)]/[1+\exp(\cdot)]$) and a single linear output node. The models fitted to the smaller and larger data sets were designated by $M_{FF1}(\theta_1)$ and $M_{FF2}(\theta_2)$ respectively. The parameter vectors, θ_1 and θ_2 were both estimated with the Levenberg-

Marquardt algorithm. In each case, the optimal model order (i.e. number of nodes in the hidden layer) was determined via cross-validation on the test data. For conciseness, $M_{FF\theta 1}$ and $M_{FF\theta 2}$ will indicate $M_{FF1}(\theta_1)$ and $M_{FF2}(\theta_2)$, respectively.

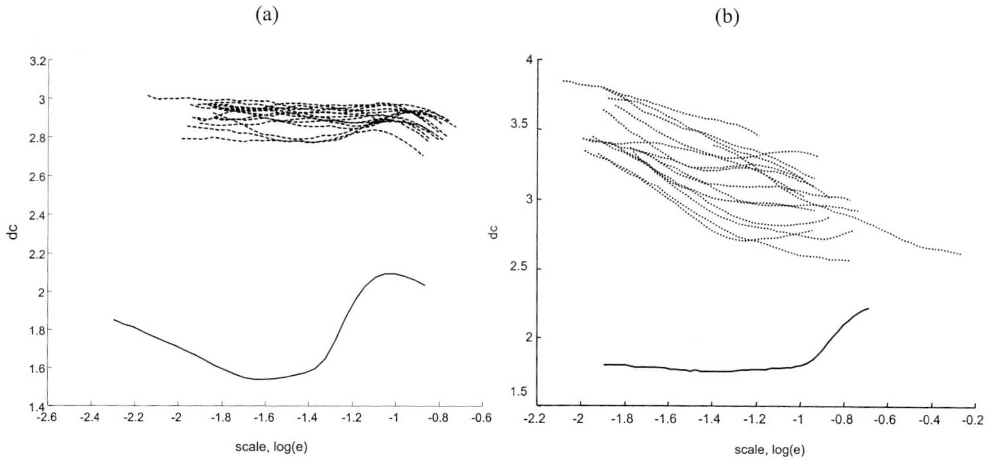

Figure 9.4. Correlation dimension (d_c) vs. scale (log(e)) for Y-state (bottom curve) of autocatalytic process and its AAFT surrogates based on (a) the smaller data set, (b) the larger set.

Both the $M_{FF\theta 1}$ and $M_{FF\theta 2}$ models were able to predict the data one-step ahead in the associated validation data set very accurately ($R^2 = 1-\Sigma(y-y_{est})^2/\Sigma(y-y_{avg})^2 = 0.935$ and 0.991 for $M_{FF\theta 1}$ and $M_{FF\theta 2}$ respectively), typically as indicated in Figure 9.5, which shows the results for the model.

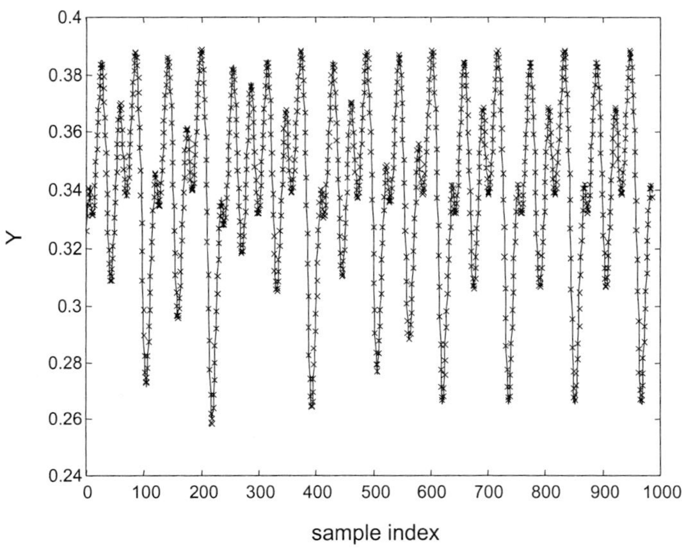

Figure 9.5. One-step prediction of autocatalytic Y-state with a multilayer perceptron model ($M_{FF\theta 1}$). The '+' marker denotes experimental observations, while the solid line denotes the predictions of the model.

The free-run predictions for the two data sets are shown in Figure 9.6(a) for $M_{FF\theta 1}$ and Figure 9.6(b) for $M_{FF\theta 2}$. As can be seen from Figure 9.6(a), $M_{FF\theta 1}$ could predict the data accurately in a free-run mode, up to about the 60th observation, after which it became unstable, following an oscillatory trajectory that grossly overestimated the actual values of the observations. $M_{FF\theta 2}$ performed significantly better, but after approximately 180 observations the predictions started to deviate significantly, as indicated in Figure 9.4(b). This result demonstrates the inability of linear statistics, such as R^2, to truly measure the performance of nonlinear models during validation. This conclusion can be extended to include cross-validation, in which R^2 plays the same role as in single-run validation.

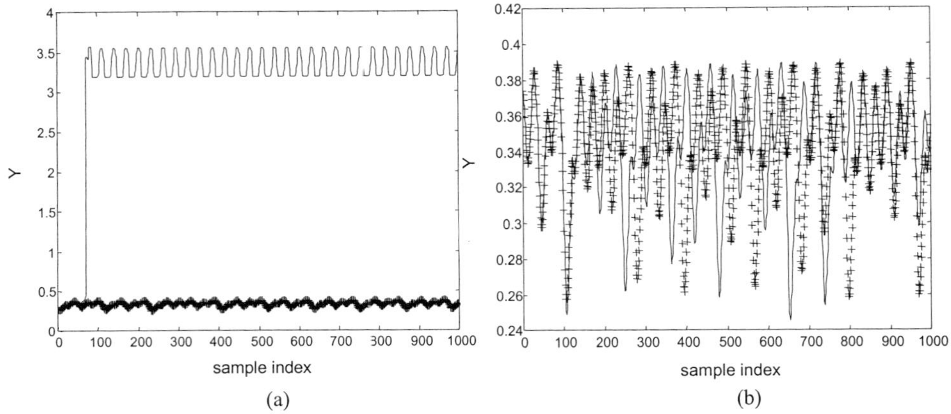

Figure 9.6. Free-run prediction of autocatalytic Y-state with a multilayer perceptron models, (a) $M_{FF\theta 1}$ and (b) $M_{FF\theta 2}$. The '+' marker denotes experimental observations, while the solid line denotes the predictions of the model.

9.2.2. Pseudo-linear radial basis function model

The set of pseudo-linear radial basis function model structures previously proposed by Small and Judd (1998a) was also fitted to the data sets by using an algorithm that optimizes model size via Rissanen's minimum description length (MDL) of the model during each iteration (Judd and Mees, 1995; Small and Judd, 1998a).

The pseudo-linear radial basis function model structure, M_{PL}, consisted of a combination of linear terms and a number of Gaussian radial basis function terms. The algorithm of Small and Judd (1998a) determines the combination and number of these terms by using minimum description length as a criterion. $M_{PL1}(\theta_1)$ based on the smaller set used 23 Gaussian kernels, while $M_{PL2}(\theta_2)$ based on the larger data set used 18 Gaussian kernels. As before, $M_{PL\theta 1}$ and $M_{PL\theta 2}$ are used instead of $M_{PL1}(\theta)$ and $M_{PL2}(\theta)$. Like the MLP network models, $M_{PL\theta 1}$ and $M_{PL\theta 2}$ were able to predict the data one-step ahead in the validation data associated very accurately ($R^2 = 0.999$), as indicated in the Figure 9.7 for $M_{PL\theta 1}$.

The free-run predictions are shown in Figure 9.8(a), using $M_{PL\theta 1}$ and Figure 9.8(b), using $M_{PL\theta 2}$. As can be seen from these figures, $M_{PL\theta 1}$ and $M_{PL\theta 2}$ could predict the data more accurately in free-run mode than $M_{FF\theta 1}$ or $M_{FF\theta 2}$. This was especially so for the smaller data set, although $M_{PL\theta 1}$ left the attractor after 150 observations. $M_{PL\theta 2}$ kept up with the attractor until after observation 200. This was at the expense of a higher processing cost owing to the larger data length. Both models managed to approximately follow the trend of the attractor, but could not quite reach the bottom of the troughs and were also partially out of phase with the observed *Y* state.

As before, it is clear that analyses of one-step ahead predictions are comparatively poor indicators of the quality of the models and that the free-run predictions provide a better idea of the adequacy of the models representing the dynamics of the system. These analyses can be formalised by comparing the surrogate data derived from the models with the actual data. The results for $M_{FF\theta2}$ are shown in Figure 9.9. The results pertaining to $M_{FF\theta1}$ are not shown, owing to the obviously poor quality of the model shown in Figure 9.8(a).

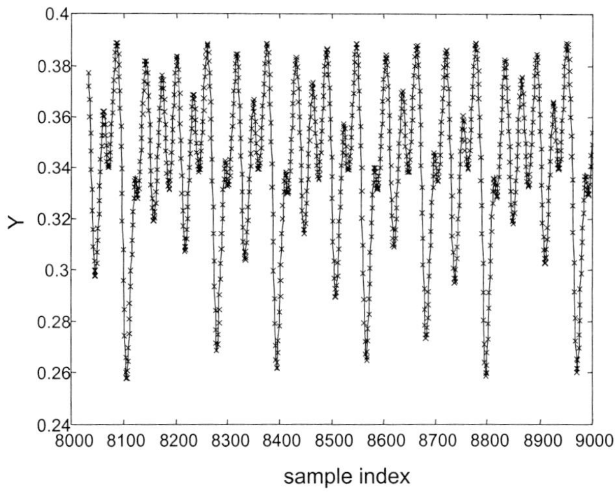

Figure 9.7. One-step prediction of autocatalytic Y-state with a pseudolinear radial basis function neural network model ($M_{PL\theta1}$). The '+' marker denotes experimental observations, while the solid line denotes the predictions of the model.

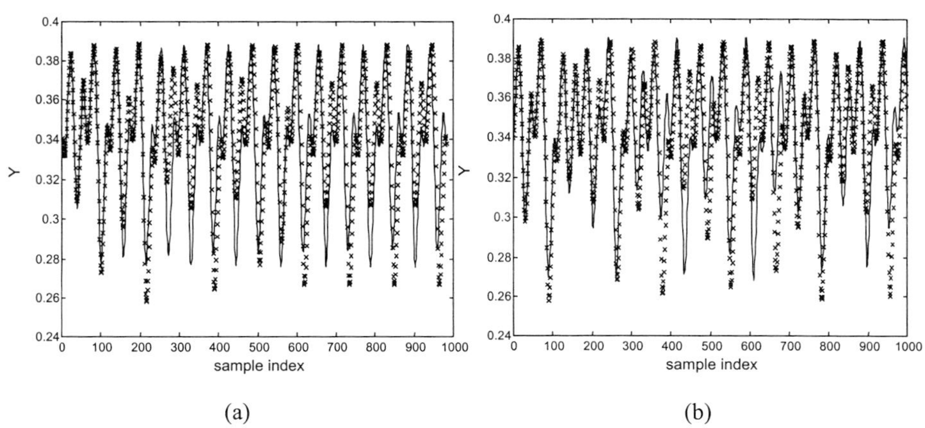

Figure 9.8. Free-run prediction of the autocatalytic Y-state with (a) $M_{PL\theta1}$ and (b) $M_{PL\theta2}$. The '+' marker denotes experimental observations, while the solid line denotes the predictions of the models.

From Figure 9.7 it is clear, judged by the broken curve at the bottom that represents the nonlinear surrogate data generated by $M_{FF\theta 2}$, that the model has not captured the structure of the data completely, except in the large-scale region of the dynamic attractor ($\log \varepsilon_0 > -0.9$). The peculiar dip in the bottom curve just after $\log \varepsilon_0 = -0.9$ can be attributed to numerical instability in the correlation dimension algorithm. In Figure 9.8(a), it can be seen that $M_{PL\theta 1}$ has captured most of the large-scale structure ($\log \varepsilon_0 > -1.3$) of the dynamic attractor. Likewise, $M_{PL\theta 2}$ has evidently captured most of the dynamic structure of the data, as indicated by Figure 9.8(b).

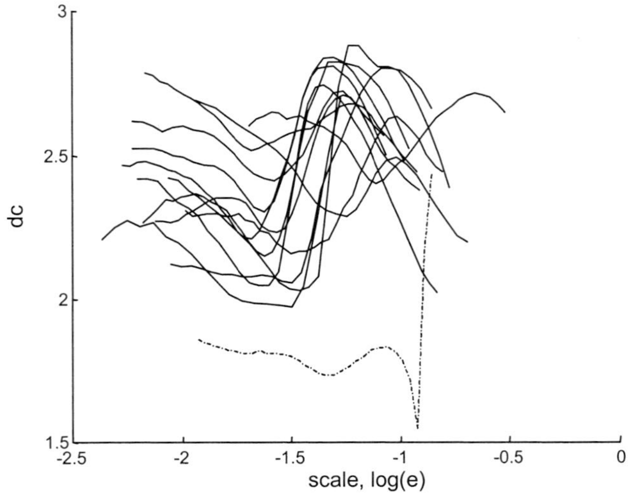

Figure 9.9. Correlation dimension curves of nonlinear surrogates of $M_{FF\theta 2}$ and that of the model residuals (broken line, bottom) from the larger data set.

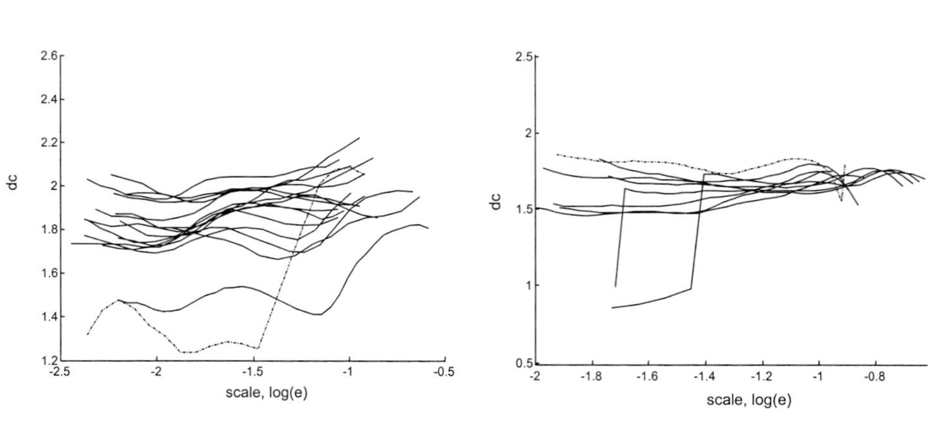

Figure 9.10. Correlation dimension curves of nonlinear surrogates and that of the observed data (broken line, bottom), for (a) $M_{PL\theta 1}$ and (b) $M_{PL\theta 2}$.

Reconstructions of the dynamic attractor of the data, based on the actual data and the free-run predicted data ($M_{PL\theta 2}$) are shown in Figure 9.11(a) and Figure 9.11(b). As can be seen from these figures, the two attractors are remarkably similar in appearance to each other and also to the attractor constructed from the *X*, *Y* and *Z* states, shown in Figure 9.1. This is confirmed by the position of the correlation dimension curve for observed data amongst the cluster of nonlinear surrogates in Figure 9.11(b).

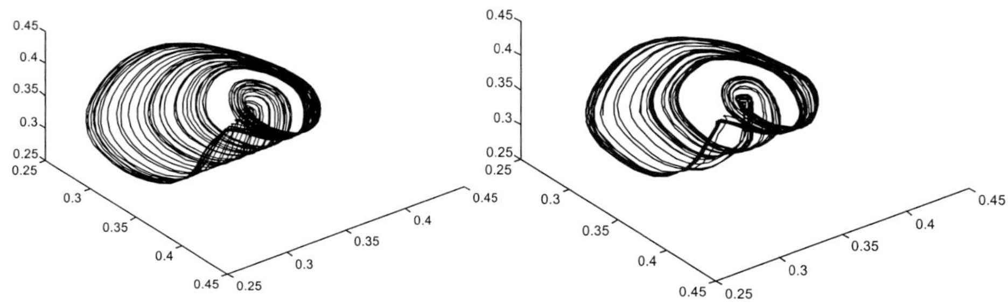

(a) (b)

Figure 9.11. Dynamic attractor of autocatalytic process reconstructed from (a) the Y-state, and (b) the $M_{PL\theta 2}$ free-run model of the Y-state.

9.3. EFFECT OF MEASUREMENT AND DYNAMIC NOISE ON THE IDENTIFICATION OF AN AUTOCATALYTIC PROCESS

9.3.1. Example 1: Effect of noise on autocatalytic data

To test the effectiveness of the identification method on noisy data, Gaussian measurement noise, as well as dynamic noise were added to the autocatalytic process data. The noise level was set at 0.1σ (10% of the sample standard deviation of the training data set) for measurement noise.

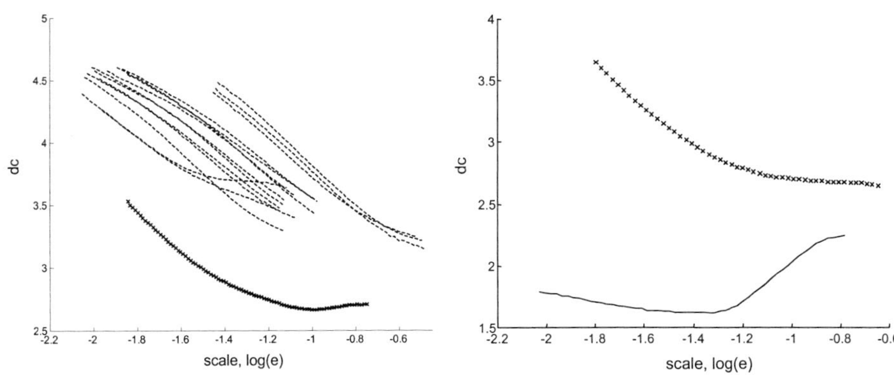

(a) (b)

Figure 9.12. Correlation dimension curves, (a) for autocatalytic Y-state with noise (crosses, bottom) and its Type 2 surrogates, and (b) for Y-state with noise (top) and Y-state without noise (bottom). The bottom curve in (a) and the top curve in (b) are the same.

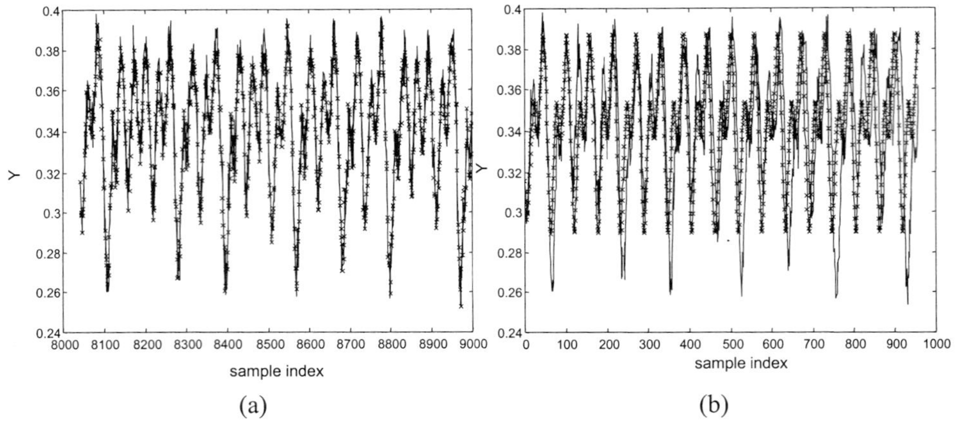

Figure 9.13. One-step ahead predictions of the autocatalytic Y-state with $M_{PL\theta3}$ in the presence of measurement and dynamic noise (a), and the free-run prediction of the same data with $M_{PL\theta3}$ (b). Experimental observations are denoted by a 'x' marker, while the solid lines show the predictions of the models.

Dynamic noise was added by a modified one-step prediction of the training data with $M_{PL\theta2}$. Noise of 0.1σ (10% of the sample standard deviation of the training data set) was added to the *i*'th point, which was then included in the embedding for prediction of the (*i*+1)'th point.

In Figure 9.12(a), the correlation dimension curves of the noisy data and their associated surrogates are shown, while the correlation dimension curves of the data with, and without noise are shown in Figure 9.12(b). The curve at the bottom of Figure 9.12(a) is the same as the curve at the top of Figure 9.12(b). It is interesting to note the relative positions of the small and large scale sections of the correlation dimension curves in Figure 9.12(b). For the data with noise, the whole correlation dimension curve has higher values than for the data without noise. Moreover, the correlation dimension curve for the noisy data converges on that for the noiseless data at larger scales.

This trend is typical of data containing measurement and dynamic noise, as opposed to noiseless data with the same dynamics. A higher correlation dimension at small scales indicates more intricate microstructure in the attractor, which is typical for data containing measurement noise. A higher correlation dimension at large scales indicates more complex macrostructure in the attractor or higher dimensionality, which is typical of data containing dynamic noise (or alternatively, more complex dynamics).

Another model structure from the set M^*_{PL} was fitted to the noisy data set, resulting in the model $M_{PL\theta3}$. This model used 18 Gaussian kernels. Figure 9.13(a) shows the ability of $M_{PL\theta3}$ to predict the data with measurement and dynamic noise one-step ahead, while Figure 9.13(b) shows the free-run predictions of $M_{PL\theta3}$. These figures indicate that though the model was able to make acceptable one-step predictions, it failed to capture the small-scale dynamics adequately during free-run predictions by not following the peaks and troughs in the observed data.

The effect of the larger data set on the identification process was significant with regard to the free-run prediction. The reason for this is that the data set was likely not to be stationary in the nonlinear sense, that is, the model structure has observed only part of the attractor. Still, for the smaller data set, the pseudolinear radial basis model managed best of the family of nonlinear models in this investigation. The pseudolinear model fit relatively little noise, since the complexity of the model was regulated by the minimum description length criterion.

Although a multilayer perceptron, as well as a pseudolinear radial basis function model were capable of similar, accurate one-step ahead prediction of the chaotic autocatalytic process, the pseudolinear radial basis function model was better able to capture the underlying dynamics of the system.

9.4. IDENTIFICATION OF AN INDUSTRIAL PLATINUM FLOTATION PLANT BY USE OF SINGULAR SPECTRUM ANALYSIS AND DELAY COORDINATES

As mentioned previously, the identification of the underlying dynamics of froth flotation systems from experimental data is complicated by the fact that erratic fluctuations in the observed behaviour are typically derived from a mixture of various plant disturbances that can be notoriously difficult to disentangle. Standard control cycles are comprised of the fixing of initial set points, a settling period for transient dynamics to subside, a period of measurement and evaluation, and a final estimate of appropriate set points. Some studies have indicated that operators often tend to make the periods for settling and measurement too small.

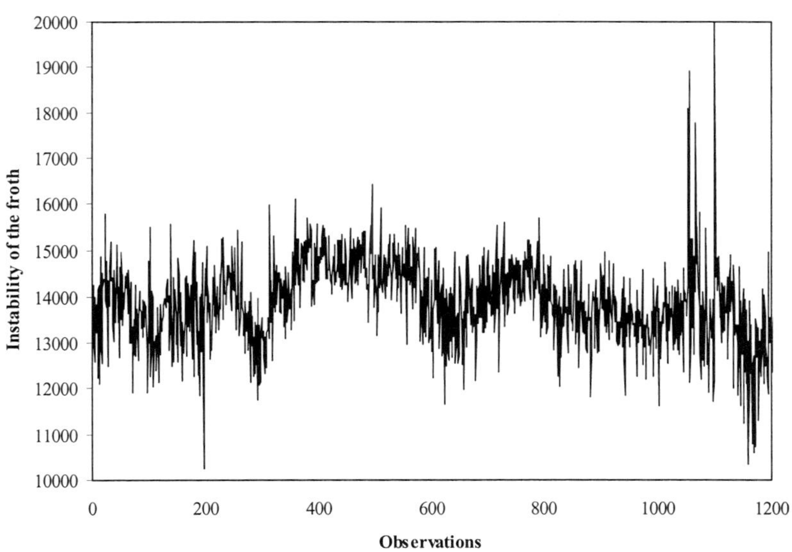

Figure 9.14. Instability measurements of froth on a platinum flotation plant. Samples were collected with intervals of one minute over a period of 20 hours.

Apart from these aspects, the inexperience or inability of the operator can have a further significant impact on the control of the plant. Consequently, optimal control is not usually maintained, especially where incipient erratic behaviour in the plant is difficult to detect. In this example, the instability measurements of a flotation froth on a platinum plant shown in Figure 9.14 are considered. The lag of k = 4 was calculated based on average mutual information critera, while the embedding dimension of m = 4 was determined by the false nearest neighbour algorithm. With a lag of k = 4 and an embedding dimension of m = 4, the attractor of the time series could be reconstructed, as shown in Figure 9.15. As can be seen from Figure 9.15, the attractor exhibited some large-scale structure, although noise appears to be quite significant.

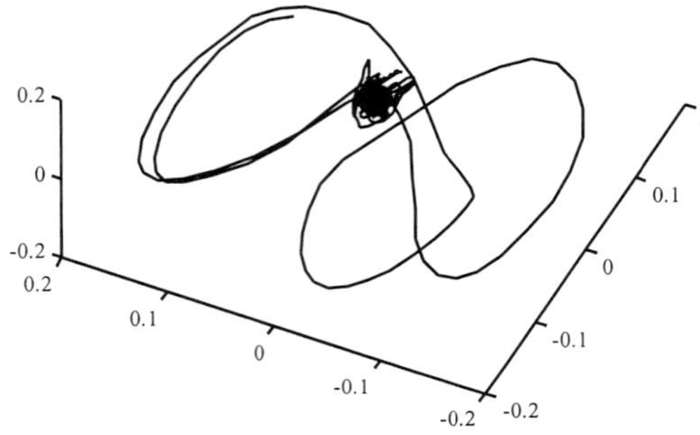

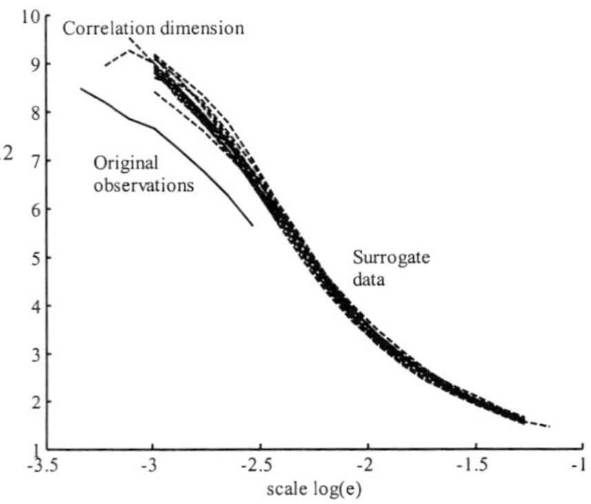

Figure 9.15(a). 3-D reconstruction of the attractor of the time series shown in Figure 9.12 and (b) the corresponding surrogate data analysis, indicating a high degree of stochasiticity (noise) in the observations.

The results of a Type 2 (AAFT) surrogate data analysis are shown in Figure 9.16. Based on these results, it can be concluded that the data are mostly stochastic (note the high correlation dimension of the curves), although some degree of determinism is indicated by the separation between the correlation dimension curve of the original time series and those of its surrogates. Identification of a reliable predictive model from the data portrayed in Figure 9.14 is therefore unlikely in this case.

9.5. IDENTIFICATION OF A HYDROMETALLURGICAL PROCESS CIRCUIT

In the third case study, historic data from a leaching circuit were considered. This data set represented the daily average concentrations of a metal downstream in the circuit and is used by the operator to control the metal concentration by adjusted the addition of acid and other reagents upstream from the measuring point. In practice the addition of these reagents depends almost entirely on the experience of the operator. Owing to inexperienced operators, and different operating strategies by different operators, this leads to a significant amount of variation in the circuit. By constructing a predictive model of the metal concentration in the circuit, a more systematic control strategy could in principle be developed.

The PL algorithm with variable embedding (with maximum $m = 4$ and fixed $k = 10$) was used to build the model. As the concentration of the metal is measured at four different points in the circuit, yielding four highly correlated time series of metal concentrations, the model was based on the first principal component of the four time series. The correlation dimension was calculated for this first principal component and compared to the correlation dimension curves for the AAFT surrogates based on the original data, as indicated in Figure 9.16.

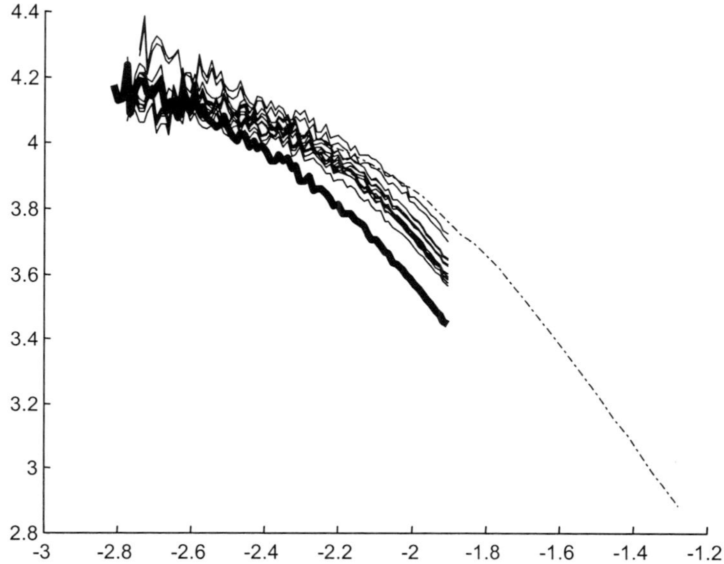

Figure 9.16. Correlation curves for the hydrometallurgical process circuit (Example 3), its first principal component and its AAFT surrogates.

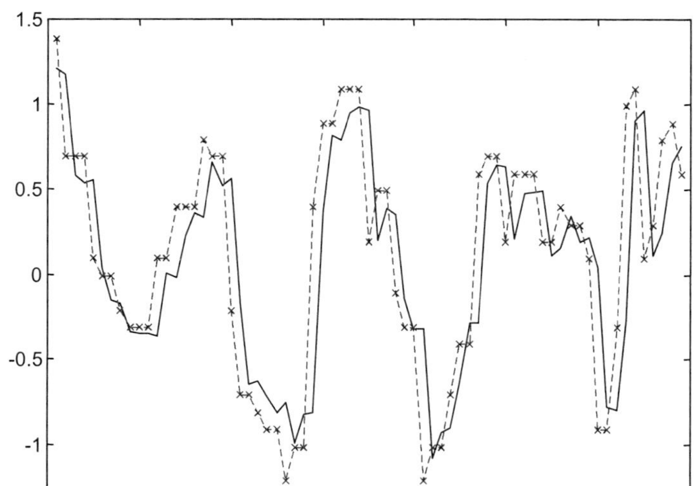

Figure 9.17. One-step ahead prediction with pseudolinear radial basis model of manganese process time series based on the first principal component.

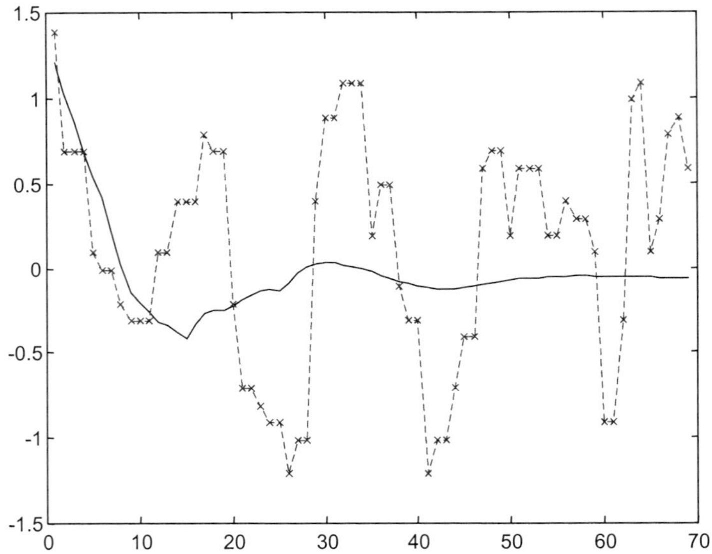

Figure 9.18. Free-run prediction with pseudolinear radial basis model of fourth observed component of manganese extraction process (x marker) vs. observed data with dynamic and observation noise.

As can be seen from this figure, the time series is very noisy, although some determinism can be discerned at larger scales. Pseudolinear radial basis function neural network models similar to those discussed in the first example, were consequently fitted to the data, which yielded models that could predict the data ($R^2 = 0.629$) as shown in Figures 9.17 and 9.18. As indicated in Figure 9.18, the model could predict the metal concentration reasonably accurately approximately 10 time steps ahead. On this basis a model predictive controller could be designed for the plant.

Chapter 10
Embedding of Multivariate Dynamic Process Systems

C. Aldrich and J. P. Barnard*

Although multivariate time series can be treated as a set of individual time series by considering each variable in isolation, this is only efficient if all the variables are independent or uncorrelated. This is often not the case. Even worse, it is not always possible to predict the time evolution of a system state from only a single observed variable. For example, the well-known Lorenz system has three state variables, x, y and z, but dx/dt = f(x,y), while dz/dt = f(x,y,z). As a consequence, z cannot be predicted properly from x and/or y only. For empirical systems the situation is usually more complicated, since the observed variables required for a predictive model and the dynamic interdependencies may not be obvious.

Under these circumstances, it would be desirable to use a strategy of embedding that exploits the redundancy between the variables. This can be accomplished by combining Takens' embedding theory (1981) with singular value decomposition or independent component analysis (Hyvärinen, 1999) to transform individual embeddings of multidimensional process observations into a vector space of uncorrelated or statistically independent vectors (state variables). The method is demonstrated on the autocatalytic process discussed previously, as well as a case study on air pollution data.

10.1. EMBEDDING OF MULTIVARIATE OBSERVATIONS

Cao et al. (1998) have proposed the embedding of all components of multi-dimensional observations by using an optimal Takens embedding for each component (time series). The optimal values of the embedding dimension for each component were found by minimizing the prediction error of a nearest neighbour, locally constant predictor. Unfortunately Cao et al. (1998) did not indicate how to optimize the embedding lag, which is crucial in reconstructing a representative attractor for practical systems. Optimization of the embedding lag is especially difficult if noise is present in the observations (Lai and Lerner, 1998).

A different approach is therefore proposed to embed multidimensional observations, which avoids both linear approximations in finding embedding dimensions and potentially sub-optimal embedding lags. With this approach, each component of an observation space, $\mathbf{Y} \in \Re^{p}$, is treated as a one-dimensional time-series. Each component is embedded individually to generate a set of subspaces. These subspaces are combined to form a first approximation of the attractor in $\Re^{\Lambda} = [\Re^{m1}\, \Re^{m2}\, \Re^{m3} \ldots \Re^{mp}]$. Finally, the lag variables are separated and the structure of the attractor optimized. This results in a reconstructed dynamic attractor based on the observation space. In addition the resultant attractor structure is an optimal projection of the original embedding variables.

10.2. MULTIDIMENSIONAL EMBEDDING METHODOLOGY

System parameterization by the multidimensional embedding method as proposed above, consists of the following steps (the method is depicted diagrammatically in Figure 10.1)

* Department of Chemical Engineering, University of Stellenbosch, Private Bag X1, Matieland, 7602, Stellenbosch, South Africa.

i. Optimal embedding of individual observed components, using Taken's embedding (Takens, 1981).

ii. Calculation of the optimal non-Gaussian linear separation matrix, using Hyvärinen's method (Hyvärinen, 1999). Alternatively, use singular spectrum analysis to calculate the separation matrix.

iii. Optimal projection of the state space onto statistically independent state variables.

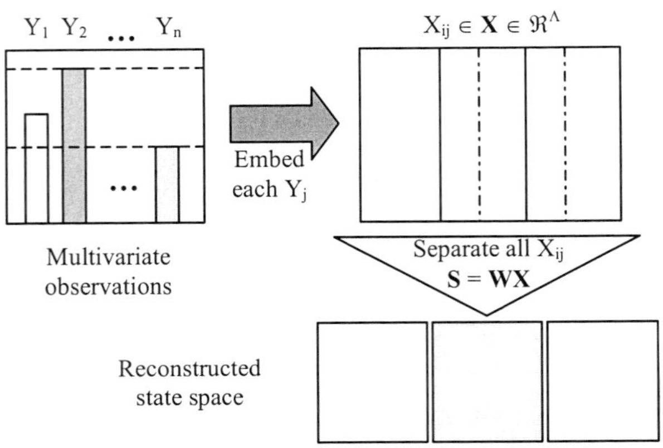

Figure 10.1. Schematic representation of embedding and separation concept to reconstruct a system state space from multivariate time series.

10.2.1 Optimal embedding of individual components

Each component of the multidimensional observation (time-series) is embedded, using the method of delay coordinates[56]. The embedding delay is determined by the minimum average mutual information (AMI) criterion proposed by Frazer and Swinney (1986) and the embedding dimension by the false nearest neighbours algorithm of Kennel et al. (1992).

Let $\mathbf{Y} = [\mathbf{Y}_1, \mathbf{Y}_2, \ldots \mathbf{Y}_n]^T$ be the array of p-dimensional observations, where $Y_i = [y_{i1}, y_{i2}, \ldots y_{ip}]$, i = 1, 2, 3,…, n. Multidimensional embedding of **Y** results in the following embedding matrix

$$\mathbf{X} = \begin{bmatrix} x_{11} & x_{12} & \cdots & x_{1p} \\ x_{21} & x_{22} & \cdots & x_{2p} \\ \vdots & \vdots & \ddots & \vdots \\ x_{n-L,1} & x_{n-L,2} & \cdots & x_{n-L,p} \end{bmatrix}, \tag{10.1}$$

where

[56] Other methods for embedding, such as singular spectrum analysis can be used as well.

$$x_{ij} = \left[y_{i+k_j(m_j-1),j} \quad y_{i+k_j(m_j-2)_j,j} \quad ,\cdots, \quad y_{i,j} \right]$$
$$i = 1 \ldots n, j = 1 \ldots p \tag{10.2}$$

and $L = \max_{1 \le j \le p} k_j(m_j - 1) + 1$.

And k_j and m_j are the optimal lag and embedding dimension for the j'th time series **Yj**, respectively. The maximum value of L forces all the columns of **X** to be of the same length.

10.2.2 Optimal projection of initial embedding

Projections of the embedded **X** onto directions that optimize the structure of the attractor and statistically separate the embedding variables are achieved by the following linear transformation

$$\mathbf{S} = \mathbf{WX} \tag{10.3}$$

where **S** is the optimal projection of the original embedding **X** and **W** is the separating matrix. The dimension of **S** may be lower than that of **X**. As discussed previously, similar results can be obtained by use of principal component analysis, where the scores are also calculated via projection of the data onto the principal components, i.e. **T** = **XP**. Thus one may achieve optimal projection, reduction of dimensionality, as well as independence of embedding variables.

a) Optimal projection by singular spectrum analysis

When applying singular spectrum analysis to embed the multiple channel observations, each observation channel is embedded as before. Instead of using average mutual information and the false nearest neighbours algorithm to obtain the embedding parameters k and m, the first stage of singular spectrum analysis is applied to each observation channel. That is, each channel is embedded using a unary delay (k = 1) in an embedding space of dimension equal to the index of the decorrelation point of that observation channel. Thereafter, the resultant embedding spaces are concatenated, as shown in the previous section. The combined initial embedding space **X** is then subjected to the second stage of singular spectrum analysis, i.e. linear separation by principal component analysis of the combined initial embedding space. In mathematical terms, we find the linear transformation matrix, **W**, consisting of the first q significant eigenvectors of the covariance matrix of **X**.

In similar style to the above application of singular spectrum analysis, it is also possible to construct an initial embedding using some other set of values for the embedding parameters, *k* and *m*. The resultant initial combined embedding space can then again be separated by principal component analysis as described above.

b) Optimal projection by linear independent component analysis

If the data are not normally distributed, the use of principal component analysis will not result in optimal separation of the components. Under these circumstances, to find **W**, the use of independent component analysis would be better, in which case one has to maximize the negentropy J_G of **X**, which is equivalent to minimizing the cross mutual information amongst components of **X**, under the constraint of decorrelation of components. Formally

$$\max \sum_{i=1}^{p} J_G(\mathbf{w}_i) \text{ with respect to } \mathbf{w}_i \tag{10.4}$$

subject to

$$E\{w_k^T w_j\} = \delta_{jk} \tag{10.5}$$

where

$$J_G(\mathbf{w}) = [E\{G(\mathbf{w}^T\mathbf{x})\text{-}E\{G(\mathbf{v})\}]^2 \tag{10.6}$$

with G some non-quadratic, contrast function that estimates the probability density function of an independent component, *c* some constant, and **v** a standardized Gaussian variable. Each vector $\mathbf{w}_i$ is a row of matrix **W**.

Four practical choices of contrast functions (with their respective derivatives, indicated by lower-case function symbols) are listed below (Hyvärinen and Oja, 1997; Hyvärinen, 1999)

- tanh-function, $G_1(u) = a_1^{-1}\log\{\cosh(a_1u)\}$, $g_1(u) = \tanh(a_1u)$ (10.7)
- Gaussian function, $G_2(u) = -a_2^{-1}\exp(-a_2u^2/2)$, $g_2(u) = u\exp(-a_2u^2/2)$ (10.8)
- power 3 function, $G_3(u) = \frac{1}{4}u^4$, $g_3(u) = u^3$ (10.9)
- skew function, $G_4(u) = u^3/3$, $g_4(u) = u^2$ (10.10)

A specific choice of contrast function serves to optimize the performance of the independent component analysis algorithm. G_1 is optimal for most data sets, G_2 for highly super-Gaussian (leptokurtotic) data, as well as when robustness against outliers is important, G_3 for sub-Gaussian (platykurtotic) data with no outliers, and G_4 for data with a skew distribution.

The maximization of equation 10.4 can be accomplished by a fixed-point algorithm, as proposed by Hyvärinen (1999). Using any contrast function G, the algorithm converges at least quadratically for practically any non-Gaussian distribution, and is roughly equivalent to projection pursuit, because it estimates independent components sequentially by deflation.

c) *Selection of a suitable model structure*

After embedding the observations, a suitable associated model structure is selected and fitted to the data set, **Z** = [**S**,**Y**]. In the case studies discussed below, the models fitted to the multidimensional embeddings were tested and validated by one-step, as well as free-run predictions. The R^2 statistic for one-step prediction versus observations was selected as a linear criterion for embedding quality and model fitness. In addition, free-run predictions were compared by way of inspection and the correlation dimension statistic was used to evaluate the nonlinear dynamic equivalence of different embedding strategies. The prediction stability boundary was inferred from the estimation of the first global Lyapunov exponent and indicated the reasonable free-run prediction horizon that could be expected.

A prediction procedure for the above multi-channel embedding strategy can be formulated as follows

i. Embed points $y_{i,j}$ through $y_{i+k(m-1),j}$ over p observation components, to form x_i, by applying equations (10.1) and (10.2).

ii. Scale embedded components to unit vectors.

iii. Separate embedded variables to form state variables, by applying equation (10.3), using **W** calculated from the data set used for estimation of the model parameters.

iv. Predict $y_{i+k(m-1)+s,j}$ as $y^{est}_{i+k(m-1)+s,j}$, where s is the step size (normally one sampling unit).

v. For free-run predictions, use the prediction $y^{est}_{i+k(m-1)+s,j}$ in the next embedding x_{i+1}, instead of the observation $y_{i+k(m-1)+s,j}$.

vi. Repeat from step iii for the length of the prediction run.

10.3 APPLICATION OF THE EMBEDDING METHOD

In the first example, the autocatalytic process system described in the previous chapter, is revisited. However, in this case, the three time series corresponding to the X, Y and Z states are considered simultaneously (and not just the Y state as before in section 9.2). Multilayer perceptron neural network model structures with single hidden layers and nodes with bipolar sigmoidal activation functions were used to develop predictive models of the autocatalytic process. The output layers of the networks were linear and the Levenberg-Marquardt algorithm was used for parameter estimation.

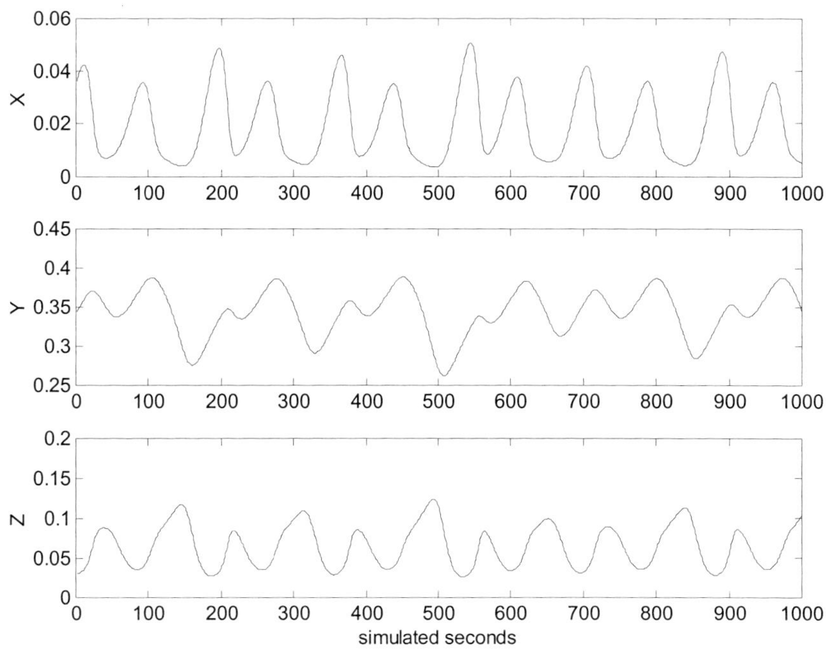

Figure 10.2. Autocatalytic data, showing X, Y, and Z states resulting from a numerical solution for the process state equation.

The X, Y and Z states of the autocatalytic process (Lynch, 1992) discussed in the previous chapter are shown below in Figure 10.2, together with the attractor of the system in Figure 10.3, which is shown again for for convenience.

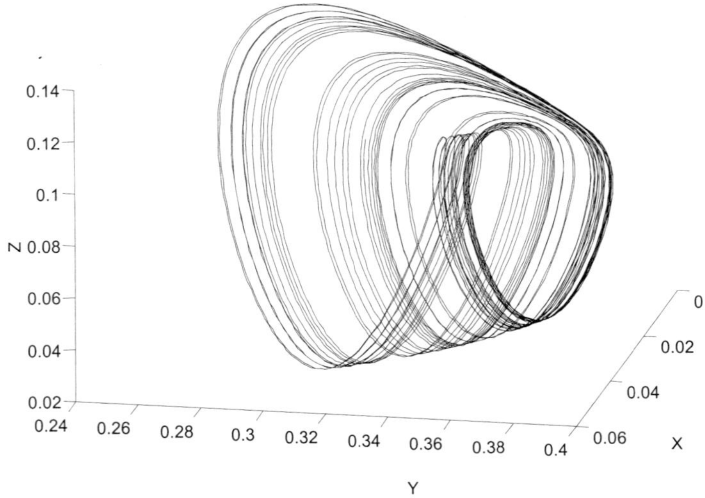

Figure 10.3. Dynamic attractor of autocatalytic process, constructed from X_t, Y_t and Z_t, resulting from a numerical solution for the process state equation.

The objective in this case was to predict the Z state from the other two states (X,Y) with a feedforward neural network, using a suitable parameterization. The Z-state is a function of all three states of the system, hence a feedforward neural network, designated M^*_{FF}, was used to predict Z_{t+1} in terms of $[X,Y]_t$.

The system was parameterized by individual embeddings of X and Y. Average mutual information and false nearest neighbour calculations for each variable resulted in an embedding dimension set of {3, 3} and lag set of {21, 26} for X and Y respectively. The large embedding lags could be attributed to the high sampling rate, which resulted in high linear correlation between successive observations. The individual embeddings were combined into $\Lambda_0 \in \Re^9$ and separated as $S_0 = W_0\Lambda_0$, using independent component analysis with the G_2 contrast function (equation 10.8). The following model structure was fitted

$$M_0\text{: } S_0(t) \rightarrow Z(t+1),\ M_0 \in M^*_{FF} \tag{10.11}$$

The optimal model topology had a hidden layer of 8 bipolar sigmoidal nodes. The effect of separation by independent component analysis was investigated in terms of a second parameterization consisting of the same embedding as above, but without the explicit separation of embedding variables. This model structure was called M_1 and used the same model topology as M_0.

After parameter estimation, the models were tested on records 25001 to 26000 and validated on records 26001 to 27000 of the dataset. The prediction of Z by $M_0(\theta_0)$ operating on the validation data appears in Figure 10.4 and had an R^2 statistic of 0.9993. The prediction of Z

by $M_1(\theta_1)$ was very similar with an R^2 of 0.9963. No free-run prediction was performed, because it is not defined for mappings of the form f: $X, Y \rightarrow Z$.

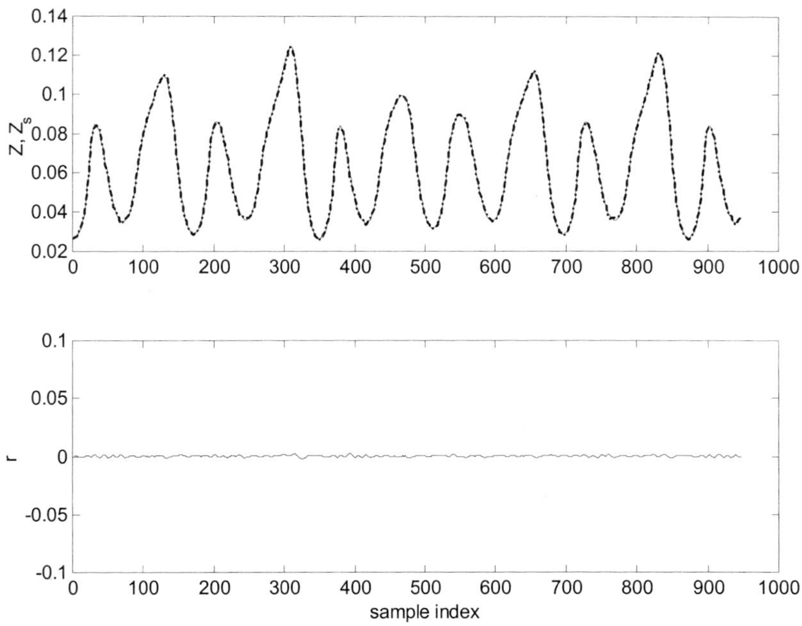

Figure 10.4. Prediction of the *Z*-state from the *X*- and *Y*-states taken from the validation data set for the autocatalysis process. Z_s is the prediction (broken line) and *Z* the observation, while *r* denotes the prediction residuals.

10.4 MODELLING OF NO_X-FORMATION

The method is demonstrated via application to the prediction of atmospheric NO_2 concentrations in the metropolitan area of Cape Town in South Africa. Nitrogen oxide (NO_x) air pollution is often a serious problem in metropolitan areas. NO and NO_2, collectively known as NO_x, are products of high-temperature, aerobic combustion, such as spark and compression ignition engines, oil-fueled power plants and the burning of automobile tyres as a means of waste disposal. Once these pollutants are released, they are further affected by solar radiation and these photochemical reactions result in the formation of NO_2 from NO (Grobliki et al., 1981). Meteorological factors cause the NO_x pollutant to be transported and dispersed in the atmosphere. Not only are these pollutants a health hazard, but this reaction cycle is also one of the precursors to photochemical smog (Dzubay, 1982).

There are essentially three approaches to atmospheric dispersion modelling, viz. the Eulerian approach, the Lagrangian approach and the statistical approach. The Eulerian approach uses a continuity equation to develop a description of the physical and chemical processes that govern the relationships between emissions and the resulting concentrations. It is a rigorous model that addresses physical and chemical processes from first principles and requires spatial and time resolved information of emission sources. Needless to say, it is complex and requires large quantities of input data and processing power.

The Lagrangian approach is applied most frequently. The motion of the pollutant particles in the atmosphere is modelled using a probabilistic description and this in turn is used to derive expressions for pollutant concentrations. The most commonly used probabilistic description is the Gaussian plume model. This model has a few simplifying assumptions that are restrictive. It assumes steady state conditions, e.g. a constant rate of emission, the probability of the wind velocity being independent of time and location, normal distribution of pollutant concentrations, etc.

With the statistical approach, statistical techniques are used to establish relationships between pollutant emissions, meteorological conditions and pollutant concentrations. The time series of the pollution data is correlated with synchronised emission and weather data using various multivariate statistical techniques. Since most of these statistical models are linear, they often fail to identify important underlying nonlinear dynamics in pollution models.

Nonetheless, during the past two decades, considerable progress has been made with the development of climatological models based on nonlinear time series analysis. These models of the formation and distribution of pollutants are often constructed in conjunction with synchronised observations of the chemical constituents and contributing environmental variables. The two main contributing environmental variables are usually taken to be ambient temperature and solar radiation.

A data set containing 8664 records of synchronised, hourly mean concentrations of NO, NO_2, ambient temperature and solar radiation for the city centre of Cape Town, observed during 1996, was used as the basis for the development of the model. NO_2 concentration (Z_{NO2}) was chosen as the dependent variable with the aim to make acceptable one-step and free-run predictions in terms of NO concentration (X_{NO}) and solar radiation (X_{Es}). The first 6000 records were selected for fitting a nonlinear model. Figure 10.5 shows the historic record of NO_2 concentrations.

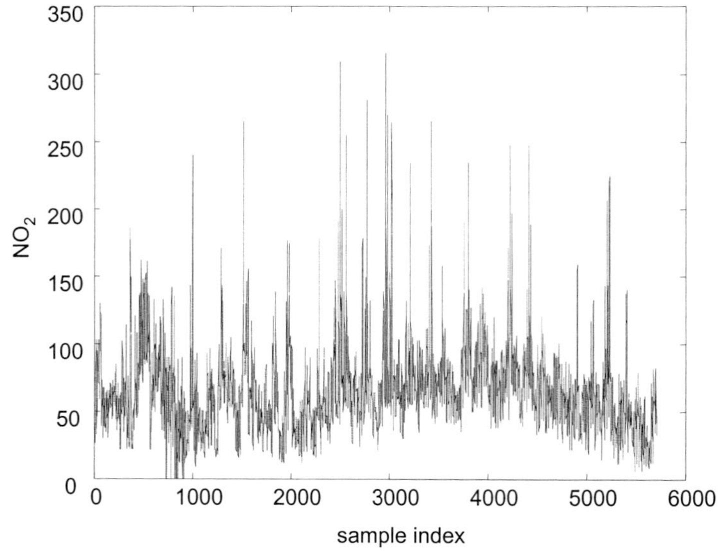

Figure 10.5. NO_2 data used for fitting a nonlinear MLP model. No filtering was applied, but outliers were removed.

After removing outliers, the data were classified as by use of the surrogate data method described in Chapter 8. This entailed the generation of surrogate data from the Z_{NO2} data by randomizing the sample index and the Fourier spectrum. Correlation dimension curves were calculated for both data and surrogates, using the algorithm by Judd (1992), and mutually compared. The data appeared to have a high degree of stochasticity from inspection of the correlation dimension curves (Figure 10.6), but some dynamics could also be discerned from the consistently lower correlation dimension values for Z_{NO2}. In the absence of a known noise model, and given the risks of noise reduction in nonlinear systems no explicit pre-filtering was applied to the data. The largest Lyapunov exponent was estimated to be 0.974.

Chaotic nonlinear deterministic systems can have a finite prediction horizon regarding free-run prediction. Starting from any initial state within the basin of attraction, any free-run prediction for such a system will become ultimately unstable outside the basin of attraction that contains the dynamic attractor (Abarbanel, 1994). This rule applies unless the dynamics have been properly reconstructed from data observed in several adjacent basins of attraction. Even within a specific basin the highly nonlinear character limits the free-run prediction horizon.

Starting a prediction from index i and predicting in L steps until i+L, an upper bound on L, called the instability horizon (Abarbanel, 1996), is approximately given as

$$t_L = \tau_s/\lambda_1 \tag{10.12}$$

where τ_s is the sampling period of the observer function and λ_1 the first Lyapunov exponent. The initial error at index i scales as a function of the number of iterations L to predict from index i to i+Lk and can be expressed in terms of the largest Lyapunov exponent estimated from

$$\varepsilon_L = \exp(L\tau_s\lambda_1) \tag{10.13}$$

In this case, $t_L = \tau_s/\lambda_1 = 1/0.974 = 1.02$. This result implied a prediction horizon of 1.02 h, which meant that any model could expect difficulty under free-run prediction. Consequently, reliable one-step predictions were the best that could be hoped for in the absence of some noise reduction scheme.

Table 10.1 lists the loading of each principal component with respect to each variable in the data set, viz. NO_2, NO, temperature and solar radiation. In addition, the column headings of the table include the eigenvalues as λ_i, as well as the percentage of variance explained by each principal component. From inspection of the table it is clear that environmental temperature played a small role in terms of the first, second and third principal components, with loadings of 0.1040, 0.0677 and 0.1265 respectively. Cumulatively, the first three principal components explained 98% of the total variance in the data.

The system was therefore redefined in terms of the three remaining observed variables: X_{NO}, X_{Es} and Z_{NO2}. Optimal parameterization of the system was investigated in terms of two alternative embedding strategies and associated model structures. Parameterization according to the first strategy implied embedding all observation components using optimal individual embeddings, as summarized in Table 10.2. In terms of the second strategy only Z_{NO2} was embedded and the observed X_{NO} and X_{Es} were included as one-dimensional time series (Table 10.3). The motivation behind the first embedding strategy was that the three variables were all dependent observations of the microclimatic system and thus allowed an embedding of each variable. On the other hand, the second strategy treated X_{NO} and X_{Es} as control variables and thus independent, which did not justify embedding them. For conciseness, $M_{2\theta}$ is used instead of $M_2(\theta_2)$, etc.

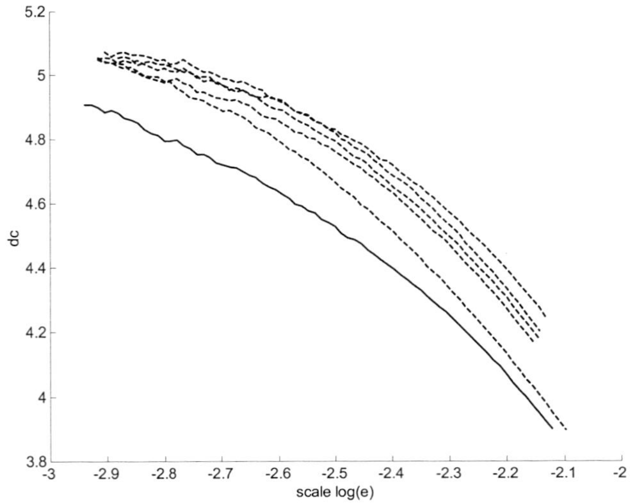

Figure 10.6. Correlation dimension curves for NO_2 concentration (solid line) and nonlinear transformed random surrogate data (dashed line) based on NO_2 concentration.

Table 10.1. Results from principal component analysis of air pollution data.

	PC_1 λ_1: 9.065e-5 (51% variance)	PC_2 λ_2: 6.798e-5 (38% variance)	PC_3 λ_3: 1.604e-5 (9% variance)	PC_4 λ_4: 2.903e-6 (2% variance)
Z_{NO2}	0.3163	-0.3527	0.8722	-0.1213
X_{NO}	0.4135	-0.7841	-0.4575	0.0691
X_T	0.1040	0.0677	0.1265	0.9841
X_{Es}	0.8473	0.5060	-0.1178	-0.1093

In accordance with the first embedding strategy, the individual embeddings were combined into $\Lambda_2 \in \Re^{17}$ and separated as $S_2 = W_2\Lambda_2$ by independent component analysis with the G_3 contrast function. The following model structure was fitted to the data

$$M_1: S_1(t) \rightarrow Z_{NO2}(t+1), M_1 \in M^*_{FF} \tag{10.14}$$

In accordance with the second embedding strategy, individual embeddings were combined as $\Lambda_3 \in \Re^7$ and separated as $S_3 = W_3\Lambda_3$. The associated model structure was

$$M_2: S_2(t) \rightarrow Z_{NO2}(t+1), M_2 \in M^*_{FF} \tag{10.15}$$

To test the influence of independent component analysis, the system was again parameterized in accordance with the first embedding strategy, but without separation by independent component analysis, resulting in the following model structure

$$M_3: \Lambda_3(t) \rightarrow Z_{NO2}(t+1),\ M_3 \in M^*_{FF} \tag{10.16}$$

In addition, the possible advantage of embedding by specifying a fourth parameterization, without embedding of any variable was also investigated. The associated model structure was

$$M_4: [X(t), Y(t)] \rightarrow Z_{NO2}(t+1),\ M_4 \in M^*_{FF} \tag{10.17}$$

MLP networks with single hidden layers and linear output layers were selected as model structures in all cases. Each hidden layer initially consisted of 16 bipolar sigmoidal nodes and training proceeded via the Levenberg-Marquardt algorithm. A more elaborate evaluation of model performance by surrogate techniques was impossible, owing to the high noise content of the data and the short prediction horizon calculated earlier. These conditions disabled free-run predictions over sufficiently long periods to preclude application of nonlinear surrogate techniques (which require at least 1000 predicted points for sufficiently accurate estimation of the correlation dimension). The optimal topologies for the other neural network models ($M_{1\theta}$ - $M_{4\theta}$) were determined in a similar fashion and are listed in Table 10.4.

Table 10.2. Embedding parameters for air pollution data: Strategy 1 (M_1).

	NO_2	NO	E_s
m	5	5	7
k	9	14	9

Table 10.3. Embedding parameters for air pollution data: Strategy 2 (M_2).

	NO_2	NO	E_s
m	5	1	1
k	9	0	0

Table 10.4. Optimal MLP network topologies for various parameterizations of the NO_x system.

	$M_{1\theta}$	$M_{2\theta}$	$M_{3\theta}$	$M_{4\theta}$
Hidden Nodes	8	14	14	8

All models were tested by doing one-step and free-run predictions on observations 6001 to 7000 and validated on observations 7001 to 8000. These data sets were embedded and separated using the respective embedding parameters and separation matrices as calculated from the training data. The validation results for one-step and free-run predictions according to the first embedding strategy ($M_{1\theta}$) are shown in Table 10.5 and Figures 10.7 and 10.8, and the results according to the second strategy in Table 10.5 only.

Table 10.5. R^2 statistics for one-step prediction of Z_{NO2}, X_{NO}, X_{Es}

	$M_{1\theta}$	$M_{2\theta}$	$M_{3\theta}$	$M_{4\theta}$
Z_{NO2}	0.7937	0.7936	0.8244	0.7374
X_{NO}	0.1989	0.2346	0.1741	0.1632
X_{Es}	0.9315	0.9143	0.9267	0.8621

One-step ahead predictions did not reveal significant differences between the embedding strategies, as can be seen from the R^2 statistics in Table 10.5. For example, Z_{NO2} was predicted best by $M_{3\theta}$, X_{NO} by $M_{2\theta}$, and X_{Es} by $M_{1\theta}$. However, parameterization without embedding appeared to lead to a less reliable model of the system, as indicated by $M_{4\theta}$ in Table 10.5. Separation of the embedding variables by independent component analysis appeared to have a slightly negative influence on the Z_{NO2}-model.

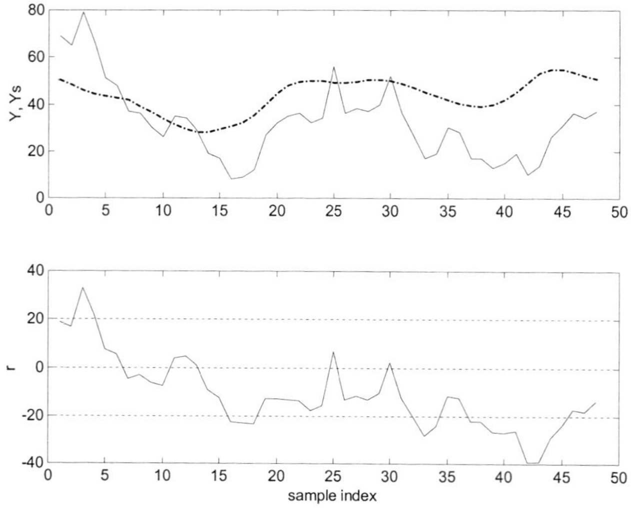

Figure 10.7. Validation of a MLP network model by simultaneous free-run prediction of NO_2, NO and E_s, using embedding strategy 1. The result for NO_2 is shown here for the first 48 hours. Y_s and Y are predicted values (dashed line) and observed values (solid line) respectively, while r denotes the prediction residuals.

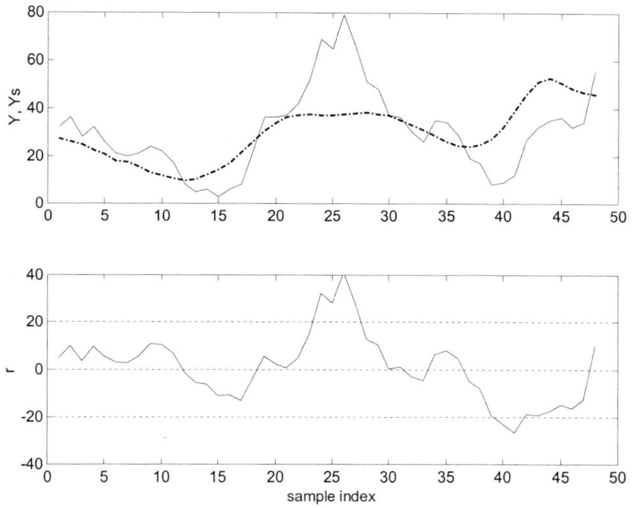

Figure 10.8. Validation of a MLP network model by simultaneous free-run prediction of NO_2, NO and E_s, using embedding strategy 1. The result for NO_2 is shown here for the second 48 hour period. Y_s and Y are predicted values (dashed line) and observed values (solid line) respectively, while r denotes the prediction residuals.

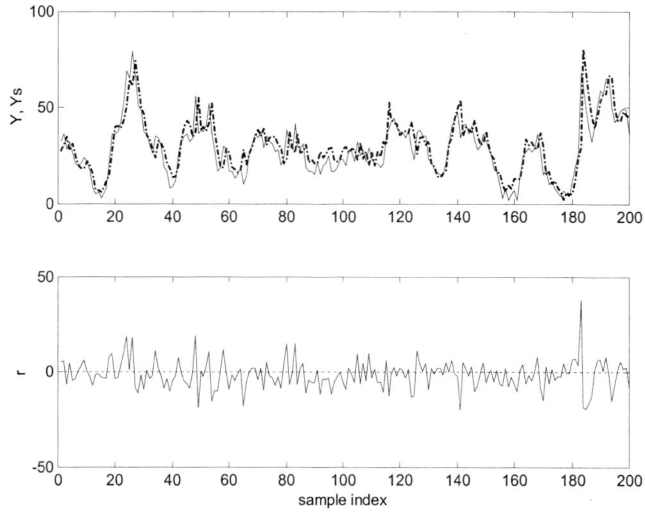

Figure 10.9. Validation of a MLP network model by simultaneous one-step prediction of NO_2, NO and E_s, using embedding strategy 1. The result for NO_2 is shown here for the first 200 hours. Y_s and Y are predicted values (dashed line) and observed values (solid line) respectively, while r denotes the prediction residuals.

Further diagnostics, such as the Hinich test (Hinich, 1982) indicated that the residuals had a Gaussian distribution and that the model had described all the inherent dynamics in the observations. Consequently, the independent components of $\mathbf{S}_2$ were tested for randomness. $S_{2,7}$, $S_{2,16}$ and $S_{2,17}$ were found to be Gaussian with more than 70% confidence. On the other hand, none of the components of the Λ_3 was Gaussian, which implied that the noise was mixed with the signals. This suggests that separation by independent component analysis after embedding led to separation of the noise components from the deterministic components of the time series. This implied that inclusion of explicit noise components in the input space had probably resulted in sub-optimal parameter estimation, compared with estimates when the noise was spread over all the input components, as was the case for $M_{3\theta}$. Data were not reconstructed without the noise components, but this could be a sensible future investigation.

Iterative free-run prediction was attempted up to 48 steps ahead. Afterwards, the input to the model $M_{2\theta}$ was reset with the embedding of Z_{NO2}, X_{NO} and X_{Es} at that index and another 48 steps were predicted. Results were poor, as could be expected from the short stability horizon, as shown in Figure 10.7. Consequently, no surrogate analysis was performed on the free-run prediction results. A more advanced model class and a suitable noise-reduction scheme could improve the results. The results for one-step prediction of Z_{NO2} by $M_{1\theta}$ over the first 200 hours of the validation data set appear in Figure 10.8. The mean prediction error normalized with the mean observation of Z_{NO2} was 0.2481. It was concluded from these results that the model was capable of sufficiently accurate one-step ahead predictions. The ability of the model to follow the peaks of the observed NO_2 concentration is particularly significant, since the maximum expected levels of NO_2 play an important role in environmental management and planning.

In summary, the results indicate that multidimensional embedding of the data in the presence of noise clearly afforded an advantage over univariate embeddings or strategies that did not involve embedding of the data at all.

Chapter 11
From Exploratory Data Analysis to Decision Support and Process Control

11.1. BACKGROUND

Although the exploratory analysis of plant data can provide valuable insights into process behaviour, which can be used to optimize plant operation, it is also an essential tool in the development of higher-level decision support and automatic control systems. As mentioned in Chapter 6, cluster analysis and subsequent classification may be used on historic plant data to develop a model for fault diagnosis or product classification. Likewise, although successful identification of the dynamic behaviour of a plant can yield useful understanding in its own right, the plant model is usually better used directly as the basis for automatic process control. In this chapter, the exploitation of process data in the development of decision support and control systems is considered briefly, by way of two examples, viz. the development of a expert system for the diagnosis of corrosion problems, as well as automatic control of a mill circuit with neural networks.

11.2. ANATOMY OF A KNOWLEDGE-BASED SYSTEM

As indicated in Figure 11.1, the two most important elements of a knowledge-based system are the knowledge base itself, as well as the inference engine. The knowledge base and inference engine are further supported by several subsystems, such as required for knowledge acquisition, interfacing with users, explanation of decisions and working memory. As was briefly discussed in Chapter 6, the facts and relationships in knowledge-based systems can be constructed and maintained with the aid of historic or other process data. This can be done by considering the decisions to be made as a classification problem, such as whether the plant is in control or not, or whether is specific action has to be taken, etc.

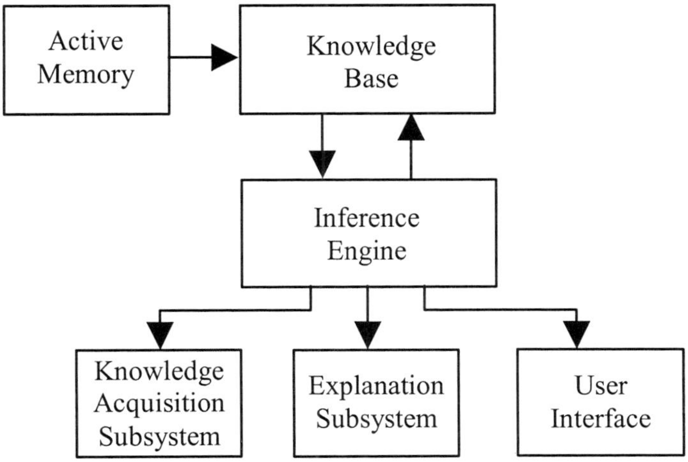

Figure 11.1. General structure of a knowledge base system.

11.2.1. Knowledge-base

The domain knowledge that is used to solve a problem can be represented in different ways. The specific mode of representation is usually determined by the type of knowledge available, as well as the inference strategies that will be used to navigate the knowledge base. The most important modes of representation comprize the use of IF-THEN rules, semantic networks, object-attribute-value triplets and frames.

Production systems are the simplest class of systems making use of rule-based approaches. IF-THEN or production rules are the most popular way of representing knowledge[57], and consist of premises (IF) and conclusions (THEN). Each premise can consist of one or more expressions or clauses connected by logical operators such as OR or AND, while the conclusions could also contain one or more clauses, e.g.

IF	the temperature is higher than 900 °C AND
	the ore is limestone AND
THEN	the residence time in the kiln is more than 3 hours
	calcination is complete

Small systems contain up to a few hundred such rules, while large systems typically contain a few thousand to tens of thousands of such rules. Some of the largest systems today cover around 180 000 facts concerning some 30 000 concepts.

One advantage of production systems is that the form and manner used to represent and manipulate knowledge is natural and plausible. Another advantage is the modularity of the system. This modularity resides throughout the system's architecture, e.g. in the structural independence of the rules of the system, and the separation of control and domain knowledge. This facilitates construction, debugging and maintenance of these systems.

By making use of classification trees, such as those considered in Chapter 4, rule-based representations of knowledge bases can be prototyped very rapidly. Neural networks are more important in this regard in the setting up of the classification problem in the first place, through cluster analysis, for example. Secondly, by using neural networks as discussed in Chapter 7 and elsewhere (Zurada and Cloete, 2000), more accurate or complex knowledge bases can be developed

11.2.2. Inference engine and search strategies

Two primary problems have to be addressed by the control portion of the inference engine. First, it has to decide on a starting point in the static knowledge base in order to initialize the reasoning process. Second, if conditions occur in which more than one production rule is ready to fire, the inference engine has to decide on the precedence of these alternative lines of reasoning.

The search strategies employed during navigation of the knowledge base can be categorized as either backward or forward chaining, each category of which can be further subdivided into breadth-first or depth-first searches, as shown in Figure 11.2. Backward chaining (or goal driven) strategies start off at the goals or conclusions in the knowledge base, and work their

[57] An estimated 80% of knowledge-based systems make use of IF-THEN rules.

way back to the initial conditions supplied by the user. These strategies are very efficient when the possible outcomes are known and reasonably small in number.

If the number of possible outcomes of a search is large or unknown, forward chaining strategies have to be used. In forward chaining (or data driven) systems, the premises of rules are examined to see whether they are true or not, given the information provided initially. If true, the conclusions of the rules are added to the known facts, and the rules are reexamined, until a conclusion or a dead-end is reached. Sophisticated systems usually accommodate both approaches in order to exploit the advantages of both strategies when needed.

Depth-first or breadth-first searches can be used in addition to backward and forward chaining strategies. In depth-first searches, the system searches for detail first and strives to reach subgoals first. In diagnosing the malfunctioning of a plant or operating unit for example, the system might ask whether the temperature had fluctuated during processing, what the frequency or duration of the fluctuations was, how severe they were, etc. Only after exhausting the possibilities associated with the particular topic will the system move on to the next issue.

In breadth-first searches the system sweeps through all the premises first, before going into more detail. Questions during the investigation of the malfunctioning of the plant might appear saltatory, e.g. were there any temperature fluctuations?, what was the pressure?, was the pH above a certain level?, etc. Only after jumping from one topic to another will the system return to the first and go into the next level of detail on all the topics, as illustrated in Figure 11.2.

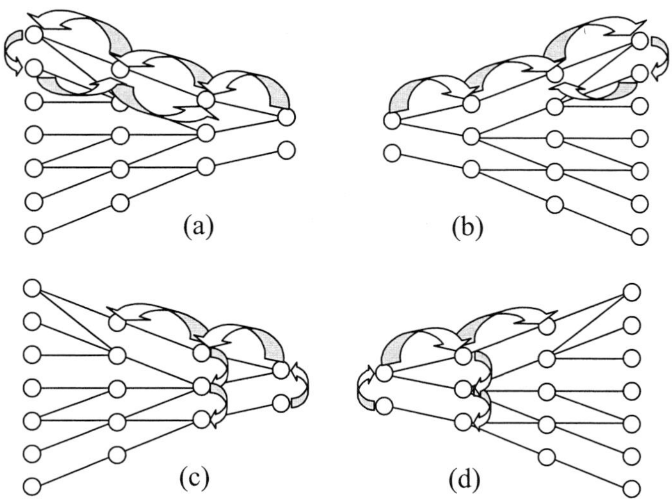

Figure 11.2. Search strategies used by inference engines, i.e. (a) depth-first backward chaining, (b) depth-first forward chaining, (c) breadth-first backward chaining and (d) breadth-first forward chaining.

Search strategies are evaluated in terms of four criteria, viz.

- *Completeness*: Guaranteeing that the search strategy will find a solution if there is one.
- *Time complexity*: The time taken to find the solution.
- *Space complexity*: The memory required to perform the search.
- *Optimality*: Finding the best solution when there are several different solutions.

If there is a solution, a breadth-first search is guaranteed to find it, i.e. it is complete. If there are multiple solutions, a breadth-first search will find the shallowest goal state first, which is optimal, provided that the path cost is a nondecreasing function of the depth of the node[58]. However, in terms of time and space complexity, breadth-first searches suffer serious drawbacks. Although the space and time complexities are both exponential, the space complexity poses a larger problem than the time complexity, so that breadth-first searches are generally good for small problems only.

In contrast, depth-first search has very modest memory requirements. It needs to store only a single path from the root node to a terminal node, along with the remaining unexpanded sibling nodes for each node on the path. This means that for a state space with a branching factor of b and a maximum depth of d_{max}, the depth-first strategy needs storage for bd_{max} nodes only, in contrast with storage requirement for b^d nodes in the case of a breadth-first search (where d is the depth of the shallowest goal). The disadvantages of the depth-first search strategies are that they are not optimal or complete.

Other search strategies include depth-limited search and iterative deepening search. In the case of the depth-limited search, the maximum depth of a path is restricted and it has time and space complexities similar to that of depth-first search strategies. The difficulty with depth-limited search is the choice of an appropriate depth limit and this is avoided by iterative deepening search, which tries all possible depth limits. It combines the benefits of depth- and breadth-first search strategies, i.e. it is optimal and complete and has modest time and space complexities. It is potentially wasteful, as so many search states are expanded multiple times, but in practice this overhead is usually small. This makes the iterative deepening search the preferred strategy when the search space is large and the depth of the solution is unknown.

11.2.3. Monotonic and non-monotonic reasoning

In monotonic reasoning all conclusions made during the reasoning process remain valid for the entire duration of the process, so that the amount of knowledge in the system increases monotonically. In non-monotonic systems conclusions identified previously as valid may be retracted during a later stage of the reasoning process. These systems are essential in the solution of planning or scheduling problems for example, where initial arrangements may have to be modified, as new information becomes available.

[58] A condition which is usually satisfied only when all the operators have the same cost.

11.3. DEVELOPMENT OF A DECISION SUPPORT SYSTEM FOR THE DIAGNOSIS OF CORROSION PROBLEMS

It is estimated that corrosion maintenance and prevention continues to consume between 2% to 4% of the gross domestic product of developed countries (Tretheway and Roberge, 1994), despite the fact that considerable knowledge is available, both with regard to fundamental aspects of corrosion and the practical application thereof. Salzberg and Watkins (1990) found that engineers often fail to use pertinent information, even if available within the company. The reasons put forward are that either they are unaware that the information is available or the cost of obtaining and translating it into a useful form is too high.

Although sophisticated and user-friendly expert systems have been developed to assist the engineer in materials selection and the understanding of corrosion (McMahon and Pitt, 1995; Smart et al., 1993), these systems are expensive and not used widely (Bamkin and Piearcey, 1990). The high cost of development can be attributed to the so-called knowledge acquisition bottleneck (Johannsen and Alty, 1991). Domain experts are often unable to formalize their domain knowledge in terms of explicit rules and exhibit cognitive biases when making decisions. An accurate representation of the expert's mental model of corrosion in a rule base is therefore difficult to achieve. This has meant that these expert systems have not proliferated, even though the advantages are clear.

By using methods to extract rules from data, such as the ANN-DT algorithm discussed in Chapter 7, or the methods discussed in chapter 6, a knowledge base can be constructed, which can then be used in a predefined expert system shell. However, for illustrative purposes, a simple classification tree was to develop a corrosion expert system, as detailed below.

11.3.1. Expert System

The expert system, developed with the assistance of the CLIPS expert system tool (Giarratano and Riley, 1994), has three primary components, viz. a *knowledge base*, *inference engine* and a *user interface*. The engine is forward-chaining and uses the Rete algorithm (Forgy, 1982) for efficient pattern matching. It makes inferences by deciding which rules in the knowledge base are satisfied by the user-supplied information and the satisfied rule with the highest priority is executed. The *user interface* allows the user to communicate with the expert system. It includes an explanation facility that shows the reasoning path followed by the expert system and a knowledge acquisition facility that allows the user to enter knowledge into the knowledge base, rather than by having a knowledge engineer explicitly code the information in the system.

The information used by the expert system is stored in two databases (Gouws and Aldrich, 1999). The *rule base* consists of the classification rules obtained through conversion of a decision tree or derived from a neural network, as well as supplementary rules supplied by domain experts. The latter includes fundamental corrosion aspects, as well as heuristic rules obtained through experience. Rules linking the most applicable corrosion case studies contained in the second database to the classification rules are also included. The *example database* contains a set of illustrative corrosion case studies, applicable to the domain under consideration. Each case study features a particular corrosion problem. Details such as visual appearance, material involved, corrosive environment (chemical, temperature, etc.), type of corrosion, time to failure and a suggested remedy for the specific problem are provided.

The performance of the system is illustrated by means of three examples, based on data obtained from the literature. In these examples, the corrosion resistance of common materials of construction, miscellaneous metals and stainless steels in seawater is highlighted. Table 11.1 summarizes the most important features of these corrosion data sets.

Table 11.1. Data set statistics.

Attributes	Construction materials	Miscellaneous metals	Stainless steels in seawater
No. of exemplars	1261	8820	487
No. of features	6	4	8
No. of classes	3	3	3
% exemplars in each class	32, 27, 41	7, 63, 30	18, 22, 60

11.3.2. Examples

a) Example 1: Corrosion of construction materials

The classification problem for the first set of data, detailing the corrosion resistance of common construction materials, was to find the decision tree, and in turn the heuristic set of DNF rules, that would be best able to predict the corrosion rate of a construction material, given a set of applicable features. The data were obtained from Perry and Green (1984). 35 common industrial construction materials, ranging from polyethylene to aluminium, were each contacted with 8 different industrial acids, including acetic acid and hydrochloric acid, at various temperatures (–20°C to 50°C) and concentrations (0 to 100%). *MaterialType* and *ChemicalType* features were added to the original data set, explicitly stating whether the particular material was made of metal or not, and whether the chemical was organic or not. The corrosion rate was discretized into three classes, labelled *Satisfactory* (< 0.51 mm p.a.), *Intermediate* (0.51 mm p.a. – 1.27 mm p.a.) and *Unsatisfactory* (> 1.27 mm p.a.). Both temperature and concentration were treated as continuous features.

b) Example 2: Miscellaneous metal corrosion

The second example is based on a data set (Sinnott, 1993), which provides information on the performance of various metals in a large number of chemical environments. 21 metals and 138 industrial and household chemicals, including beer, sulphuric acid and yeast were considered. To aid the induction algorithm in determining the optimum decision tree, a further discrete *ChemicalType* feature was included in the original set of exemplars, explicitly distinguishing between organic and inorganic chemicals. A temperature feature indicated the maximum temperature at which the metal remained resistant to corrosion. The suitability of the metals was classed as R, resistant to corrosion up to the temperature indicated, ND, where no data were available, and U, where the metal-chemical combination is unsuitable. C4.5 was used to find the optimum relationships or rules, linking the corrosion resistance of a given metal to the abovementioned features.

c) Example 3: Seawater corrosion of stainless steels

The data set used in the third example (Schumacher, 1979) concerns the corrosion rate performance of 18 stainless steel alloys in seawater, exposed over various periods at different depths under the sea surface. Together with these two features, the environment (with values

W indicating total exposure to seawater; and S indicating embedding of lower portions of the specimen in bottom sediments), maximum corrosion pit depth, crevice depth, maximum tunnel length and type of corrosion (15 possible types). All, except the environment- and corrosion-type features were continuous properties. The corrosion rate was split into three discrete groups, viz. *Low* (0.0 mm p.a. - 0.05 mm p.a.), *Intermediate* (0.05 mm p.a. - 0.6 mm p.a.) and *High* (> 0.6 mm p.a). An induction algorithm was used to infer the decision tree that would be best able to classify corrosion rate performance in terms of environmental conditions, type of corrosion, corrosion effects, etc.

11.3.3. Experiments and results

The average classification accuracy on the test data, both before and after tree pruning, as well as the accuracy predicted by the C4.5 algorithm for unseen exemplars, are compared in Figure 11.3. The complexity of the decision trees, measured in terms of the number of nodes (before and after pruning) of each decision tree is compared in Figure 11.4.

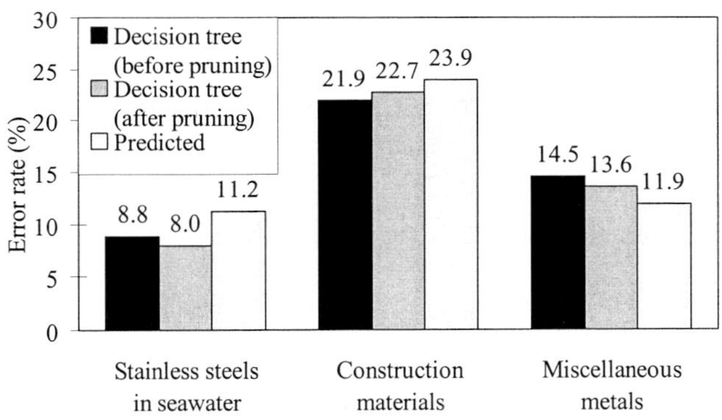

Figure 11.3. Performance of the classification tree models.

Although no direct quantitative comparison can be made to the classification accuracy that a domain expert would achieve on these particular corrosion classification problems, numerous studies have shown that machine learning techniques, such as inductive learning, are able to provide comparable, and in most cases better results than those achieved with explicit knowledge acquisition techniques (Ben-David and Mandel, 1995; Quinlan et al., 1987; Shaw and Gentry, 1990; Silverman, 1994; Smets and Bogaerts, 1992).

From Figure 11.4 it is clear that C4.5 is able to produce relatively small decision trees, especially after pruning. Fairly low error rates were attained for both the seawater corrosion and miscellaneous metals corrosion problems. Good generalization was achieved for these two problems, as the predicted error rates (11.2% and 11.9% respectively) for unseen exemplars were very similar to those obtained using the given data sets. Optimum results were obtained when C4.5 was allowed to group feature values, allowing the underlying patterns in the data to be expressed in the decision trees.

The comparatively poor performance in terms of tree size on the construction material and miscellaneous metal data sets is attributable to the *problem of small disjuncts* (Holte et al., 1989). Terminal nodes in a decision tree that are only able to cover a small number of training exemplars tend to have a much higher error rate than those that cover a large number of exemplars. In both data sets, on average less than five exemplars were available for each material-chemical combination. As a result C4.5 was forced to create a large number of terminal nodes, each of which covered only a few exemplars, in order to cover the entire training set. This is substantiated by the significant decrease in the number of tree nodes after pruning has occurred. To remedy this problem, more comprehensive data about the same number of materials and chemicals should be supplied to the induction algorithm during training.

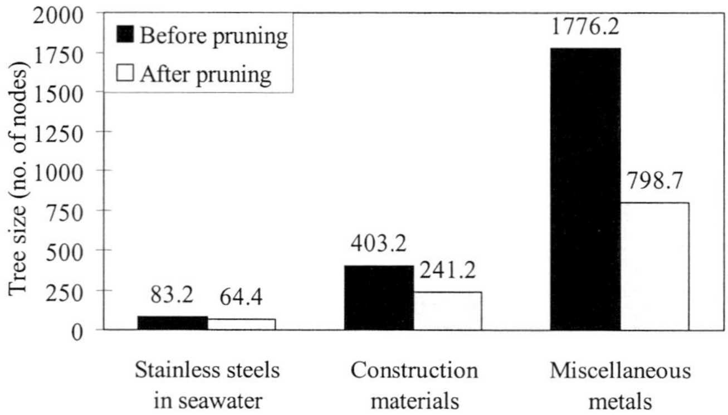

Figure 11.4. Complexity of classification trees before and after pruning.

Both Silverman (1994), and Smets and Bogaerts (1992) used neural networks to extract information from case histories to aid corrosion performance prediction. Although good performance results were obtained in both studies, the neural networks were unable to explicitly relate the information gleaned from the case histories to the user. The neural networks were treated as so-called black box models, packaged in an expert system to aid user friendliness.

11.4. ADVANCED PROCESS CONTROL WITH NEURAL NETWORKS

In the last few decades, the development and applications of nonlinear control methodologies have attracted a growing interest from academia and industry alike. One of the chief barriers to the more widespread use of advanced modelling and control techniques in the process industries is the cost associated with the development and validation of models. It is not uncommon for modelling costs to account for more than 75% of the overall cost of the design of an advanced controller.

Since the use of neural networks provides a potentially cost effective approach to the construction of accurate process models, their application in process control systems has been investigated extensively to date. Despite these efforts, neurocontrol systems have not been as successful as other applications of neural networks in the process industries yet, such as fault diagnosis, data validation or the interpretation of sensor data. The reason for this is not so much the failure of neural networks with regard to system identification and modelling, but has more to do with the maintenance of these models in the face of continuously changing plant conditions. This is especially the case in the mineral process industries, where the ore type and quality has a major effect on process operations and can vary substantially from day to day.

The majority of neural networks used in control applications are multilayer perceptrons and radial basis function neural networks and to a lesser extent recurrent or learning vector quantization neural networks. Three major categories of control applications can be distinguished, viz. predictive control, inverse model-based control and adaptive control.

11.4.1 Predictive neurocontrol schemes

Predictive control schemes are the most commonly used and in these schemes the controller determines a profile of manipulated variables that optimize some open loop performance criterion on a time interval, i.e. from the current time up to a certain prediction horizon.

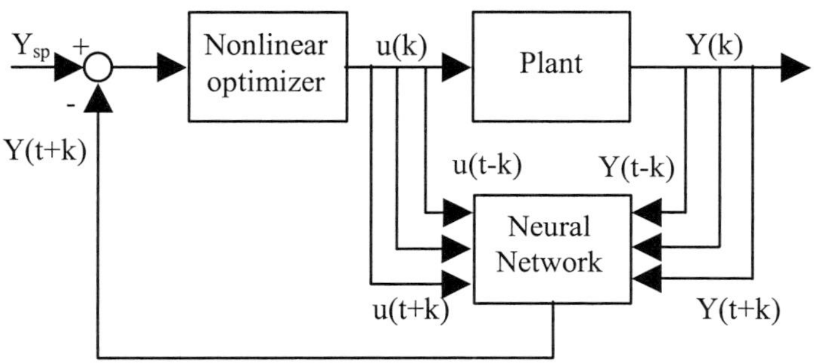

Figure 11.5. Structure of a predictive neurocontroller.

In nonlinear model predictive control systems, the model, performance criterion and constraints are nonlinear functions of the system variables. In neurocontrollers, neural networks are used to represent these functions, as indicated in Figure 11.5. In model predictive control algorithms, future output deviations from the set point are minimized, while taking into account the necessary control sequence to achieve this objective. The (neural network) model predicts the outputs for some k time steps ahead, y(t+k), while a cost function J is used to obtain the optimal sequence of control actions, i.e.

$$J = \sum_{k=N(1)}^{N(2)}[e(t+k)]^2 + \sum_{i=1}^{N(3)}[p_i\delta u(t+i)]^2 \quad (11.1)$$

where N(1) and N(2) define the prediction horizon, N(3) defines the control horizon, e is the error between the predicted and the target output, p_i is the i'th weighting factor and δu_i is the i'th change in the control sequence. The first term in equation 11.1 is a measure of the distance between the desired future trajectory and the trajectory predicted by the neural network, while the second term penalizes excessive movement of the manipulated variables. This approach is very similar to the classical optimal control strategy, except that a forward neural network model is used in the equality constraint equations, instead of the normal plant model.

11.4.2. Inverse model-based neurocontrol

Neural networks can be used in direct inverse and internal model control schemes, as indicated in Figure 11.6. In the former, the inverse model is cascaded with the plant and controls it without any feedback. The neurocontroller is trained to specify the appropriate control parameters, given the desired targets (set points) as inputs, as well as past plant inputs and outputs. The set points constitute the desired targets at the output layer.

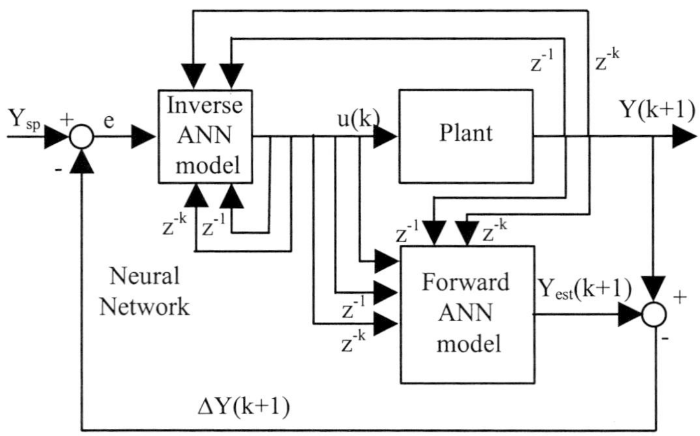

Figure 11.6. Neural networks in an internal model control strategy.

Although nonlinear internal model control is similar to direct inverse control, it is more robust and stable. It differs from the direct inverse approach in that it has a forward neural network model placed in parallel with the plant to account for model mismatches. In addition, the error between the plant output and the forward neural network model is subtracted from the set point before being presented to the inverse model.

In most cases the necessary control signals from the neural network inverse model is computed by numerically inverting the forward neural network model at each interval by Newton's method or substitution methods based on the contraction mapping theorem. The computational cost associated with these numerical methods is high and the results can be influenced considerably by initial estimates.

11.4.3. Adaptive neurocontrol systems

In direct adaptive control schemes, no explicit attempt is made to model the plant. Instead, controller parameters are adjusted directly on-line to track and stabilize the closed loop system. With neural networks, the weights of the neurocontrollers are adjusted on-line to control the plant by minimizing some appropriate cost function of the plant output and the desired response. This is similar to the direct inverse modelling control scheme, except that controller adjustment takes place on-line based on a model inference signal.

11.5. SYMBIOTIC ADAPTIVE NEURO-EVOLUTION (SANE)

For operation in complex nonlinear environments it is desirable to design controllers that operate with as little dependence on human interaction as possible (Gupta and Rao, 1994).

Although evolutionary methods are becoming increasingly popular as a means to develop such control schemes for complicated systems, the canonical approach is not particularly suitable for real-world problems (Grefenstette et al., 1990; Whitley et al., 1993; Moriarty and Miikkulainen, 1996). For example, assessment of solutions (fitness functions) may not be readily available. Moreover, eventual convergence of the population to a single best solution hinders adaptation in dynamic environments.

These problems can be avoided by use of *cooperative* evolutionary algorithms (Horn et al., 1994) used in conjunction with reinforcement learning. As indicated in the Figure 11.7(a), in conventional evolutionary algorithms, each individual codes a complete solution to the problem (neurocontroller). In contrast, in cooperative evolutionary algorithms, each individual represents a partial solution to the problem, so that a complete solution can only be obtained by appropriate groupings of individuals, as indicated in Figure 11.7(b). In the case of a neurocontroller, each individual would code a node in the neural network, rather than a set of nodes completely defining the network. Since the individuals have to cooperate to present a solution to the problem at hand, convergence is not an issue and therefore the algorithm can deal with changes in the environment in a natural way.

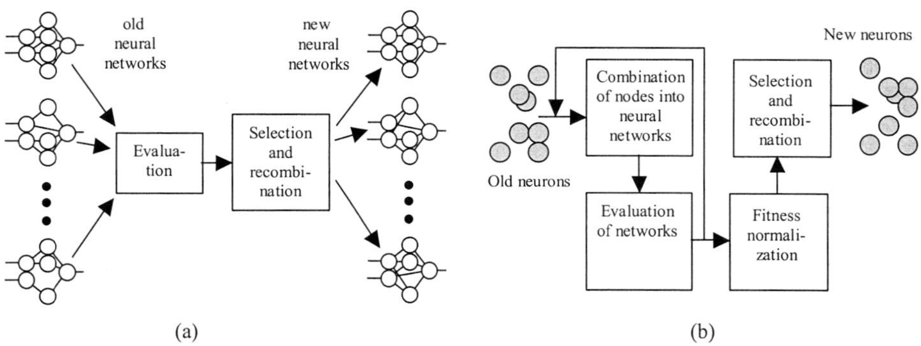

Figure 11.7(a) Conventional evolution of neurocontrollers as opposed to (b) neurocontroller evolution performed by SANE.

Reinforcement learning involves a search for a particular controller structure that executes an appropriate set of actions that yield the highest possible reward (Sutton and Barto, 1998). Its approach is different from other (supervised) learning techniques in that the emphasis is on learning from direct interaction with the environment, *without* exemplary supervision or even complete models of the environment. A learning process is typically initialized to a randomly structured controller that is unfamiliar with which behavioural pattern will lead to the successful completion of the proposed control task.

As indicated in Figure 11.8(a), a learning controller interacts with a discrete dynamic system in an iterative fashion. Initially the system may be at an arbitrary state, s_t. The controller perceives a partial or full state of the environment and selects a control action a_t based on the available environmental information. The discrete dynamic environment subsequently attains a new state, s_{t+1}, as a result of the control action. A reward r_t is assigned to the controller based on the contribution of the new state in reaching the final goal or completing the given task. The input signals from the environment do not explicitly direct the controller towards attaining a given control task. Learning progresses purely on how much reward can be amassed in a given learning trial. Learning may therefore require experimenting with behavioural patterns that occasionally produce unacceptable results in a real world environment. Simulated systems are thus often used in lieu of real world systems for learning control strategies (Grefenstette et al., 1990).

Evolutionary reinforcement learning entails searching a population of possible controller structures (behavioural patterns or policies) in order to find, for example, a neurocontroller structure that encompasses effective control actions in the environment. Neurocontrollers are comprised of collections of neurons, with each neuron specifying connections and their associated weights to the input layer (environmental states) and output layer (control actions). In the genetic algorithm framework, effective neurocontrollers are allowed to produce offspring, which promotes the propagation of effective neurons (genetic material) in the population. Moriarty and Miikkulainen (1998) developed a state of the art genetic algorithm, viz. symbiotic adaptive neuro-evolution (SANE), that uses implicit fitness sharing to ensure genetic diversity in the controller population, while allowing for an aggressive search in the solution space. A robust search for the global optimum is thus ensured.

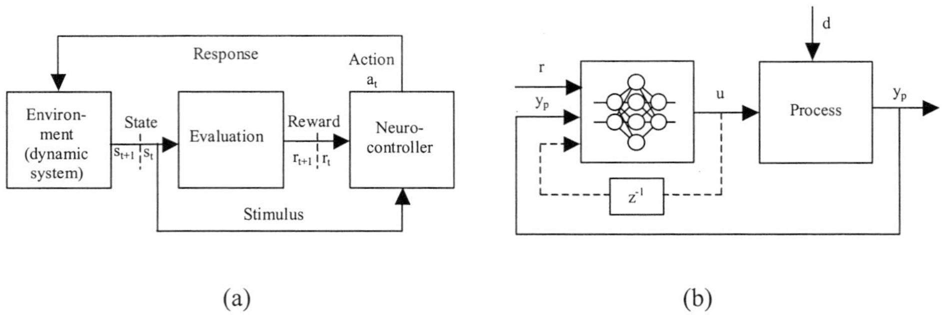

Figure 11.8(a) Controller-environment interaction and (b) closed loop architecture of the neuro-controller.

In the SANE framework, the neurocontroller is implemented in a closed loop as illustrated in Figure 11.8(b). The development of neurocontrollers for mineral processing applications using SANE for nonlinear controller design, is demonstrated for a grinding circuit case study, as discussed in more detail below.

11.6. CASE STUDY: NEUROCONTROL OF A BALL MILL GRINDING CIRCUIT

Ball mill grinding circuits are important unit operations in the production of metals from ore, as it precedes concentration operations such as flotation, gravity or dense media separation, etc. For effective concentration or subsequent mineral liberation, grinding circuits have to provide optimal particle size distributions. Another operational concern lies in the energy inefficiency of ore size reduction. With less than 10% of the electrical power input typically contributing to grinding of the ore, the total operating cost of a grinding facility may constitute a significant fraction of the overall cost of mineral processing. It is therefore essential that grinding circuits be subject to effective control in which both an optimal particle size distribution is provided to downstream operations and optimal utilisation of electrical energy is assured (Rajamani and Herbst, 1991).

The majority of grinding circuits employ the principle of a closed loop consisting of a mill and a classifier (hydrocyclone). Ore is partly milled and fed to the classifier where the finer material is split off and the coarser material is recycled to the mill for further milling. In this case, the slurry sump, which accepts the mill discharge, acts as a buffer to fluctuations in the incoming flow from the mill.

Disturbances occur frequently during grinding circuit operation, viz. changes in ore hardness, ore feed rate and feed particle size. Variations in ore hardness and feed size may reduce the mill throughput drastically by causing the mass flow rate of the hydrocyclone overflow and particle size distribution to experience continuous perturbations. In order to counteract the effects of these disturbances, a basic circuit usually has a minimum of two manipulated variables, viz. the fresh solids feed rate to the circuit and the dilution water to the sump. The most common process-manipulated variable pairings are illustrated in Table 11.1 (Rajamani and Herbst, 1991).

Table 11.2. Common Single-Input-Single-Output (SISO) control loop pairings.

Pairing	*Controlled variable*	*Manipulated variable*
I	Particle size hydrocyclone overflow	Sump water dilution rate
	Hydrocyclone feed rate	Fresh solids feed (and dilution rate)
II	Particle size hydrocyclone overflow	Fresh solids feed (and dilution rate)
	Hydrocyclone feed rate	Sump water dilution rate

A high degree of controller interaction between the control pairings in Table 11.1 is evident. For pairing I, the dynamic response to an upward step in the product size set point illustrates this interaction. The positive set point change initially causes the first control loop to increase the sump dilution rate. An increase in the dilution rate to the sump causes an increase in the flow rate and a decrease in the solids concentration of the feed to the cyclone, both of which have the effect of causing an initially finer classification. Although resulting in a finer classification, an increase in the cyclone feed rate at an approximately constant solids concentration, implies that a larger portion of the solids report to the underflow. The consequently larger mass flow rate of coarse material to the mill reduces the mill's ability to grind the material as fine as before the step change. Hence the discharge from the mill slowly becomes coarser. Once this coarser particle distribution reaches the cyclone, the cyclone overflow particle size distribution becomes coarser and the particle size distribution to the overflow may return to a similar state as to before the step change in the dilution rate. This dynamic response is illustrated in Figure 11.9 in the open loop. Although the product specification initially increases, it more or less returns to the initial value with a step change in sump dilution rate (Barker and Hulbert, 1983).

The increase in the hydrocyclone feed rate initiated by the first control loop causes the second control loop in pairing I to decrease the fresh feed to the mill (reducing the mill production rate). This is primarily a response to the mill's inability to cope with the large amount of coarse particles in the mill. Therefore the long-term change in product size (Figure 11.10) is largely the result of a change in the fresh solids feed rate, but this arrangement (pairing I) results in an upset to the cyclone in order to obtain the change. Continuous interaction between the two control loops may lead to prolonged upsets to the product specification, before the new set point is finally reached. Pairing II similarly results in significant controller interaction (Barker and Hulbert, 1983).

The time constant of the hydrocyclone classification is negligible compared to any of the other time constants in the grinding circuit, necessitating precise hydrocyclone control to prevent short-term fluctuations from affecting the product unnecessarily. The sump discharge flow rate should thus be as steady as possible, because the feed flow to the cyclone has a significant effect on the behaviour of the cyclone, and thus on the behaviour of the rest of the circuit. As particle size instrumentation is prone to time delay, any form of control based on the measurement of particle size is prone to be ineffective with regard to high frequency disturbances to the hydrocyclone. Many control strategies have regarded the sump level as a separate entity from the mill control, in that the sump level is controlled separately by a single-input-single-output (SISO) controller (to ensure that the remainder of the circuit is open loop stable). High performance sump level control, which allows the sump level control to contribute to the control objectives of the entire circuit, may ensure a steady flow to the cyclone and reduce the likelihood of short-term fluctuations in the product size distribution. As a minimum requirement, a decoupled SISO arrangement or a multi-input-multi-output (MIMO) controller design is thus required to eliminate unwanted controller interaction (Barker and Hulbert, 1983).

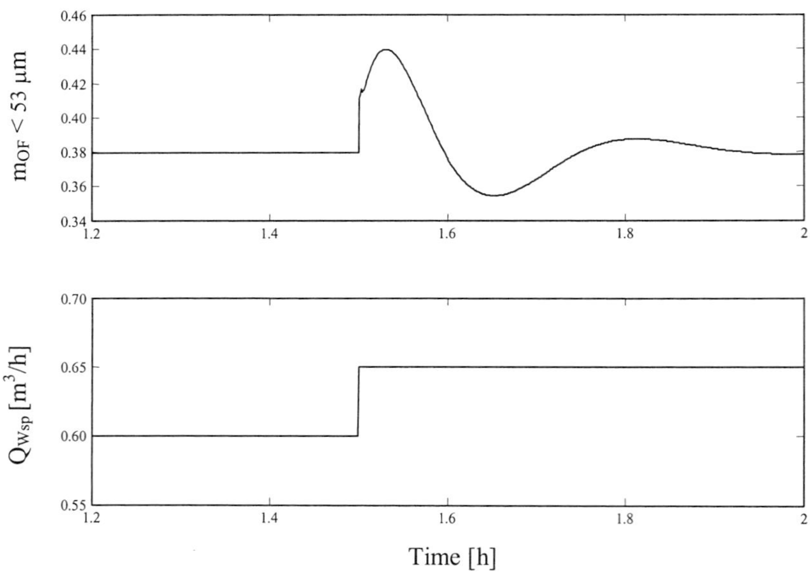

Figure 11.9. Particle size response (m_{OF}, top) in hydrocyclone overflow in response to an open loop step change in dilution rate (Q_{Wsp}, bottom).

In view of the severe interaction between controlled and manipulated variables (Table 11.1), better control may be achieved by incorporating mill rotation speed into the control system. Mill speed has been proposed as a less interactive manipulated variable and may be considered ideal as it directly affects the grinding kinetics of the mill and therefore should eliminate the interactions caused by control through flow rate manipulation. The basic premise is that a build-up of coarse material in the mill should result in an increase in mill speed. Large variable speed drive motors would allow for the continuous adjustment in mill speed.

Although grinding circuits exhibit nonlinear dynamic behaviour, controller design has largely been investigated from a linear controller perspective (i.e. PID control). Linear controller design techniques (modern and classical) require the use of linear process models. As rigorous grinding circuit models are necessarily nonlinear, these nonlinear model need to be linearized and control limited to this linearized region of the state space. The robustness of the linear controller cannot be guaranteed, should operation stray from the linearized region. In contrast, the use of nonlinear controllers offers substantial opportunities for operation over a wider region of the state space.

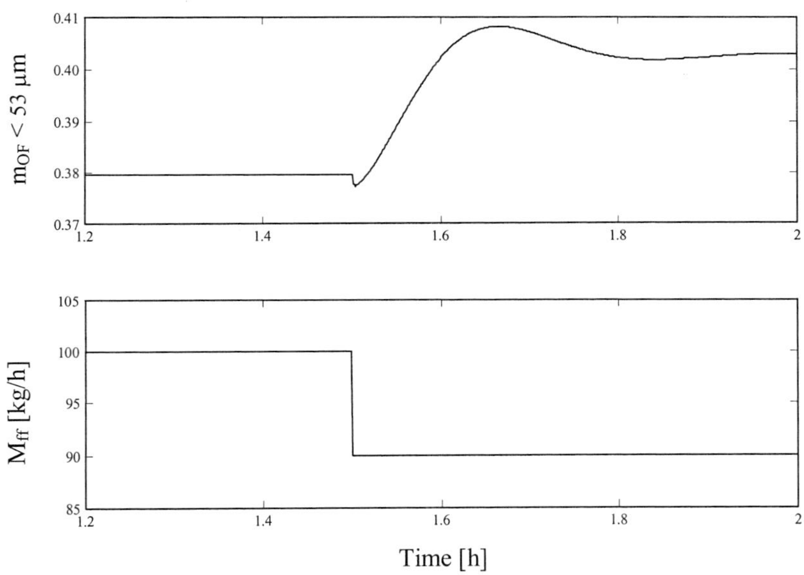

Figure 11.10. Particle size response (m_{OF}, top) in the hydrocyclone overflow, after an open loop step change in fresh solids feed rate (M_{FF}, bottom).

In this case study the potential of evolutionary reinforcement learning, in particular the SANE algorithm (Moriarty and Miikkulainen, 1998), for the development of nonlinear (neuro)-controllers for ball mill grinding circuits is demonstrated. The high performance of the neurocontroller is demonstrated for disturbances in the feed particle size distributions. The novel approach to the design of plant wide nonlinear controllers can incorporate all process and manipulated variables into an optimal control law (neural network), which effectively deals with the control challenges posed by ball mill grinding circuits. The flow diagram of the grinding circuit is shown in Figure 11.11.

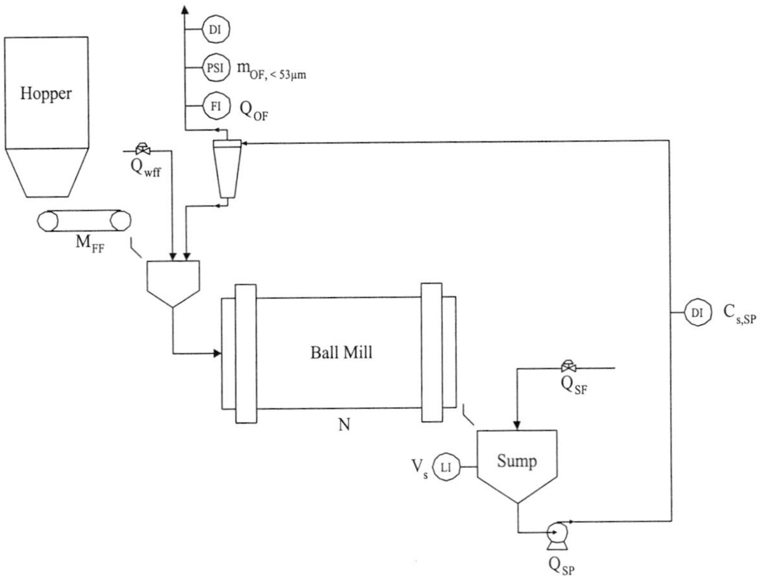

Figure 11.11. Grinding circuit flow diagram as modelled by Rajamani and Herbst (1991).

11.7. NEUROCONTROLLER DEVELOPMENT AND PERFORMANCE

11.7.1. SANE implementation

The SANE algorithm (Moriarty and Miikkulainen, 1998) was used to develop a feedforward neurocontroller with 6 input nodes (state variables), 12 hidden nodes and 5 output nodes (manipulated variables). The variables associated with the 6 inputs nodes and the 5 output nodes are listed in Table 11.3.

Table 11.3. Description of neurocontroller inputs and outputs.

Input nodes	*Output nodes*
Product mass fraction < 53 μm	Solids fresh feed mass flow rate
Product mass fraction < 53 μm set point	Fresh feed water dilution addition rate
Sump volume	Sump water dilution addition rate
Product volumetric flow rate	Sump volumetric discharge rate
Sump solids concentration	Fraction of mill critical speed
Product solids concentration	

SANE was required to minimize the fitness function in equation 11.19 (maximize reward). A neurocontroller is thus rewarded based on the effectiveness with which it tracks the set point, while maintaining the highest possible production rate.

$$\text{Fitness} = |f_{OF,53\,\mu m} - f_{setpoint}|\cdot|1 - C_{s,OF}\cdot Q_{OF}/1200| \qquad (11.19)$$

Table 11.4. Operating ranges of manipulated variables.

Manipulated variable (Output nodes)	*Operating range constraint*	
Solids fresh feed mass flow rate	90 - 136	[$kg \cdot h^{-1}$]
Fresh feed water dilution addition rate	0.1 - 2.0	[$m^3 \cdot h^{-1}$]
Sump water dilution addition rate	0.1 - 3.0	[$m^3 \cdot h^{-1}$]
Sump volumetric discharge rate	0.1 - 2.0	[$m^3 \cdot h^{-1}$]
Fraction of mill critical speed	0.3 - 0.8	[-]

During each learning trial the state variables are initialized with a Gaussian distribution around the mean values listed in Table 11.4. Initialization serves as a starting point for the dynamic optimization conducted via the SANE algorithm.

11.7.2. Set point changes

With the reinforcement learning methodology, learning occurs without explicitly showing the neurocontroller *how* the task is to be performed. Instead, the reinforcement learning framework is concerned with *what* is to be accomplished, allowing the learning algorithm (SANE) to explore the simulated environment and to discover implicitly *how* the task should be performed.

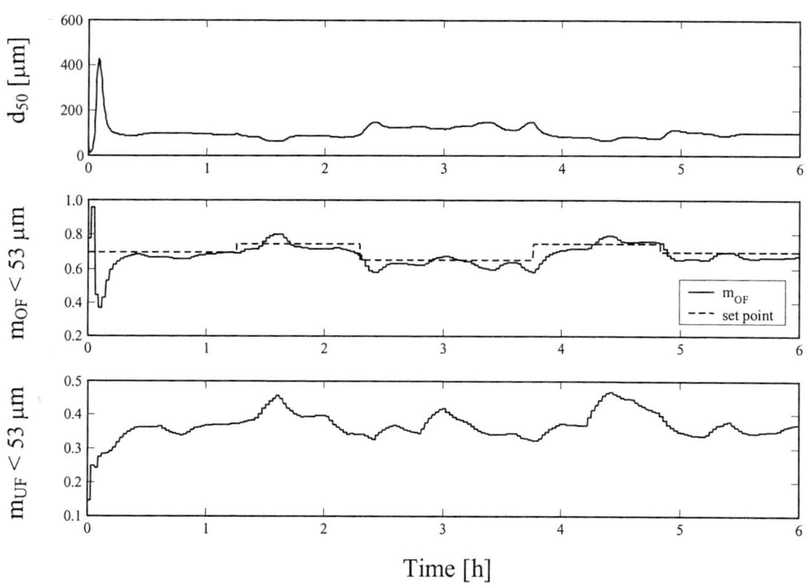

Figure 11.12. Set point change responses of the hydrocyclone process variables in the presence of particle size disturbances in the fresh solids feed.

No prior analysis of the simulated environment, for example, possible controller pairings or likely control strategies were provided to SANE. The learned behaviour is purely an indication of what could be gauged from cause-effect interactions with the simulated environment.

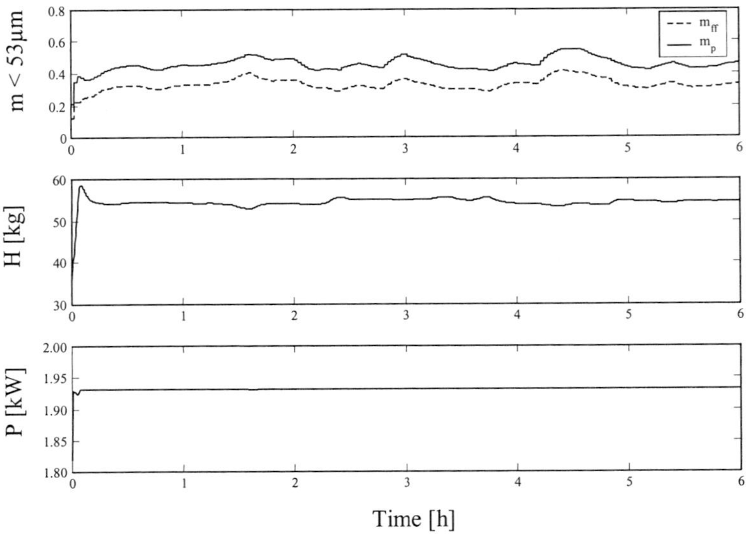

Figure 11.13. Set point change responses of the ball mill process variables in the presence of particle size disturbances in the fresh solids feed.

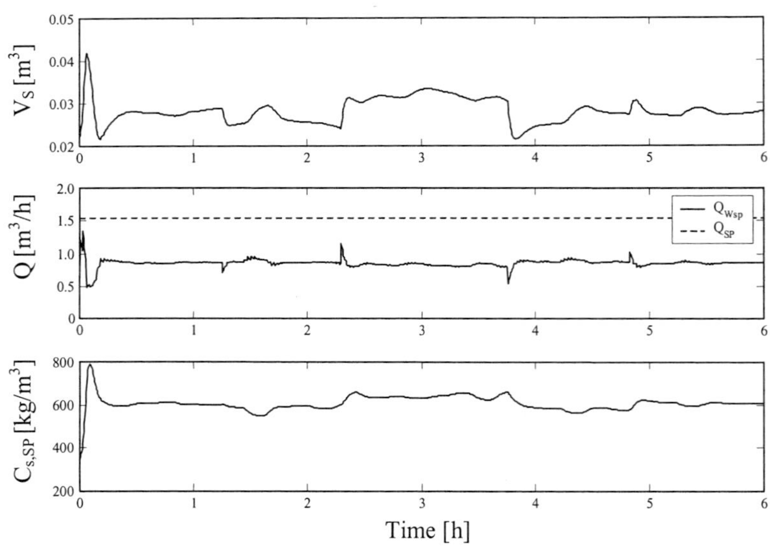

Figure 11.14. Set point change responses of the sump process variables and the associated control actions (with particle size disturbances in fresh solids feed).

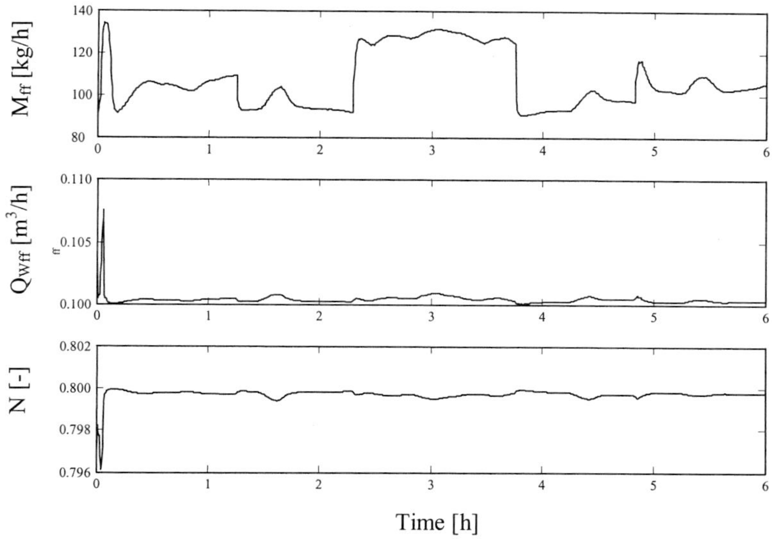

Figure 11.15. Control actions of the manipulated variables of the ball mill based on the observed inputs to the neurocontroller (with particle size disturbances in fresh solids feed).

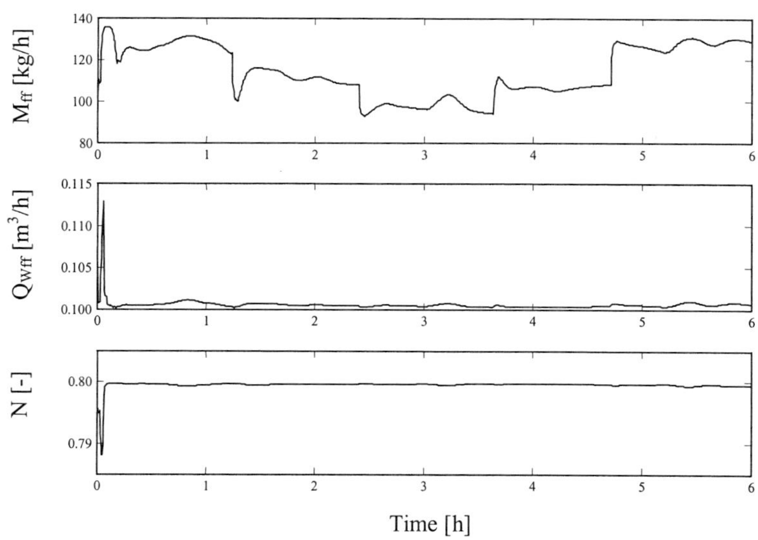

Figure 11.16. Control actions of the manipulated variables of the ball mill based on the observed inputs to the neurocontroller (with particle size disturbances in fresh solids feed).

Figure 11.12 (hydrocyclone), Figure 11.13 (ball mill) and Figure 11.14 (sump) illustrate the setpoint changes initiated by the neurocontroller in the presence of disturbances in the feed particle size. As can be seen from the top graph in Figure 11.12, the disturbance is introduced as a twofold increase in the mass fractions of the four largest size intervals in the fresh mill feed, after which it changes randomly every 12 minutes.

11.7.3. Particle size disturbances in the feed

Figures 11.15 and 11.16 show the control actions of the manipulated variables of the ball mill based on the observed inputs to the neurocontroller with particle size disturbances in fresh solids feed. It is clear that the controller has learnt to manipulate the fresh solids feed rate more than the other manipulated variables.

11.8. CONCLUSIONS

This, and other studies have indicated that evolutionary reinforcement learning offers significant opportunities for the development of nonlinear controllers (neurocontrollers) for mineral processing plants. For the ball mill grinding circuit case study the SANE algorithm was able to implicitly learn to eliminate controller interactions. The robust and highly autonomous control provided by the neurocontroller was demonstrated for particle size disturbances in the fresh mill feed. Similar results were obtained for ore hardness variation (Conradie and Aldrich, 2001), which suggest that the SANE algorithm is a promising approach to the generation of robust neurocontrollers.

References

Abarbanel, H. D. I. 1994. *Nonlinearity and Chaos in Engineering Dynamics*, Wiley and Sons.

Abarbanel, H. D. I. 1996. *Analysis of Observed Chaotic Data*, Springer-Verlag, New York, NY, USA.

Abarbanel, H. D. I., Brown, R. and Kennel, M.B. 1991. Lyapunov exponents in chaotic systems: Their importance and their evaluation using observed data. *International Journal of Modern Physics B*, 5, 1347-1375.

Abarbanel, H. D. I., Gills, Z., Liu, C. and Roy, R. 1996. *Physical Review A*, 53, 440.

Addison, P. S. 1997. *Fractals and Chaos*. Institute of Physics Publishing, Bristol and Philadelphia, USA.

Akaike, H. 1970. Statistical predictor identification. *Annals of the Institute for Statistical Mathematics*, 22, 203-217.

Akaike, H. 1973. Information theory and an extension of the maximum likelihood principle. In *Proceedings of the Second International Symposium on Information Theory*, 267-281, Akademia Kiado, Budapest, Hungary.

Akaike, H. 1974. A new look at the statistical model identification. *IEEE Transactions on Automatic Control*, AC-19(6), 716-723.

Akaike, H. 1977. On entropy maximization principle. In: Krishnaiah, P. (ed.). *Applications of Statistics*, 27-41, North-Holland Publishing Company.

Aldrich C. and van Deventer J. S. J. 1995. Modelling of Induced Aeration in Turbine Aerators by Use of Radial Basis Function Networks, *The Canadian Journal of Chemical Engineering*, 73, 808-816.

Aldrich, C. 2000. Cluster analysis of mineral process data with autoassociative neural networks. *Chemical Engineering Communications,* 177, 121-137.

Aldrich, C. and Reuter, M.A. 1999. Monitoring of metallurgical reactors by the use of topographic mapping of process data. *Minerals Engineering*, 12(11), 1301-1312.

Aldrich, C., Moolman, D. W., Eksteen, J. J. and Van Deventer, J. S. J. 1995. Characterization of flotation processes with self-organizing neural nets. *Chemical Engineering Communications*, 139, 25-39.

Alecsic, Z. 1991. Estimating the embedding dimension. *Physica D*, 52, 362-368.

Alpern, B. and Carter, L. 1991. Hyperbox. In: Nielson, G.M. and Rosenblum, L.R. (eds). *Proceedings of IEEE Visualization '91*, 133-139, San Diego, CA, USA.

Amari, S. I. 1977. Neural theory of association and concept formation. *Biological Cybernetics*, 26, 175-185.

Amari, S. I. 1990. Mathematical foundations of neurocomputing. *Proceedings of the IEEE*, 78(9), 1443-1463.

Amirthalingam, R. and Lee, J. H. 1997. Subspace identification based inferential control of a continuous pulp digester. *Computers and Chemical Engineering*, 21, Supplement, S1143-S1148.

Anderberg, M. R. 1973. *Cluster Analysis for Applications*, Academic Press, Inc., New York, NY, USA.

Anderson, E. 1954. Efficient and inefficient methods of measuring specific differences. In: Kempthorne, O. (ed) *Statistics and Mathematics in Biology*, 98-107, Ames, Iowa State College Press, USA.

Anderson, E. 1957. A semi-graphocal method for the analysis of complex problems. *Proceedings of the National Academy of Sciences, USA*, 43, 923-927 (Reprinted in Technometrics, 1960, 2, 387-392).

Anderson, J. A., Silverstein, J. W., Ritz, S. A. and Jones, R. S. 1977. Distinctive features, categorical perception, and probability learning: Some applications of a neural model. *Psychological Review*, 84, 413-451.

Andrews, D. F. 1972. Plots of high-dimensional data. *Biometrics*, 28, 125-136.

Andrews, R., Diederich, J. and Tickle, A. B. 1995. Survey and critique of techniques for extracting rules from trained artificial neural networks. *Knowledge-Based Systems*, 8(6), 373-383.

Angeline, P. J. 1997. Evolving basis function with dynamic receptive fields. *Proceedings of the 1997 IEEE International Conference on Systems, Man, and Cybernetics*, 5, 4109-4114, IEEE Press, Piscataway, NJ, USA.

Anon. 1963. Acoustic controller for improving grinding mill output. *Measurement and Control*, 2(11), 444-446.

Antonisse, J. 1989. A new interpretation of schema notation that overturns the binary encoding constraint, 86-91. In: Schaffer, J. D. (ed), *Proceedings of the 3rd International Conference on Genetic Algorithms and their Applications*, Morgan Kaufmann, San Mateo, CA, USA.

Arnoldi, W. E. 1951. *The principle of minimized iteration in the solution of the matrix eigenvalue problem.* Quarterly Applied Mathematics, 9, 17-29.

Ash, T. 1989. Dynamic node creation in backpropagation networks. *Connection Science*, 1(4), 365-375.

Azimi-Sadjadi, M. R., Shhedvash, S. and Trujillo, F. O. 1993. Recursive dynamic node creation in multilayer neural networks. *IEEE Transactions on Neural Networks*, 4(2), 242-256.

Bäck, T. and Schwefel, H.-P. 1993. An overview of evolutionary algorithms for parameter optimization. *Evolutionary Computation*, 1(1), 1-23.

Bäck, T., Hammel, U. and Schwefel, H.-P. 1997. Evolutionary computation: Comments on the history and current state. *IEEE Transactions on Evolutionary Computation*, 1(1), 3-17.

Bamkin, R. J. and Piearcey, B. J. 1990. Knowledge-based material selection in design. *Materials and Design,* 11(1), 25-29.

Barber, C. B., Dobkin D. P., and Huhdanpaa, H. T. Dec. 1996. The quickhull algorithm for convex hulls. *ACM Transactions on Mathematical Software*, 22(4), 469-483.

Barker, I. J. and Hulbert, D. G. 1983. Dynamic behaviour in the control of milling circuits. In: Westerlund, T. (ed), *Proceedings of the 4th IFAC Symposium on Automation in Mining, Mineral and Metal Processing*, 139-152, Pergamon Press, Oxford, UK.

Barnard, J. P., Aldrich, C. and Gerber, M. 2001. Identification of dynamic process models with surrogate data. *AIChE Journal*, 47(9), 2064-2075.

Barron, A. 1984. Predicted square error: A criterion for automatic model selection. In: Farlow, S. (ed). *Self-Organizing Methods in Modeling*, Marcel Dekker, New York, NY, USA.

Bartlett, E. B. 1994. Dynamic node architecture learning: An information-theoretic approach. *Neural Networks*, 7(1), 129-140.

Bartlett, P. L. 1997. For valid generalization, the size of the weights is more important than the size of the network. *Advances in Neural Information Processing Systems*, 9, 134-140, MIT Press, Cambridge, MA, USA.

Battiti, R. and Tecchiolli, G. 1994. The reactive tabu search. *ORSA Journal on Computing*, 6(2),126-140.

Bauer, H.-U., Herrmann, M., and Villmann, T. 1999. Neural maps and topographic vector quantization. *Neural Networks*, 12, 659-676.

Baxter, J. 1992. The evolution of learning algorithms for artificial neural networks. In: Green, D. and Bossomaier, T. (eds) *Complex Systems*, 313-326, IOS Press, Amsterdam, Netherlands.

Belew, R. K., McInerney, J. and Schraudolph, N. N. 1991. Evolving networks: Using genetic algorithms with connectionist learning. Technical Report #CS90-174 (Revised), Computer Science and Engineering Department, University of California at San Diego, La Jolla, CA, USA.

Ben-David, A. and Mandel, J. 1995. Classification accuracy: Machine learning vs. explicit knowledge acquisition. *Machine Learning*, 18, 109-114.

Benettin, G., Galgani, L., Giorgilli, A. and Strelcyn, J.-M. 1980. Lyapunov characteristic exponents for smooth dynamical systems and for hamiltonian systems; a method for computing all of them. Part II: Numerical application. *Meccanica*, 15, 21-30.

Bezdek, J. C. and Pal, N. R. 1995. An index of topological preservation for feature extraction. *Pattern Recognition*, 28, 381-391.

Bishop, C. M. 1995. *Neural Networks for Pattern Recognition*. Clarendon Press, Oxford, OX2 6DP, United Kingdom.

Bishop, C. M., Svensén, M. and Williams, C. K. I. 1997. GTM: A principled alternative to the self-organizing map. In Mozer, M.C., Jordan, M.I. and T, P.T. (eds), *Advances in Neural Information Processing Systems*, 9, MIT Press.

Bishop, C. M., Williams, C. K. I. and Svensén, M. 1998. GTM: The generative topographic mapping. *Neural Computation*, 10, 215-234.

Boese, K. D. and Kahng, A. B. 1993. Simulated annealing of neural networks: The 'cooling strategy' reconsidered. *Proceedings of the IEEE International Symposium on Circuits and Systems*, 2572-2575.

Bohachevsky, I. O., Johnson, M. E. and Stein, M. L. 1986. Generalized simulated annealing for function optimization. *Technometrics*, 28(3), 209-217.

Bonifazi, G. and Massacci, P. 1995. Ore liberation modelling by minerals topological evaluation. *Minerals Engineering*, 8(6), 649-658.

Breiman, L. and Friedman, J. H. 1985. Estimating optimal transformations for multiple regression and correlation. (with discussion). *Journal of the American Statistical Association*, 80, 580-619.

Breiman, L., Friedman, J. H., Olshen, R. A. and Stone, C. J. 1984. *Classification and Regression Trees*, Chapman and Hall, 115 5th Avenue New York, NY 10003, USA.

Broomhead, D. S. and King, G. P. 1986. Extracting qualitative dynamics from experimental data, *Physica D*, 20, 217-236.

Brown, R., Bryant, P. and Abarbanel, H. D. I. 1991. Computing the Lyapunov spectrum of a dynamical system from observed time series. *Physical Review A*, 43, 2787-2806.

Bryant, P., Brown, R. and Abarbanel, H. D. I. 1990. Lyapunov exponents from observed time series. *Physics Letters A*, 151, 27-32.

Bryson, A. E. Jr and Ho, Y. C. 1969. *Applied Optimal Control*. Blaisdell, Hemisphere Publishing, Washington, DC, USA.

Burnham, A. J., MacGregor, J. F. and Viveros, R. 1999. Latent variable multivariate regression modelling. *Chemometrics and Intelligent Laboratory Systems*, 48, 167-180.

Buzug, T. and Pfister, G. 1992. Optimal delay time and embedding dimension for delay-time coordinates by analysis of the global static and local dynamical behavior of strange attractors. *Physical Review A*, 45, 7073-7084.

Cao, L. 1997. Practical method for determining the minimum embedding dimension of a scalar time series. *Physica D*, 110, 43-50.

Cao, L., Mees, A. and Judd, K. 1998. Dynamics from multivariate time series. *Physica D*, 121, 75-88.

Carpenter, G. A. and Grossberg, S. 1987. ART2: Self-organization of stable category recognition codes for analog input patterns. In: Proceedings of the *IEEE International Conference on Neural Networks*, II, 727-736. San Diego, CA, USA.

Carpenter, G. A. and Grossberg, S. 1990. ART3: Hierarchical search using chemical transmitters in self-organizing pattern recognition architectures. *Neural Networks*, 3, 129-152.

Carpenter, G. A. and Grossberg, S. 1995. Adaptive resonance theory, In: Arbib, M.A. (ed) *The Handbook of Brain Theory and Neural Networks*, MIT Press, Cambridge, MA, USA.

Casdagli, M. 1989. Nonlinear prediction of chaotic time series. *Physica D*, 35, 335-356.

Caudell, T. P. and Dolan, C. P. 1989. Parametric connectivity: Training of constrained networks using genetic algorithms. In: Schaffer, J. D. (ed), *Proceedings of the 3rd International Conference on Genetic Algorithms and their Applications*, 370-374, Morgan Kaufmann, San Mateo, CA, USA.

Chang, C. L. and Lee, R. C. T. 1973. A heuristic relaxation method for nonlinear mapping in cluster analysis. *IEEE Transactions on Systems, Man, and Cybernetics*, SMC-3, 197-200.

Chang, K.-Y. and Ghosh, J. 1998. Principal curves for nonlinear feature extraction and classification. *SPIE: Applications of Artificial Neural Networks in Image Processing III*, 3307, 120-129.

Chang, K.-Y. and Ghosh, J. 2001. A unified model for probabilistic principal surfaces. *IEEE Transactions on Pattern Analysis and Machine Intelligence*, 23(1), 22-41.

Chatterjee, S. and Price, B. 1977. *Regression Analysis by Example.* John Wiley, New York.

Chauvin, Y. 1989. A back-propagation algorithm with optimal use of hidden units. In: Touretzky, D. S. (ed), *Advances in Neural Information Processing Systems*, 1, 519-526, Morgan Kaufman, San Mateo, CA, USA.

Cheeseman, P. and Stutz, J. 1996. Bayesian classification (AutoClass): Theory and results. In: Fayyad, U.M., Piatetsky-Shapiro, G., Smyth, P. and Uthurusamy, R. (eds) *Advances in Knowledge Discovery and Data Mining.* AAAI Press, MIT.

Cheeseman, P., Kelly, J., Self, M., Stutz, J., Taylor, W. and Freeman, D. 1988. AutoClass: A Bayesian classification system. In: *Proceedings of the 5th International Conference on Machine Learning.* Ann Arbor, MI, USA, Morgan Kaufmann, San Francisco, CA, USA.

Chemaly, T. P. and Aldrich, C. 2001. Data visualization by use of genetic programming. *Computers and Chemical Engineering,* 25(9-10), 1341-1349.

Chen, Z.-P., Jiang, J.-H., Li, Y. and Yu, R.-Q. 1999. Nonlinear mapping using real-valued genetic algorithm. *Chemometrics and Intelligent Laboratory Systems*, 45, 409-418.

Cheng, Y., Karjala, T. W. and Himmelblau, D. M. 1995. Identification of nonlinear dynamic processes with variable dead time using an internal recurrent neural network. *Industrial and Engineering Chemistry Research*, 34, 1735-1742.

Chernoff, H. 1973. Using faces to represent points in k-dimensional space graphically. *Journal of the American Statistical Association*, 68, 361-368.

Choudhury, S. and Murti, M. N. 1990. A divisive scheme for constructing minimal spanning trees in coordinate space. *Pattern Recognition Letters*, 11(6), 385-389.

Chung, T. L. and Lee, T. 1995. Network-growth approach to design of feedforward neural networks. *IEE Proceedings, Control Theory and Applications*, 142(5), 486-492.

Cook, R. D. and Weisberg, S. 1982. *Residuals and Influence in Regression*. Chapman and Hall, London.

Craven, M. W. and Shavlik, J. W. 1994. Using sampling and queries to extract rules from trained neural networks. *Proceedings of the 11th International Conference on Machine Learning*, San Francisco, CA, USA.

Craven, M. W. and Shavlik, J. W. 1996. Extracting Tree-Structured Representations of Trained Networks. *Advances in Neural Information Processing Systems* 8: Touretzky, D.S., Mozer, M.C. and Hasselmo, M.E. (eds.), Morgan Kaufman, MIT Press, London, UK.

Craven, P. and Wahba, G. 1979. Smooting noisy data with spline functions: Estimating the correct degree of smoothing with generalized cross-validation. *Numerische Mathematik*, 31, 377-403.

Crosier, R. B. 1988. Multivariate generalizations of cumulative sum quality-control schemes. *Technometrics*, 30(3), 291-303.

Crouzeix, M., Philippe, B. and Sadkane, M. 1994. The Davidson method. *SIAM Journal of Scientific Computing*, 15, 62-75.

Cutler, C. D. 1993. A review of the theory and estimation of fractal dimension. In: Tong, H. (ed), *Dimension Estimation and Models*, Nonlinear Time Series and Chaos, 1, World Scientific, Singapore.

Datta, B. N. 1995. *Numerical Linear Algebra*, Brookes/Cole Publishing Company, New York, NY, USA.

Davidson, E. R. 1975. The iterative calculation of a few of the lowest eigenvalues and corresponding eigenvectors of large real-symmetric matrices. *Journal of Computational Physics*, 17, 87-94.

Davies, P. T. and Tso, M. K.-S. 1982. Procedures for reduced rank regression. *Applied Statistics*, 31, 244-255.

Day, W. H. E. 1992. Complexity theory: An introduction for practitioners of classification. In: Arabie, P. and Hubert, M. (eds), *Clustering and Classification*, World Scientific Publishing Co., Inc., River Edge, NJ, USA.

De Garis, H. 1991. GenNets: Genetically programmed neural nets – using the genetic algorithms to train neural nets whose inputs and/or outputs vary in time. *Proceedings of the 1991 IEEE International Joint Conference on Neural Networks (IJCNN 91)*, 2, 1391-1396, IEEE Press, New York, NY, USA.

De Ridder, D. and Duin R. P. W. 1997. Sammon's mapping using neural networks: A comparison. *Pattern Recognition Letters*, 18, 1307-1316.

De Ruijter, M. A. 1989. The Leslie shales. *The Role of the Practical Metallurgist Symposium*, Johannesburg, South Africa.

De Veaux, R. D., Schumi, J., Schweinsberg, J. and Ungar, L. H. 1998. Prediction intervals for neural networks via nonlinear regression. *Technometrics*, 40(4), 273-282.

De Waal, S. A. 1975. *The mineralogy, chemistry, and certain aspects of reactivity of chromitite from the Bushveld Igneous Complex*, Report 1709, National Institute of Metallurgy, Johannesburg, South Africa.

Dempster, A. P. Laird, N. M. and Rubin, D. B. 1977. Maximum likelihood from incomplete data via the EM algorithm. *Journal of the Royal Statistical Society*, B 39(1), 1-38.

Diaconis, P. and Shahshahani, M. 1984. On nonlinear functions of linear combinations. *SIAM Journal of Scientific and Statistical Computing*, 5(1), 175-191.

Dickerson, J. A. and Kosko B. 1993. Fuzzy Function Learning with Covariance Ellipsoids. *Proceedings of the IEEE International Conference on Neural Networks* (IEEE ICNN-93) San Francisco, CA, USA, 1162-1167

Dillon, W. R. and Goldstein, M. 1984. *Multivariate Analysis - Methods and Applications*. John Wiley and Sons, Inc., New York, USA.

Domingos, P. 1994. The RISE system: Conquering without separating. In *Proceedings of the 6th IEEE International Conference on Tools with Artificial Intelligence*, 704-707, New Orleans, Los Angeles. USA, IEEE Computer Society Press.

Donaldson, J. R. and Schnabel, R. B. 1987. Computation experience with confidence regions and confidence intervals for nonlinear least squares. *Technometrics*, 29, 67-82.

Dong, D. and McAvoy, T. J. 1996. Non-linear principal component analysis – based on principal curves and neural networks. *Computers and Chemical Engineering*, 20, 65-78.

Draper, N. R. and Smith, H. 1981. *Applied Regression Analysis* (2nd ed). John Wiley, New York.

Dreyfus, S. 1962. The numerical solution of variational problems. *Mathematical Analysis and Applications*, 5(1), 30-45.

Duda, R. O. and Hart, P. E. 1973. *Pattern Classification and Scene Analysis*. Wiley, New York.

Durbin, R. and Mitchison, G. 1990. A dimension reduction framework for understanding cortical maps. *Nature*, 343, 644-647.

Dzubay, T. G. 1982. Visibility and aerosol composition in Houston, Texas. *Environmental Science and Technology,* 16, 514-525.

Eckmann, J.-P. and Ruelle, D. 1985. Ergodic theory of chaos and strange attractors. *Review of Modern Physics*, 57, 617-656.

Eckmann, J.-P. and Ruelle, D. 1992. Fundamental limitations for estimating dimensions and Lyapunov exponents in dynamical systems. *Physica D*, 56, 185-187.

Eckmann, J.-P., Kamphorst, S. O., Ruelle, D. and Ciliberto, S. 1986. Lyapunov exponents from time series. *Physical Review A*, 34, 4971-4979.

Efron, B. and Tibshirani, R. J. 1993. *An Introduction to the Bootstrap*, 57, Monographs on Statistics and Applied Probability. Chapman and Hall, New York, NY, USA.

Ellner, S., Gallant, A. R., McCaffrey, D. and Nychka, D. 1991. Convergence rates and data requirements for Jacobian-based estimates of Lyapunov exponents from data. *Physics Letters A*, 153, 357-363.

Elman, J. L. 1990. Finding structure in time. *Cognitive Science*, 14, 179-211.

Elsner, J. B. and Tsonis, A. A. 1996. *Singular Spectrum Analysis: A New Tool In Time Series Analysis*, Plenum Press, New York, NY, USA.

Eltoft, T. and deFigueiredo, R. J. P. 1998. A new network for cluster-detection-and-labeling. *IEEE Transactions on Neural Networks*, 9(5), 1021-1035.

Evgeniou T., Pontil M. and Poggio T. 2000. Regularization networks and support vector machines. *Advances in Computational Mathematics*, 13(1), 1-50.

Fahlman, S. E. and Lebiere, C. 1990. *The cascade correlation learning architecture*. Technical Report CMU-CS-90-100, Carnegie-Mellon University.

Farmer, J. D. and Sidorowich, J. J. 1987. Predicting chaotic time series, *Physical Review Letters*, 34, 845-848.

Fogel, D. B. 1994. An introduction to simulated evolutionary optimization. *IEEE Transactions on Neural Networks*, 5(1), 3-14.

Fogel, D. B., Fogel, L. J. and Porto, V. W. 1990. Evolving neural networks. *Biological Cybernetics*, 63, 487-493.

Forgy, G. 1982. Rete: A fast algorithm for the many pattern / many object pattern match problem. *Artificial Intelligence*, 19, 17-37.

Frank, I. E. 1995. Modern non-linear regression methods. *Chemometrics and Intelligent Laboratory Systems*, 27, 1-19.

Frank, I. E. and Lanteri, S. 1988. ACE: A non-linear regression model. *Chemometrics and Intelligent Laboratory Systems*, 3, 301-313.

Frazer, A. M. 1989a. Reconstructing attractors from scalar time series: A comparison of singular system and redundancy criteria. *Physica D*, 34, 391-404.

Frazer, A. M. 1989b. Information and entropy in strange attractors. *IEEE Transactions on Information Theory*, 35, 245-262.

Frazer, A. M. and Swinney, H. L. 1986. Independent coordinates for strange attractors. *Physical Review A*, 33, 1134-1140.

Fredkin, D. R. and Rice, J. A. 1995. Method of false nearest neighbors: A cautionary note. *Physical Review E*, 51(4), 2950-2954.

Friedman, J. H. 1987. Exploratory projection pursuit. *Journal of the American Statistical Association*, 82(397), 249-266.

Friedman, J. H. 1991. Multivariate adaptive regression splines (with discussion). *Annals of Statistics*, 19, 301-313.

Friedman, J. H. and Stuetzle, W. 1981. Projection pursuit regression. *Journal of the American Statistical Association*, 76, 817-823.

Friedman, J. H. and Tukey, J. W. 1974. A projection pursuit algorithm for exploratory data nalysis. *Transactions on Computers*, 23C(9), 881-889.

Fritzke, B. 1992. Growing cell structures – a self-organising network in *k* dimensions. In *Artificial Neural Networks*, 2, 1051-1056, North-Holland, Amsterdam, Netherlands.

Fritzke, B. 1994a. Fast learning with incremental RBF networks. *Neural Processing Letters*, 1(1), 2-5.

Fritzke, B. 1994b. Growing cell structures – a self-organizing network for unsupervised and supervised learning. *Neural Networks*, 7(9), 1441-1460.

Fritzke, B. 1995. A growing neural gas network learns topologies. In: Tesauro, G., Touretzky, D.S. and Leen, T.K. (eds.) *Advances in Neural Information Processing Systems*, 7, 625-632, MIT Press, Cambridge, MA, USA.

Fu, L. M. 1991. Rule learning by searching on adapted nets. *Proceedings of the Ninth National Conference on Artificial Intelligence*, Anaheim, CA, USA, 590-595.

Fukushima, K. 1980. Neocognitron: A self-organizing neural network model for a mechanism of pattern recognition unaffected by shift in position. *Biological Cybernetics*, 36, 193-202.

Furnival, G. M. and Wilson W. W. Jr. 1974. Regression by Leaps and Bounds. *Technometrics*, 16(4), 499-511

Gallant, S.I. 1993. *Neural Network Learning and Expert Systems*. MIT Press, Cambridge, MA, USA.

Galton, F. 1885. Regression towards mediocrity in hereditary stature. *Journal of the Anthropological Institute*, 15, 246-263.

Gao, J. and Zheng, Z. 1993. Local exponential divergence plot and optimal embedding of a chaotic time series. *Physical Letters A*, 181, 153-158.

Gao, J. and Zheng, Z. 1994. Direct dynamical test for deterministic chaos and optimal embedding of a chaotic time series. *Physical Review E*, 49, 3807-3814.

Geist, K., Parlitz, U. and Lauterborn, W. 1990. Comparison of different methods for computing Lyapunov exponents. *Progress in Theoretical Physics*, 83, 875-893.

Geladi, P. and Kowalski, B. R. 1986. PLS tutorial. *Analytica Chimica Acta*, 185(1), 1-17.

Gencay, R. and Dechert, W. D. 1992. An algorithm for the n Lyapunov exponents of an n-dimensional unknown dynamical system. *Physica D*, 59, 142-157.

Ghez, J. M. and Vaienti, S. 1992. Integrated wavelets on fractal sets: II The generalized dimensions. *Nonlinearity*, 5, 777-790.

Gianetto, A. and P. L. Silverston. 1986. *Multiphase Chemical Reactors-Theory Design and Scale-Up*. Hemisphere Publishing Corp., Washington, DC, USA.

Giarratano, J. and Riley, G. 1994. *Expert Systems*. 2nd Edition, PWS Publishing Company, Boston, MA, USA.

Ginsberg, D. W. and Whiten, W. J. 1991. Cluster analysis for mineral processing applications. *Transactions of the Institute of Mining and Metallurgy (Section C; Mineral Processing and Extractive Metallurgy)* 100, C139-C146

Golub, G. H. and van Loan, C. F. 1996. *Matrix computations*, 3rd ed, The Johns Hopkins University Press, Baltimore, London.

Golyandina, N., Nekrutkin, V. and Zhiglyavsky, A. 2001. *Analysis of Time Series Structure – SSA and Related Techniques*. Chapman and Hall/CRC Press, Boca Raton, FL, USA.

Gouws, F. S. 1999. *Induction of Fuzzy Rules from Chemical Process Data using Growing Neural Gas and Reactive Tabu Search Methods.* Ph.D. Dissertation, Department of Chemical Engineering, University of Stellenbosch, Stellenbosch, South Africa.

Gouws, F. S. and Aldrich, C. 1999. Development of a corrosion expert system with inductive and genetic algorithms. *Proceedings of the 14th International Conference on Corrosion (CD-ROM)*, Cape Town, South Africa, 27 September – 1 October.

Gowda, K. C. and Krishna, G. 1977. Agglomerative clustering using the concept of mutual nearest neighborhood. *Pattern Recognition*, 10, 105-112.

Gower, J. C. 1971. A general coefficient of similarity and some of its properties. *Biometrics*, 27, 857-872.

Grassberger, P. and Procaccia, I. 1983. Characterization of strange attractors. *Physical Review Letters,* 50, 346-349.

Grassberger, P., Hegger, R., Kantz., H., Schaffrath, C., and Schreiber, T. 1993. On noise reduction methods for chaotic data, *Chaos,* 3, 127-140.

Grassberger, P., Schreiber, T. and Schaffrath, C. 1991. Nonlinear time sequence analysis. *International Journal of Bifurcation and Chaos,* 1(3), 521-547.

Gray, P. and Scott, S. K. 1983. Autocatalytic reactions in the isothermal, continuous stirred tank reactor. Isolas and other forms of multi-stability, *Chemical Engineering Science*, 38, 29-43.

Gray, P. and Scott, S. K. 1984. Autocatalytic reactions in the isothermal, continuous stirred tank reactor. Oscillations and instabilities in the system A + 2B → 3B, B → C. *Chemical Engineering Science*, 39, 1087-1097.

Grefenstette, J. J., Ramsey, C. L., and Schultz A. C. 1990. Learning Sequential Decision Rules Using Simulation Models and Competition. *Machine Learning*, 5, 355-381.

Grobliki, P. J., Wolff, G. T., and Countess, R., J. 1981. Visibility reducing species in the Denver "Brown Cloud". *Atmospheric Environment*, 15(12), 2473-2484.

Grossberg, S. 1976. Adaptive pattern classification and universal recoding: Part 1. Parallel development and coding of neural feature detectors. *Biological Cybernetics*, 23, 121-134.

Grossberg, S. 1982. *Studies of Mind and Brain*. Reidel, Boston, MA, USA.

Gupta, M. M. and Rao, D. H. 1994. Neuro-Control Systems: A Tutorial. In: Gupta, M. M. and Rao, D. H. (eds) *NeuroControl Systems: Theory and Application*, IEEE Press, 1-43.

Hair, J. F., Anderson, R. E., Tatham, R. L. and Black, W. C. 1995. *Multivariate Data Analysis with Readings*, 4th ed., Prentice-Hall Inc., Upper Saddle River, N.J., USA.

Hamilton, J. D. 1989. A new approach to the economic analysis of nonstationary time series and the business cycle. *Econometrics*, 2, 357-384.

Hancock, P. J. 1992. Genetic algorithms and permutation problems: A comparison of recombination operators for neural net structure specification. In: Whitley, D. and Schaffer, J. D (eds), *Proceedings of the International Workshop on Combinations of Genetic Algorithms and Neural Networks (COGANN 92)*, 108-122, IEEE Computer Society Press, Los Alamitos, CA, USA.

Hanson, S. J. and Pratt, L. Y. 1989. Comparing biases for minimal network construction with back-propagation. In: Touretzky, D. S. (ed), *Advances in Neural Information Processing*, 1, 177-185. Morgan Kaufmann, San Mateo, CA, USA.

Haralick, R. M., 1979, Statistical and Structural Approaches to Texture, *Proceedings of the IEEE*, 67(5), 786-803.

Haralick, R. M., Shanmugan, K. and Dinstein, I. 1973, Textural features for image classification, *IEEE Transactions on Systems, Man, and Cybernetics*, SMC-3(6), November, 610-621.

Hartigan, J. A. 1975. *Clustering Algorithms*, Wiley, New York, USA.

Hartman, E. and Keeler, J. D. 1991. Predicting the future: Advantages of semi-local unts. *Neural Computation*, 3, 566-578.

Hartman, E., Keeler, J. D, and Kowalski, J. 1990. Layered neural networks with gaussian hidden units as universal approximators. *Neural Computation*, 2, 210-215.

Hassibi, B., Stork, D. G. and Wolff, G. J. 1992. Optimal brain surgeon and general network pruning. *IEEE International Conference on Neural Networks*, 1, 293-299, San Francisco, CA, USA.

Hastie, T. J. and Stuetzle, W. 1989. Principal curves. *Journal of the American Statistical Association*, 84, 502-516.

Haykin, S. 1999. *Neural Networks: A Comprehensive Foundation.* Macmillan College Publishing Company, 866 3rd Avenue, New York, NY 10022, USA.

Hays, W. L. 1988. *Statistics*, Holt, Rinehart and Winston, Inc. Orlando, FL 32887 USA.

Healy, J. D. 1987. A note on multivariate CUSUM procedures. *Technometrics*, 29(4), 409-416.

Hebb, D. O. 1949. *The Organization of Behaviour*. John Wiley & Sons, New York, NY, USA.

Heistermann, J. and Eckardt, H. 1989. Parallel algorithms for learning in neural networks with evolution strategy. In: Evans, D. J., Joubert, G. R. and Peters, F. J. (eds), *Proceedings of Parallel Computing 89*, 275-280, Elsevier Science Publishers, B.V. Amnsterdam, Netherlands.

Herbst, J. A., Robertson, K. and Rajamani, K. 1983. Mill speed as a manipulated variable for ball mill grinding control. In: Westerlund, T. (ed), *Proceedings of the 4th IFAC Symposium on Automation in Mining, Mineral and Metal Processing*, 153-160, Pergamon Press, Oxford, UK.

Hertz, J., Krogh, A. and Palmer, R. G. 1991. *Introduction to the Theory of Neural Computation*. Santa Fe Institute Studies in the Sciences of Complexity lecture notes. Addison-Wesley Longman Pub. Co., Inc., Reading, MA, USA.

Hinich, M. J. 1982. Testing for Gaussianity and linearity in a stationary time series. *Journal of Time Series Analysis*, 3, 169-176.

Hinton, G. E. 1989. Connectionist learning procedures. *Artificial Intelligence*, 40(1), 195-234.

Hirose, Y., Yamashita, K. and Hijiya, S. 1991. Back-propagation algorithm which varies the number of hidden units. *Neural Networks*, 4, 61-66.

Hoerl, A. E. 1962. Application of ridge analysis to regression problems. *Chemical Engineering Progress*, 58, 54-59.

Hoerl, A. E. and Kennard, R. W. 1968. On regression analysis and biased estimation. *Technometrics*, 10, 422-423.

Hoerl, A. E. and Kennard, R. W. 2000. Ridge regression: Biased estimation for non-orthogonal problems. *Technometrics*, 42(1), 80-86.

Hoerl, A. E., Kennard, R. W. and Baldwin, K, F. 1975. Ridge regression: Some simulations. *Communications in Statistics*, 4, 105-123.

Holte, R. C., Acker, L. E., and Porter, B. W. 1989. Concept Learning and the Problem of Small Disjuncts. *Proceedings of the International Joint Conference on AI,* 813-819.

Holzfuss, J. and Lauterborn, W. 1989. Liapunov exponents from a time series of acoustic chaos. *Physics Review A*, 39, 2146-2152.

Holzfuss, J. and Parlitz, U. 1991. Lyapunov exponents from time series. *Proceedings of the Conference Lyapunov exponents*, Oberwolfach, eds: Arnold, L., Crauel, H. and Eckmann, J.-P., in *Lecture Notes in Mathematics*, Springer-Verlag, Berlin.

Hopfield, J. J. 1982. Neural networks and physical systems with emergent properties collective computational abilities. *Proceedings of the National Academy of Sciences*, USA, 79, 2554-2558.

Hopfield, J. J. 1984. Neurons with graded response have collective computational properties like those of two-state neurons. *Proceedings of the National Academy of Sciences*, USA, 81, 3088-3092.

Horn, J. Goldberg, D. E. and Deb, K. 1994. Implicit niching in a learning classifier system: Nature's way. *Evolutionary Computation*, 2(1), 37-66.

Hotelling, H. 1947. Multivariate quality control, illustrated by the air testing of sample bombsights. In: Eisenhart, C., Hastay, M. W. and Wallis, W. A. (eds), *Techniques of Statistical Analysis*, McGraw-Hill, New York, NY, USA.

Huang, Y., McAvoy, T. J. and Gertler, J. 2000. Fault isolation in nonlinear systems with structured principal component analysis and clustering analysis. *Canadian Journal of Chemical Engineering*, 78, 569-577.

Huber, P. J. 1985. Projection pursuit. *Annals of Statistics*, 13(2), 435-475.

Huerta, R., Santa Cruz, C., Dorronsore, J. R. and Lòpez, V. 1995. Local state space reconstruction using averaged scalar products of dynamical-system flow vectors. *Europhysical Letters*, 29, 13-18.

Hurst, H. E. 1951. Long-term storage of reservoirs. *Transactions of the American Society of Civil Engineers*, 770-779.

Hwang, J. T. G. and Ding, A. A. 1997. Prediction intervals for artificial neural networks. *Journal of the American Statistical Association*, 92, 748-757.

Hwang, J.-N., Lay, S.-R., Maechler, M., Martin, R. D. and Schimert, J. 1994. Regression modeling in back-propagation and projection pursuit learning. *IEEE Transactions on Neural Networks*, 5(3), 342-353.

Hwang, J.-N., Li, H., Maechler, M., Martin, R. D. and Schimert, J. 1992. A comparison of projection pursuit and neural network regression modelling. In: *Advances in Neural Information Processing Systems*, 4, Moody, J.E., Hanson, S.J. and Lippmann, R.P. (eds), 1159-1166. MIT Press.

Hwang, J.-N., Li, H., Martin, R. D. and Schimert, J. 1991. The learning parsimony of projection pursuit and back-propagation networks. *Conference Record of the 25th Asilomar Conference on Signals, Systems, and Computers*, 1, 491-495, Pacific Grove, CA, USA.

Hyvärinen, A. 1999. Survey of independent component analysis. *Neural Computing Surveys*, 2, 94-128.

Hyvärinen, A. and Oja, E. 1997. A fast fixed-point algorithm for independent component analysis. *Neural Computation*, 9(7), 1483-1492.

In: Hewitt, G.F., Delhaye, J.M. and Zuber, N. (eds) *Multiphase Science and Technology*, 6, Hemisphere Publishing Corporation, New York, NY, USA.

Inselberg, A. and Dimsdale, B. 1990. Parallel coordinates: A tool for visualizing multi-dimensional geometry. *Proceedings of the IEEE Conference on Visualization*. 361 - 378.

Ishikawa, M. 1996. Structural learning with forgetting. *Neural Networks*, 9, 509-521.

Isliker, H. 1992. A scaling test for correlation dimensions. *Physical Letters A*, 169, 313-322.

Izenman, A. J. 1975. Reduced-rank regresssion for the multivariate linear model. *Journal of Multivariate Analysis*, 5, 248-264.

J. 1989. Fast learning in networks of locally-tuned processing units. *Neural Computation,* 1(4), 281-294.

Jackson, J. E. 1991. *A User's Guide to Principal Components*, John Wiley and Sons, Inc., New York, NY, USA.

Jaeckle, C. and MacGregor, J. F. 1998. Product design through multivariate statistical analysis of process data. *AIChE Journal*, 44(5), 1105-1118.

Jain, A. K., Murty, M. N. and Flynn, P. J. 1999. Data clustering: A review. *ACM Computing Surveys*, 31(3), 264-319.

Jang, J.-S. R. and Sun, C.-T. 1993. Functional equivalence between radial basis function networks and fuzzy inference systems. *IEEE Transactions on Neural Networks*, 4(1), 156-159.

Jang, J.-S. R., Sun, C.-T. and Mizutani, E. 1997. *Neuro-Fuzzy and Soft Computing*, Prentice-Hall, New Jersey, USA.

Janson, D. J. and Frenzel, J. F. 1992. Application of genetic algorithms to the training of higher order neural networks. *Journal of Systems Engineering*, 2, 272-276.

Janson, D. J. and Frenzel, J. F. 1993. Training product unit neural networks with genetic algorithms. *IEEE Expert*, 8(5), 26-33.

Jia, F., Martin, E. B. and Morris, A. J. 1998. Non-linear principal component analysis for process fault detection. *Computers and Chemical Engineering*, 22, Supplement, S851-S854.

Johannsen, G. and Alty, J. L. 1991. Knowledge engineering for industrial expert systems. *Automatica*, 27(1), 97-114.

Jones, M. C. and Sibson, R. 1987. What is projection pursuit? *Journal of the Royal Statistical Society, ser A*, 150, 1-36.

Joseph, A. and Bryson, N. 1998. Parametric linear programming and cluster analysis. *European Journal of Operational Research*, 111, 582-588.

Judd, K. 1992. An improved estimator of dimension and some comments on providing confidence intervals. *Physica D*, 56, 216-228.

Judd, K. 1994. Estimating dimension from small samples. *Physica D*, 71, 421-429.

Judd, K. and Mees, A. 1995. On selecting models for non-linear time series. *Physica D*, 82, 426-444.

Kadtke, J. B., Brush, J. and Holzfuss, J. 1993. Global dynamical equations and Lyapunov exponents from noisy chaotic time series. *International Journal of Bifurcation and Chaos*, 3, 607-616.

Kahng, A. B. 1992. Exploiting fractalness in error surfaces: New methods for neural network learning. *Proceedings of the IEEE International Symposium on Circuits and Systems*, 41-44.

Kambhatla, N. and Leen, T. K. 1997. Dimension reduction by local principal component analysis. *Neural Computation*, 9, 1493-1516.

Kantz, H. and Schreiber, T. 1997. *Nonlinear Time Series Analysis*: Cambridge Nonlinear Time Series 7, Cambridge University Press, Cambridge, CB2 2RU, United Kingdom.

Kaplan, D. 1994. Exceptional events as evidence for determinism. *Physica D*, 73, 38-48.

Kaye, B. H. 1997. Characterizing the flowability of apowder using the concepts of fractal geometry and chaos theory. *Particles and Particle Systems Characterization*, 14, 53-66.

Kaye, B. H., Gratton-Liimatainen, J. and Faddis, N. 1995. Studying the avalanching behaviour of a powder in a rotating disc. *Particles and Particle Systems Characterization*, 12, 232-236.

Kelley, H. J. 1969. Gradient theory of optimal flight path. *ARS Journal*, 30(10), 947-954.

Kember, G. and Fowler, A. C. 1993. A correlation function for choosing time delays in phase portrait reconstructions. *Physical Letters A*, 179, 72-80.

Kennel, M. B. 1997. Statistical test for dynamical nonstationarity in observed time-series data. *Physical Review E*, 56(1), 316-321.

Kennel, M. B., Brown, R. and Abarbanel, H. D. I. 1992. Determining minimum embedding dimension using a geometrical construction. *Physica Review A,* 45, 3403-3411.

Khorasani, K. and Weng, W. 1994. Structure adaptation in feedforward neural networks. *Proceedings of the IEEE Conference on Neural Networks*, 3, 1403-1408, Orlando, FL, USA.

Kirkpatrick, S. Gelatt, C. D. and Vecchi, M. P. 1983. Optimization by simulated annealing. *Science*, 220, 671-680.

Kirsch, A. 1996. *An Introduction to the Mathematical Theory of Inverse Problems*. Springer-Verlag, New York, NY, USA.

Köhn, P. 1984. *Combining Genetic Algorithms and Neural Nnetworks: The Encoding Problem*. Master's Thesis, University of Tennessee, Knoxville, TE, USA.

Kohonen, T. 1977. *Associative Memory: A System-Theoretical Approach*. Springer-Verlag, Berlin.

Kohonen, T. 1982. Self-organized formation of topologically correct feature maps. *Biological Cybernetics*, 43, 59-69.

Kohonen, T. 1984. *Self-organization and Associative Memory*. Springer-Verlag, Berlin

Kohonen, T. 1988. The 'neural' phonetic typewriter. *IEEE Computer*, 27(3), 11-22.

Kohonen, T. 1990. The self-organizing map. *Proceedings of the IEEE*, 78(9), 1464-1480.

Kohonen, T. 1995. *Self-organizing Maps*. Springer-Verlag, Berlin.

Kosanovich, K., Dahl, K. S. and Piovoso, M. J. 1996. Improved process understanding using multiway principal component analysis. *Industrial and Engineering Chemistry Research*, 35, 138-146.

Kostelich, E. J. and Yorke, J. A. 1988. Noise reduction in dynamical systems. *Physical Review A,* 38, 1649.

Kourti, T. and MacGregor, J. F. 1995. Process analysis, monitoring and diagnosis, using multivariate projection methods. *Chemometrics and Intelligent Laboratory Systems*, 28, 3-21.

Kramer, M. A. 1991. Nonlinear principal component analysis using autoassociative neural networks. *AIChE Journal*, 37(2), 233-243.

Kramer, M. A. 1992. Autoassociative neural networks. *Computers and Chemical Engineering*, 16(4), 313-328.

Kresta, J., MacGregor, J. F. and Marlin, T. E. 1991. Multivariate statistical monitoring of process operating performance. *Canadian Journal of Chemical Engineering*, 69, 35-47.

Kruel, T. M., Eiswirth, M. and Schneider, F. W. 1993. Computation of Lyapunov spectra: Effect of interactive nosie and application to a chemical oscillator. *Physica D*, 63, 117-137.

Kruger, U., Chen, Q., Sandoz, D. J. and McFarlane, R. C. 2001. Extended PLS approach for enhanced condition minotoring of industrial processes. *AIChE Journal*, **47**(9), 2076-2091.

Kruskal, J. B. 1964. Nonmetric multidimensional scaling: A numerical method. *Psychometrica*, 29(2), 115-129.

Krzanowski, W. J. and Marriott, F. H. C. 1994. *Kendall's Library of Statistics 1, Multivariate Analysis. Part I. Distributions, Ordination and Inference*. John Wiley and Sons Inc, 605 Third Avenue, New York, NY 10158.

Krzanowski, W. J. and Marriott, F. H. C. 1995. *Kendall's Library of Statistics 2, Multivariate Analysis, Part 2. Classification, covariance structures and repeated measurements*. Arnold Publishers, London, UK.

Kubat, M. 1998. Decision trees can initialize radial basis function neural networks. *IEEE Transactions on Neural Networks*, 9, 813-821.

Kurita, T. 1991. An efficient agglomerative clustering algorithm using a heap. *Pattern Recognition*, 24(3), 205-209.

Kwok, T.-Y. and Yeung, D.-Y. 1999. Constructive algorithms for structure learning in feed-forward neural networks for regression problems. *Technical Report HKUST-CS95-43*, Department of Computer Science, Hong Kong University of Science and Technology, Clear Water Bay, Kowloon, Hong Kong, China.

Laghi, A. and Lizzanie, L. 1997. Projection pursuit regression with mixed variables. In: *Classification and Data Analysis* (Eds: Vichi, M. and Opitz, O.). Proceedings of the Biannual Meeting of the Classification Group of Societá Italiana di Statistica (SIS), Pescara, 3-4 July.

Lai, Y-C. and Lerner, D. 1998. Effective scaling region for computing the correlation dimension from chaotic time series. *Physica D*, 115, 1-18.

Laine, S., Lappalainen, H. and Jämsä-Jounela, S.-L. 1995. On-line determination of ore type using cluster analysis and neural networks. *Minerals Engineering*, 8(6), 637-648.

Landa, P. S. and Rosenblum, M. G. 1991. Time series analysis for system identification and diagnostics. *Physica D*, 48, 232-254.

Lasota, A. and MacKay, M. C. 1994. *Chaos, Fractals and Noise.* Applied Mathematical Sciences, 2nd ed, 97, Springer, New York, NY, USA.

LeCun, Y., Denker, J. S. and Solla, S. A. 1990. Optimal brain damage. *Advances in Neural Information Processing Systems*, 2, 598-605, Morgan Kaufmann, San Mateo, CA, USA.

Lee, D. H. and Park, D. 1997. An efficient algorithm for fuzzy weighted average. *Fuzzy Sets and Systems*, 87, 39-45.

Lerner, B., Guterman, H., Aladjeni, M. and Dinstein, I. 1999. A comparative study of neural network based feature extraction paradigms. *Pattern Recognition Letters*, 20(1), 7-14.

Levine, S. H. and Kreifeldt, J. G. 1989. Minimal information for the unique representation of a pattern of points. *Proceedings of the IEEE Conference on Systems, Man, and Cybernetics*, 828-830.

Levine, S. H. and Kreifeldt, J. G. 1994. Uniquely representing point patterns with minimal information. *IEEE Transactions on Systems, Man, and Cybernetics*, 24, 895-900.

Liebert, W., Pawelzik, K. and Schuster, H. G. 1991. Optimal embeddings of chaotic attractors from topological considerations. *Europhysical Letters*, 14, 521-526.

Limpo, J. L., Luis, A. and Cristina, M. C. 1995. Hydrolysis of zinc chloride in aqueous ammoniacal ammonium chloride solutions. *Hydrometallurgy*, 38, 235-243.

Liu, Y. and Yao, X. 1996. Evolutionary design of artificial neural networks with different nodes. *Proceedings of the 1996 IEEE International Conference on Evolutionary Computation* (ICEC '96) Nagoya, Japan, 670-675, IEEE Press, New York, NY, USA.

Lorber, A. Wangen, L.E. and Kowalski, B.R. 1987. A theoretical foundation for the PLS algorithm. *Journal of Chemometrics*, 1(9), 19-31.

Lorenzen L. 1995. Some guidelines to the design of a diagnostic leaching experiment. *Minerals Engineering*, 8, 247-256.

Lorenzen L. and Van Deventer J. S. J. 1994. The interrelationship between mineral liberation and leaching behaviour. *International Journal of Mineral Processing*, 41, 1-15.

Lowe, D. and Tipping, M. E. 1996. Feed-forward neural networks and topographic mappings for exploratory data analysis. *Neural Computing and Applications*, 4, 83-95.

Lowry, C. A., Woodall, W. H., Champ, C. W. and Rigdon, S. E. 1992. A multivariate exponentially weighted moving average control chart. *Technometrics*, 34, 46-53.

Luttrell, S. P. 1990. Derivation of a class of training algorithms. *IEEE Transactions on Neural Networks*, 1, 229-232.

Lynch, A. J. 1977. *Mineral Crushing and Grinding Circuits*. Elsevier Scientific Publishing Company, Amsterdam, Netherlands, 287-297.

Lynch, D. T. 1992. Chaotic behavior of reaction systems: parallel cubic autocatalators. *Chemical Engineering Science*, 47(2), 347-355.

MacGregor, J. F. and Kourti, T. 1995. Statistical process control of multivariate processes. *Control Engineering Practice*, 3(3), 403-414.

Maechler, M., Martin, D., Schimert, J., Csoppenszky, M. and Hwang, J. N. 1990. Projection pursuit learning networks for regression. *Proceedings of the 2nd International IEEE Conference on Tools for Artificial Intelligence*, 350-358, Herndon, VA, USA.

Malthouse, E. C. 1998. Limitations of nonlinear principal component analysis as performed with generic neural networks. *IEEE Transactions on Neural Networks*, 9(1), 165-173.

Mao, J. and Jain, A. K. 1995. Artificial neural networks for feature extraction and multivariate data projection. *IEEE Transactions on Neural Networks*, 6(2), 296-317.

Marengo, E. and Todeschini, R. 1991. A fast method for the calculation of partial least squares coefficients. *Chemometrics and Intelligent Laboratory Systems*, 12, 117-120.

Martin, E. B., Morris, A. J. and Kiparissides, C. 1999. Manufacturing performance enhancement through multivariate statistical process control. *Annual Reviews in Control*, 23, 35-44.

Martin, E. B., Morris, A. J. and Zhang, J. 1996. Process performance monitoring using multivariate statistical process control. *Systems Engineering for Automation*, 143(2), 132-144.

Martinetz, T. 1993. Competitive Hebbian learning rule forms perfectly topology preserving maps. *Proceedings of the International Conference on Artificial Neural Networks (ICANN 93)*, 426-438, Amsterdam, Springer, New York, NY, USA.

Martinetz, T. and Schulten, K. 1991. A "neural-gas" network learns topologies. In: Kohonen, T., Mäkisara, K., Simula, O. and Kangas, J. (eds.) *Artificial Neural Networks*, North-Holland, Amsterdam, Netherlands.

Masters, T. 1993. *Practical Neural Network Recipes in C++*. Academic Press, Inc.

McCullagh, P. and Nelder, J. A. 1989. *Generalized Linear Models*, 2nd edition, Chapman and Hall, London, UK.

McCulloch, W. S. and Pitts, W. H.. 1943. A logical calculus of the ideas imminent in nervous activity. *Bulletin of Mathematical Biophysics*, 5, 115-133.

McMahon, C. A. and Pitt, D. J. 1995. Hybrid computer database systems for materials engineering. *Materials and Design,* 16(1), 3-13.

Mees, A. I., and Judd, K. 1993. Dangers of geometric filtering. *Physica D,* 68, 427-436.

Menczer, F. and Parisi, D. 1992. Evidence of hyperplanes in the genetic learning of neural networks. *Biological Cybernetics*, 66, 283-289.

Michalski, R., Stepp, R. E. and Diday, E. 1983. Automated construction of classifications: Conecptual clustering versus numerical taxonomy. *IEEE Transactions on Pattern Analysis and Machine Intelligence*, PAMI-5, 396-409.

Mihalisin, T., Timlin, J. and Schwegler, J. 1991. Visualizing multivariate functions, data, and distributions. *IEEE Computer Graphics and Applications*, 11(3), 28-35.

Mingers, J. 1989. An Empirical Comparison of Pruning methods for Decision Tree Induction. *Machine Learning*, 4, 227-243.

Minsky, M. 1954. *Neural Nets and the Brain*. Doctoral dissertation, Princeton University, NJ, USA.

Minsky, M. and Papert, S. 1969. *Perceptrons*. MIT Press, Cambridge, MA, USA.

Mitchison, G. 1995. A type of duality between self-organizing maps and minimal wiring. *Neural Computation*, 7, 25-35.

Mjolsness, E., Sharp, D. H. and Alpert, B. K. 1989. Scaling, machine learning, and genetic neural nets. *Advances in Applied Mathematics*, 10, 137-163.

Montana, D. and Davis, L. 1989. Training feedforward neural networks using genetic algorithms. *Proceedings of the 11th International Joint Conference on Artificial Intelligence*, 762-767, Morgan Kaufmann, San Mateo, CA, USA.

Moody, J. 1994. Prediction risk and architecture selection for neural networks. In: Cherkassky, V., Friedman, J.H. and Wechsler, H. (eds), *From Statistics to Neural Networks: Theory and Pattern Recognition Applications*, 136, 147-165, NATO ASI Series F, Springer-Verlag.

Moody, J. and Darken, C.Moody, J. E. 1991. Note on generalization, regularization, and architecture selection in nonlinear learning systems. In: Juang, B.H., Kung, S.Y., Kamm,

C.A. (eds), *Neural Networks for Signal Processing*. Proceedings of the 1991 IEEE Workshop, Princeton, NJ, USA.

Moody, J. E. and Rögnvaldsson, T. 1997. Smoothing regularizers for projective basis function networks. *Advances in Neural Information Processing Systems*, 9, 585-591, Morgan Kaufmann, San Mateo, CA, USA.

Moolman, D. W. 1995. *The Interpretation of Flotation Froths by Using Computer Vision and Connectionist Networks*. Ph.D. thesis, Department of Chemical Engineering, University of Stellenbosch, Stellenbosch, South Africa.

Moolman, D. W. Aldrich, C. and Van Deventer, J. S. J. 1995a. The monitoring of froth surfaces on industrial flotation plants using connectionist image processing techniques. *Minerals Engineering*, 8(1/2), 23-30.

Moolman, D. W., Aldrich, C. and Van Deventer, J. S. J. 1995b. The interpretation of flotation froth surfaces by using digital image analysis and neural networks. *Chemical Engineering Science*, 55(22), 3501-3513.

Moolman, D. W., Aldrich, C. and Van Deventer, J. S. J. 1995c. Analysis of videographic process data with neural nets. *Acta Chimica Slovenica*, 42(1), 137-142, 1995.

Moolman, D. W., Aldrich, C., Van Deventer, J. S. J. and Stange, W. W. 1994. Digital image analysis as a tool for on-line monitoring of froth in flotation plants. *Minerals Engineering*, 7(9), 1149-1164.

Moolman, D. W., Van Deventer, J. S. J. and Aldrich, C. 1998. Chapter 8: The interrelationship between flotation variables and froth appearance. In: Laskowski J. S. & Woodburn, E. T., *Frothing and Flotation II*, 245-273. Gordon and Breach Science Publishers Ltd, New York.

Morgan, R. B. and Scott, D. S. 1993. Preconditioning the Lanczos algorithm for sparse symmetric eigenvalue problems. *SIAM Journal of Scientific Computing*, 14, 585-593.

Moriarty, D. E., and Miikkulainen, R. 1996. Efficient Reinforcement Learning Through Symbiotic Evolution. *Machine Learning*, 22, 11-32.

Moriarty, D. E., and Miikkulainen, R. 1998. Forming Neural Networks through Efficient and Adaptive Coevolution. *Evolutionary Computation,* 1998, 5(4), 1-28.

Morozov, V. A. 1993. *Regularization Methods for Ill-posed Problems*. CRC Press, Boca Raton, FL, USA.

Muhamad, A. K. and Deravi, F. 1992. Neural Networks for texture classification. *The 4th International Conference on Image Processing and its Applications*, 7-9 April, Maastricht Exhibition and Convention Centre, The Netherlands.

Myers, R. H. and Montgomery, D. C. 1995. *Response Surface Methodology: Process and Product Optimization Using Designed Experiments*. John Wiley and Sons, Inc., New York, USA.

Niemi, A. J., Ylinen, R. and Hyötyniemi, H. 1997. On characterization of pulp and froth in cells of flotation plant. *International Journal of Mineral Processing*, 51, 51-65.

Nomikos, P. and NacGregor, J. F. 1995. Multivariate SPC charts for monitoring batch processes. *Technometrics*, 37, 41-59.

Oja, E. 1996. Robust fitting by nonlinear neural units. *Neural Networks*, 9, 435-444.

Oliver, J. J. and Hand, D. 1994. *Introduction to Minimum Encoding Inference*. Technical Report 4-94, Department of Statistics, Open University, Walton Hall, United Kingdom. (Also published as Technical Report 205, Department of Computer Science, Monash University, Clayton, Victoria, Australia, 1994.)

Osborne, A. R. and Provenzale, A. 1989. Finite correlation dimension for stochastic systems with power-law spectra. *Physica D*, 35, 357-381.

Ott, E., Sauer, T. and Yorke, J. A. 1994. *Coping with Chaos*, Wiley, New York, NY, USA.

Ott, E., Withers, W. D. and Yorke, J. A. 1984. Is the dimension of chaotic attractors invariant under coordinate change? *Journal of Statistical Physics*, 36, 687-697.

Packard, N. J., Crutchfield, J. P., Farmer, J. D. and Shaw, R. S. 1980. Geometry from a time series. *Physical Review Letters*, 45, 712-716.

Pal, N. R. and Eluri, V. K. 1998. Two efficient connectionist schemes for structure preserving dimensionality reduction. *IEEE Transactions on Neural Networks*, 9(6), 1142-1154.

Palus, M., Albrecht, V and Dvorak, I. 1993. Information theoretic tests for nonlinearity in time series. *Physical Letters A*, 175, 203-209.

Park, J. and Sandberg, I.W. 1991. Universal approximation using radial basis function networks. *Neural Computation*, 3, 246-257.

Parker, R. E. and Tummala, M. 1992. Identificationof Volterra systems with a polynomial neural network. *Proceedings of the 1992 IEEE International Conference on Acoustics, Speech and Signal Processing*, 4, 561-564, San Francisco, CA, USA.

Parlett, B. N. 1980. *The Symmetric Eigenvalue Problem*. Prentice-Hall, Englewood Cliffs, N.J.

Parlitz, U. 1992. Identification of true and spurious Lyapunov exponents from time series. *International Journal of Bifurcation and Chaos*, 2, 155-165.

Parlitz, U. 1993. Lyapunov exponents from Chua's circuit. *Journal of Circuits, Systems and Computers*, 3, 507-523.

Parlitz, U. 1999. Nonlinear time-series analysis. In: *Non-linear Modelling*, Suykens, J.A.K. and VandeWalle, J. (eds), Kluwer Academic Publishers, Boston, MA, USA.

Pascui, M. I., Murciego, A., Pellitero, E., Babkine, J. and Dusausoy, Y. 1997. Sn-Ge-Cd-Cu-Fe-bearing sulfides and sulfosalts from the Barquilla deposit, Salamanca, Spain. *The Canadian Metallurgist*, 35, 39-52.

Penny, K. I. 1996. Appropriate critical values when testing for a single multivariate outlier by using the Mahalanobis distance. *Applied Statistics*, 45(1), 73-81.

Perry, R. H. and Green, D. 1984. *Perry's Chemical Engineer's Handbook.* (eds), 50th Edition, McGraw-Hill, Ontario, Canada.

Pickett, R. and Grinstein, G. 1988. Iconographic displays for visualizing multidimensional data. *Proceedings of the 1988 IEEE Conference on Systems, Man, and Cybernetics*, 514-519, Beijing and Shenyang, China.

Poggio, T. and Girosi F. 1990. Networks for approximation and learning. *Proceedings of the IEEE*, 78(9), 1481-1497.

Pop, E., Hayward, R. and Diederich, J. 1994. RULENG: Extracting rules from a trained artificial neural network by stepwise negation. *QUT NRC, December.*

Powell, M. J. D. 1985. Radial basis functions for multivariable interpolation: A review. *IMA Conference on Algorithms for the Approximation of Functions and Data*. Mason, J.C. and Cox, M.G. (eds). University Press, Oxford, UK, 143-167.

Press, W. H., Teukolsky, S. A., Vetterling, W. T. and Flannery, B. P. 1992. *Numerical Recipes in C: The Art of Scientific Computing* (2nd ed.), Cambridge University Press, Cambridge, United Kingdom.

Proakis, J. G. and Manolakis, D. M. 1996. *Digital Signal Processing.* 3rd ed. Prentice Hall International Editions, New Jersey, USA.

Provenzale, A., Smith, L.A., Vio, R. and Murante, G. 1992. Distinguishing between low-dimensional dynamics and randomness in measured time series. *Physica D*, 58, 50-76.

Pulkkinen, K., Ylinen, R., Jamsa-Jounela, S.-L. and Jarvensivu, M. 1993. Integrated expert control system for grinding and flotation. *Proceedings of XVIII International Mineral Processing Congress*, Sydney, 325-334.

Pykett, C. E. 1978. Improving the efficiency of Sammon's non-linear mapping by using clustering archetypes. *Electronic Letters*, 14, 799-800.

Qin, S. J. 1996. Neural networks for intelligent sensors and control - practical issues and some solutions. In: Elliott, D. (ed), *Neural Networks for Control*, Chapter 8., Academic Press.

Qin, S. J. and McAvoy, T. J. 1992. Nonlinear PLS modelling using neural networks. *Computers and Chemical Engineering*, 16, 379-391.

Quinlan, J. R. 1986. Induction of decision trees. *Machine Learning*, 1, 81-106.

Quinlan, J. R. 1987. Simplifying decision trees. *International Journal of Man-Machine Studies*, 27, 221-234.

Quinlan, J. R. 1988. Decision trees and multi-valued attributes. *Machine Intelligence*, 11, 305-318.

Quinlan, J. R. 1993. *C4.5: Programs for Machine Learning.* Morgan Kaufmann, San Mateo, CA, USA.

Quinlan, J. R., Compton, P. J., Horn, K. A. and Lazarus, L. 1987. Inductive knowledge acquisition: A case study. In: Quinlan, J. R. (ed.), *Applications of Expert Systems,* Addison Wesley, Sydney, Australia, 157-173.

Rabiner, L. R. 1989. A tutorial on hidden Markov models and selected applications in speech recognition. *Proceedings of the IEEE*, 2, 257-286.

Rajamani, R. K. and Herbst, J. A. 1991. Optimal control of a ball mill grinding circuit - I. Grinding circuit modelling and dynamic simulation. *Chemical Engineering Science*, 46(3), 861-870.

Rapp, P. E., Albano, A. M., Zimmerman, I. D., Jiménez-Moltaño, M. A. 1994. Phase-randomized surrogates can produce spurious identifications of non-random structure. *Physical Letters A*, 192, 27-33.

Rauf, F. 1997. *Non-linear Adaptive Filtering: A Unified Approach.* Ph.D. Thesis, Boston University, Boston, MA, USA.

Rauf, F. and Ahmed, H. 1997. New non-linear adaptive filters with applications to chaos. *International Journal of Bifurcation and Chaos*, 7(8), 1791-1809.

Reed, R. 1993. Pruning algorithms – a survey. *IEEE Transactions on Neural Networks*, 4(5), 740-747.

Reimann, J., John, H. and Seeger, W. 1992. Transition from slug to annular flow in horizontal pipes.Reklaitis, G. V., Ravindran, A. and Ragsdell, K. M. 1983. *Engineering Optimization*. John Wiley and Sons, Inc., NY, USA.

Renals, S. 1988. Radial basis function network for speech pattern classification. *Electronics Letters*, 25(7), 437-439.

Renyi, A. 1971. *Probability Theory*, North-Holland, Amsterdam, Netherlands.

Reuter, M. A. 1997. Modelling and control of ill-defined metallurgical furnaces with particulate feeds. *Proceedings of the Minerals and Materials Process Control Workshop*, 74-101, Stellenbosch, South Africa.

Rissanen, J. 1975. Modelling by shortest data description. *Automatica*, 14, 465-471.

Rissanen, J. 1989. *Stochastic Complexity in Statistical Inquiry*. World Scientific, Singapore.

Roosen, C. B. and Hastie, T. J. 1994. Automatic smoothing spline projection pursuit. *Journal of Computational and Graphical Statistics*, 3(3), 235-248.

Rosenstein, M. T., Collins, J. J. and De Luca, C. J. 1994. Reconstruction expansion as a geometry-based framework for choosing proper delay times. *Physica D*, 73, 82-98.

Rowland, C. A., 1963. Automation of grinding circuits. *Pit and Quarry*, 56(1), 176-179.

R-Roda, I., Comas, J., Poch, M., Sànchez-Marrè, M. and Cortés, U. 2001. Automatic knowledge acquisition from complex processes for the development of knowledge-based systems. *Industrial and Engineering Chemistry Research*, 40, 3353-3360.

Rubner, J. and Schulten, K. 1990. Development of feature detectors by self-organization. *Biological Cybernetics*, 62, 193-199.

Rubner, J. and Tavan, P. 1989. A self-organizing network for principal component analysis. *Europhysics Letters*, 10, 693-698.

Ruelle, D. 1980. Strange attractors. *Mathematical Intelligencer*, 2, 37-48.

Rumelhart, D. E. and McClelland, J. L. 1986. *Parallel Distribution Processing: Exploration in the Microstructure of Cognition,* 1, MIT Press, Cambridge, MA, USA.

Ryan, T. P. 1989. *Statistical Methods for Quality Improvement*, Wiley, New York, NY, USA.

S. I. 1972. Learning patterns and pattern sequences by self-organizing nets of threshold elements. *IEEE Transactions on Computers*, C-21, 1197-1206.

Saad, Y. 1992. *Numerical Methods for Large Eigenvalue Problems*. Manchester University Press, Manchester, United Kingdom.

Saha, A., Christian, J. Tang, D.S. and Wu, C.-L. 1991. Orientated non-radial basis functions for image coding and analysis. *Advances in Neural Information Processing Systems 3*, Lippmann, R.P., Moody, J.E. and Touretzky, D.S. (eds), 728-734. Morgan Kaufmann, San Mateo, CA, USA.

Saito, K. and Nakano, R. 1988. Medical diagnostic expert systems based on PDP model. *Proceedings of IEEE International Conference on Neural Networks*, 1, San Diego, CA, USA, 255-262.

Salvino, L. W. and Cawley, R. 1994. Smoothness implies determinism: A method to detect it in time series. *Physical Review Letters*, 73, 1091-1094.

Salzberg, S. and Watkins, M. 1990. Managing information for concurrent engineering: Challenges and barriers. *Research in Engineering Design* 2(1), 35-52.

Sammon, J. W. 1969. A nonlinear mapping for data structure analysis. *IEEE Transactions on Computers*, C-18, 401-409.

Sano, M. and Sawada, Y. 1985. Measurement of the Lyapunov spectrum from a chaotic time series. *Physics Letters Review A*, 55, 1082-1085.

Saravanan, N. and Fogel, D. B. 1995. Evolving neural control systems. *IEEE Expert*, 10(3), 23-27.

Sarkar, M. and Yegnanarayana, B. 1997a. Evolutionary program-based probabilistic neural network construction technique. *Proceedings of the 1997 IEEE International Conference on Neural Networks*, 1 (of 4), 456-461, IEEE Press, Piscataway, NJ, USA.

Sarkar, M. and Yegnanarayana, B. 1997b. Feedforward neural networks configuration using evolutionary programming. *Proceedings of the 1997 IEEE International Conference on Neural Networks*, 1 (of 4), 438-443, IEEE Press, Piscataway, NJ, USA.

Sauer, T., Yorke, J.A., Casdagli, M. 1991. Embedology. *Journal of Statistical Physics*, 65, 579-616.

Savit, R. and Green, M. 1991. Time series and dependent variables. *Physica D*, 50, 95-116.

Schachter, B. 1978. A non-linear mapping algorithm for large databases. *Computational Graphics and Image Processing*, 7, 271-278.

Schmitz G. P. J. and Aldrich C. 1999. Combinatorial evolution of regression nodes in feed-forward neural networks. *Neural Networks*, 12(1), 175-189.

Schmitz, G. P. J. 1999. *Training of Neural Networks with the CERN algortihm*. Ph.D. Thesis, Department of Chemical Engineering, University of Stellenbosch, Stellenbosch, Western Cape, South Africa.

Schmitz, G. P. J., Aldrich, C. and Gouws, F. S. 1996. Extracting decision trees from artificial neural networks. Proceedings of *Minerals and Materials '96*: Somerset West, South Africa, 1, 250-257.

Schmitz, G. P. J., Aldrich, C. and Gouws, F. S. 1999a. ANN-DT: An algorithm for extracting decision trees from artificial neural networks. *IEEE Transactions on Neural Networks*, 10(6), 1392-1401.

Schmitz, G. P. J., Aldrich, C. and Gouws, F. S. 1999b. Extraction of decision trees from artificial neural networks. In: *Knowledge-based Neurocomputing*, Zurada, J.A. and Cloete, I. 11, MIT Press, Cambridge, MA, USA.

Schölkopf, B., Smola, A. and Müller, K.-R. 1998. Nonlinear component analysis as a kernel eigenvalue problem, Neural Computation, 10, 1299-1313.

Schreiber, T. 1998. Constrained randomization of time series data. *Physical Review Letters*, 80(10), 2105-2108.

Schreiber, T. and Schmitz, A. 1996. Improved surrogate data for non-linearity tests. *Physical Review Letters*, 77(4), 635-638.

Schumacher, M. (ed.) *Seawater Corrosion Handbook*. Noyes Data Corporation, Park Ridge, New Jersey, U.S.A., 1979.

Schwartz, G. 1978. Estimating the dimension of a model. *The Annals of Statistics*, 6, 461-464.

Schwenker, F., Kestler, H.A. and Palm, G. 2001. Three learning phases for radial basis function networks. *Neural Networks*, 14, 439-458.

Seasholtz, M. B. 1999. Making money with chemometrics. *Chemometrics and Intelligent Laboratory Systems*, 45, 55-63.

Sebald, A. V. and Chellapilla, K. 1998. On making problems evolutionarily friendly, Part I: Evolving the most convenient representations. In: Porto, V.W., Saravanan, N., Waagen, D. and Eiben, A.E. (eds) *Evolutionary Programming VII: Prceedings of the 7th Annual Conference on Evolutionary Programming*, Vol 1447 of Lecture Notes in Computer Science, 271-280, Springer-Verlag, Berlin.

Seber, G. A. F. 1977. *Linear Regression Analysis*. Wiley, Wiley Series in Probability and Mathematical Statistics. New York, NY, USA.

Seisto, S. and Dillon, T. 1993. Automated knowledge acquisition of rules with continuously valued attributes". *Proceedings of the 12th International Conference on Expert Systems and their Applications*, Avignon, France, May 1992, 645-656.

Sethi, I.K. 1995. Neural implementation of tree classifiers. *IEEE Transactions on Systems, Man, and Cybernetics*, 25(8), 1243-1249.

Setiono, R. and Hui, L. C. K. 1995. Use of a quasi-Newton method in a feedforward neural network construction algorithm. *IEEE Transactions on Neural Networks*, 6(1), 273-277.

Setnes, M. 1995. *Fuzzy Rule-Based Simplification using Similarity Measures*. M.Sc. Thesis, Department of Electrical Engineering, Delft University of Technology, Delft, Netherlands.

Setnes, M., Babuška, R., Kaymak, U. and van Nauta Lemke, H.R. 1998. Similarity measures in fuzzy rule simplification. *IEEE Transactions on Systems, Man, and Cybernetics – Part B: Cybernetics*, 28(3), 376-386.

Shaw, M. J. and Gentry, J. A. 1990. Inductive learning for risk classification. *IEEE Expert*, 47-53.

Shepard, R. N. 1962. The analysis of proximities: Multidimensional scaling with an unknown distance function i. *Psychometrica*, 27, 125-140.

Shimada, I. And Nagashima, T. 1979. A numeric approach to ergodic problems of dissipative dynamical systems. *Progress in Theoretical Physics*, 61, 1605-1616.

Shin, Y. and Ghosh, J. 1992. Approximation of multivariate functions using ridge polynomial networks. *Proceedings of International Joint Conference on Neural Networks*, II, 380-385, Baltimore, MD, USA.

Shin, Y. and Gosh, J. 1995. Ridge polynomial networks. *IEEE Transactions on Neural Networks*, 6(3), 610-622

Siew, L. H., Hodgson, R. M., Wood, E. J. 1988. Texture measures for carpet wear assessment, *IEEE Transactions on Pattern Analysis and Machine Intelligence*, 10(1), 92-105.

Silverman, B. W. 1981. Using kernel densities to investigate multimodality. *Journal of the Royal Statistical Society*, Series B, 43, 97-99.

Silverman, D. C. 1994. Artificial Neural Network Predictions of Degradation of Nonmetallic Lining Materials from Laboratory Tests. *Corrosion Science,* 50(6), 411-418.

Sinnott, R. K. 1993. *Coulson & Richardson's Chemical Engineering*. Volume 6, 2nd Edition, Pergamon Press, Oxford, England.

Small, M. and Judd, K. 1998a. Comparison of new non-linear modelling techniques with applications to infant respiration, *Physica D*, 117, 283-298

Small, M. and Judd, K. 1998b. Correlation dimension: A pivotal statistic for non-constrained realizations of composite hypotheses in surrogate data analysis. *Physica D*, 129, 386-400.

Smart, N.R., Holmes, D.R. and Poulson, B. The Achilles Expert System on Corrosion and Protection: Consultations on Aspects of Stress Corrosion Cracking. *Corrosion Science* 1993, **35**(1-4), 403-410.

Smets, H.M.G. and Bogaerts, W.F.L. Deriving corrosion knowledge from case histories: the neural network approach. *Materials and Design* 1992, **13**(3), 149-153.

Smith, L. 1992. Identification and prediction of low-dimensional dynamics. *Physica D*, 58, 50-76.

Specht, D. F. 1991. A general regression neural network. *IEEE Transactions on Neural Networks*, 2(6), 568-576.

Srinivas, M. and Patnaik, L. M. 1991. Learning neural network weights using genetic algorithms – improving performance by search space reduction. *Proceedings of the 91 IEEE Internaitonal Joint Conference on Neural Networks (IJCNN 91)*, 2, 2331-2336, IEEE Press, New York, NY, USA.

Stefanovska, A., Strle, S. and Kroselj, P. 1997. On the overestimation of the correlation dimension. *Physical Letters A,* 235, 24-30.

Stone, G. O. 1986. An analysis of the delta rule and the learning of statistical associations. In: *Parallel Distributed Processing*, 1, 444-459. MIT Press, Cambridge, McClelland, R.D.E. and McClelland, J.L. (eds.).

Stone, M. 1974. Cross-validatory choice and assessment of statistical predictions (with discussion). *Journal of the Royal Statistical Society B*, 36, 111-147.

Stone, M. and Brooks, R. J. 1990. Continuum regression: Cross-validated sequentially constructed prediction embracing ordinary least squares, partial least squares and principal components regression. *Journal of the Royal Statistical Society B*, 52, 337-369.

Stoop, R. and Meier, P. F. 1988. Evaluation of Lyapunov exponents and scaling functions from a time series. *Journal of the Optical Society of America B*, 5, 1037-1045.

Stoop, R. and Parisi, J. 1991. Calculation of Lyapunov exponents avoiding spurious elements. *Physica D*, 50, 84-89.

Stork, D. G., Walker, S., Burns, M. and Jackson, B. 1990. Preadaptation in neural circuits. *Proceedings of the International Conference on Neural Networks,* 1, Washington, DC, USA, 202-205, Lawrence Erlbaum Associates, Hillsdale, NJ, USA.

Storn, R. and Price, K. 1995. Differential Evolution - a Simple and Efficient Adaptive Scheme for Global Optimization over Continuous Spaces. Technical Report TR-95-012. ICSI

Stremler, F. G. 1990. *Introduction to Communications Systems.* 3rd ed. Addison-Wesley Publishing Company, Massachusetts, USA.

Sun, C. and Wee, W. G. 1983. Neighbouring grey level dependence matrix for texture classification. *Computer Vision, Graphics and Image Processing*, 23, 341-352.

Sutton, R. S., and Barto, A. G., *Reinforcement Learning - An Introduction*, 1st ed. 1998, A Bradford Book, Cambridge Massachusetts.

Suykens J. A. K. and Vandewalle J. (eds.), 1998. *Nonlinear Modeling: Advanced black-box techniques,* Kluwer Academic Publishers, Boston, MA, USA.

Suykens J. A. K., Vandewalle J., De Moor B. 1996. *Artificial Neural Networks for Modelling and Control of Non-Linear systems*, Kluwer Academic Publishers, Boston, MA, USA.

Suykens, J. A. K. 2001. Nonlinear modeling and support vector machines. *Proceedings of the IEEE Instrumentation and Measurement Technology Conference*. Budapest, Hungary, May 21-23.

Takane, Y. and Young, F. W. 1977. Nonmetric individual differences multidimensional scaling: An alternating least squares method with optimal scaling features. *Psychometrica*, 42(1), 7-67.

Takens, F. 1981. Detecting strange attractors in turbulence. *Lecture Notes in Mathematics*, 898, Springer (Berlin), 366-381.

Takens, F. 1985. On the numerical determination of the dimension of an attractor. In: Braaksma, B. L., Broer, H. W. and Takens, F. (eds), *Dynamical Systems and Bifurcations* (Lecture notes in Mathematics), Groningen 1984, 1125, Springer Berlin

Takens, F. 1993. Detecting non-linearities in stationary time series. *International Journal of Bifurcations and Chaos*, 3, 241-256.

Tan, S. and Mavrovouniotis, M. L. 1995. Reducing data dimensionality through optimizing neural network inputs. *AIChE Journal*, 41(6), 1471-1480.

Tattersall G. D. and Limb P. R. Visualisation techniques for data mining, *BT Technology Journal*, 12(4), 23-30 (1994).

Tenorio, M. F. and Lee, W. T. 1990. Self-organizing network for optimum supervised learning. *IEEE Transactions on Neural Networks*, 1(1), 100-110.

Theiler, J. 1990. Statistical precision of dimension estimators. *Physics Review A*, 41, 3038-3051.

Theiler, J. 1995. On the evidence for low-dimensional chaos in an epileptic encephalogram. *Physics Letters A*, 196, 335-341.

Theiler, J. and Pritchard, D. 1996. Constrained realization Monte Carlo method for hypothesis testing, *Physica A*, 94, 221-235.

Theiler, J. and Rapp, P. E. 1996. Re-examination of the evidence for low-dimensional nonlinear structure in the human electroencephalogram. *Encephalography and Clinical Neurophysiology*, 98, 213-222.

Theiler, J., Eubank, E., Longtin, A. and Galdrikian, B. 1992a. Testing for non-linearity in time series: The method of surrogate data. *Physica D*, 58, 77-94.

Theiler, J., Galdrikian, B., Lontin, A., Eubank, S. and Farmer, J. D. 1992b. *Using surrogate data to detect nonlinearity in time series*. In: *Nonlinear Modeling and Forecasting*, eds, Casdagli, M. and Eubank, S., SFI Studies in the Sciences of Complexity, XII, 163-188, Addison-Wesley, Reading, MA.

Theron, D. A. and Aldrich, C. 2000. Acoustic monitoring of the comminution of complex sulphide ores in a laboratory ball mill. *Transactions of the Journal of the South African Institute of Mining and Metallurgy*, 100(4), 243-248.

Theron, H. 1994. *Specialization by Exclusion: An Approach to Concept Learning*. Ph.D. dissertation, Department of Computer Science, University of Stellenbosch, Stellenbosch, South Africa.

Theron, H. and Cloete, I. 1996. BEXA: A covering algorithm for learning propositional concept descriptions. *Machine Learning*, 24, 5-40.

Thrun, S. 1994. Extracting provable correct rules from artificial neural networks, *Technical Report IAI-TR-93-5*, Institut für Informatik III Universität Bonn, Germany.

Thrun, S. 1995. Extracting rules from artificial neural networks with distributed representation. *Advances in Neural Information Processing Systems*, 7, Tesauro, G., Touretzky, D. and Leen, T. (eds).

Tibshirani, R. 1988. Estimating transformations for regression via additivity and variance stabilization. *Journal of the American Statistical Association*, 83, 394-405.

Tickle, A. B., Orlowski, M. and Diederich, J. 1994. DEDEC: Decision detection by rule extraction from neural networks, *QUT NRC*, December.

Tikhonov, A. N. and Arsenin, V. Y. 1977. *Solutions of Ill-posed Problems*. W. H. Winston, Washington, DC, USA.

Torgerson, W. S. 1952. Multidimensional scaling, I: Theory and method. *Psychometrika*, 17, 401-419.

Towell, G. and Shavlik, J. W. 1993. Extracting refined rules from knowledge based neural networks. *Machine Learning*, 13, 71-101.

Trethway, K. R. and Roberge, P. R. 1994. Lifetime prediction in engineering systems: The influence of people. *Materials and Design*, 15(5), 275-285.

Tso, M. K.-S. 1981. Reduced rank regression and canonical analysis. *Journal of the Royal Statistical Society, Series B*, 43(2), 183-189.

Tsogo, L., Masson, M.-H. and Bardot, A. 2001. Recovery of the metric structure of a pattern of points using minimal information. *IEEE Transactions on Systems, Man, and Cybernetics – Part A: Systems and Humans*, 31(1), 30-42.

Vapnik, V. N. 1992. Principles of risk minimization for learning theory. *Advances in Neural Information Processing Systems*, 4, 831-838, San Mateo, CA, Morgan Kaufmann.

Vapnik, V. N. 1995. *The Nature of Statistical Learning Theory*. Springer-Verlag, New York, NY, USA.

Vapnik, V. N. 1998. *Statistical Learning Theory*. Wiley, New York, NY, USA.

Vautard, R. and Ghil, M. 1989. Singular spectrum analysis in nonlinear dynamics with applications to paleoclimatic time series. *Physica D*, 35, 395-424.

Vautard, R., Yiou, P. and Ghil, M. 1992. Singular spectrum analysis: A toolkit for short, noisy and chaotic time series. *Physica D*, 58, 95-126.

Vinod, H. D. 1978. A survey of ridge regression and related techniques for improvements over ordinary least squares. *Review of Economics and Statistics*, 60, 121-131.

Wachs, A. and Lewin, D. R. 1999. Improved principal component analysis methods for process disturbance and failure identification. *AIChE Journal*, 45(8), 1688-1700.

Wang, X. Z. 2001. Knowledge discovery through mining process operational data. In: Mujtaba, I. M. and Hussain, M. A. (eds) *Application of Neural Networks and Other Process Learning Methodologies in Process Engineering*, 287-328, Imperial College Press, London, UK.

Wang, X. Z. and McGreavy, C. 1998. Automatic classification for mining process operational data. *Industrial and Engineering Chemistry Research*, 37, 2215-2222.

Watson, J. L. 1985. An analysis of mill grinding noise. *Powder Technology*, 41, 83-89.

Watson, J. L. and Morrison, S.D. 1986. Estimation of pulp viscosity and grinding mill performance by means of mill noise measurements. *Journal of Mineral and Metallurgical Processing*, AIME-SME, 4, 216-227.

Wayland, R., Bromley, D. and Pickett, D. and Passamante, A. 1993. Recognizing determinism in a time series. *Physical Review Letters*, 70, 580-582.

Weigend, A. S. and LeBaron, B. 1994. Evaluation neural network predictors by bootstrapping. *Proceedings of International Conference on Neural Information Processing*, 2, 1207-1212, Seoul, Korea.

Weigend, A. S., Rumelhart, D. E. and Huberman, B. A. 1991. Generalization by weight elimination with application to forecasting. *Advances in Neural Information Processing Systems*, 3, 875-882, Morgan Kaufmann, San Mateo, CA, USA.

Werbos, P. J. 1974. *Beyond regression: New Tools for prediction and analysis in the behavioural science*. Ph.D. Thesis, Harvard University, Cambridge, MA, USA.

West, M. and Harrison, J. 1989. Bayesian Forecasting and Dynamic Models, Springer, New York.

White, A. P. and Liu, W. Z. 1994. Bias in information-based measures in decision tree induction. *Machine Learning*, 15, 321-329.

Whiteley, J. R. and Davis, J. F. 1994. A similarity based approach to interpretation of sensor data using adaptive resonance theory. *Computers and Chemical Engineering*, 18, 637.

Whitley, D., 1989. The GENITOR algorithm and selective pressure: Why rank-based allocation of reproductive trials is best, 116-121. In: Schaffer, J. D. (ed), *Proceedings of the 3rd International Conference on Genetic Algorithms and their Applications*, Morgan Kaufmann, San Mateo, CA, USA.

Whitley, D., Dominic, S., Das, R. and Anderson, C. W. 1993. Genetic reinforcement learning for neurocontrol problems. *Machine Learning*, 13, 259-284.

Whitley, D., Starkweather, T. and Bogart, C. 1990. Genetic algorithms and neural networks; Optimizing connections and connectivity. *Parallel Computing*, 14, 347-361.

Widrow, B. and Hoff, M.E. 1960. Adaptive switching circuits. In *IRE WESCON Convention Record*, 96-104, New York.

Wieland, A. P. 1991. Evolving neural network controllers for unstable systems. Proceedings of the 1991 *IEEE International Joint Conference on Neural Networks (IJCNN 91)*, 2, 667-673, IEEE Press, New York, NY, USA.

Wierda, S. J. 1994. Multivariate statistical process control – recent results and directions for future research. *Statistica Neerlandica*, 48(2), 147-168.

Williams, C. K. I. 1998. Prediction with Gaussian processes: from linear regression to linear prediction and beyond. *Learning and Inference in Graphical Models*, Jordan, M.I. (Ed.), Kluwer Academic Press, 599-621.

Wise, B. M. and Ricker, N. L. 1993. Identification of finite impulse response models using continuum regression. *Journal of Chemometrics*, 7(1), 1-14.

Wise, B. M. and Ricker, N. L., Veltkamp, D. F. and Kowalski, B. R. 1990. A theoretical basis for the use of principal component models for monitoring multivariate processes. *Process Control and Quality*, 1, 41-51.

Wishart, D. 1969. *Mode Analysis.* In: Cole, A. J. (ed), *Numerical Taxonomy*, Academic Press, New York, USA.

Wishart, D. 1987. *CLUSTAN User Manual*, University of St Andrews, Scotland.

Wolf, A., Swift, J. B., Swinney, L. and Vastano, J. A. 1985. Determining Lyapunov exponents from a time series. *Physica D*, 16, 285-317.

Wong, P. C. and Bergeron, R. D. 1997. 30 years of multidimensional multivariate visualization. In: Nielson, G.M., Hagan, H. and Muller, H. (eds), *Scientific Visualization - Overviews, Methodologies and Techniques*, 3-33, Los Alamitos, CA, IEEE Computer Society Press.

Wornell, G. W. 1996. *Signal Processing with fractals: A wavelet-based approach*. Prentice-Hall, Inc., Upper Saddle River, NJ, USA.

Yao, X. 1999. Evolving artificial neural networks. *Proceedings of the IEEE*, 87(9), 1423-1447.

Yao, X. and Liu. Y. 1997. Fast evolution strategies. *Control and Cybernetics*, 26(3), 467-496.

Ylinen, R., Jämsä-Jounela, S.-L. and Miettunen, J. 1992. Use of cluster analysis in process control. 12th World Congress on IFAX, Sydney, Australia.

Young, F. W., De Leeuw, J and Takane, Y. 1976. Regression with qualitative and quantitative variables: An alternating least squares method with optimal scaling features. *Psychometrika*, 41, 505-528.

Zeng, X., Eykholt, R. and Pielke, R. A. 1991. Estimating the Lyapunov exponents spectrum from a short time series of low precision. *Physics Review Letters*, 66, 3229-3232.

Zeng, X., Pielke, R. A. and Eykholt, R. 1992. Extracting Lyapunov exponents from a short time series of low precision. *Modern Physics Letters B*, 6, 55-75.

Zeng, Y. and Forssberg, K. S. E. 1996. Vibration signal measurement from mono-particle breakage. *International Journal of Mineral Processing*, 44-45, 59-69.

Zeng, Y. and Forssberg, K. S. E. 1992. Effects of operating parameters on vibration signal under laboratory scale ball grinding conditions. *International Journal of Mineral Processing*, 35, 273-290.

Zeng, Y. and Forssberg, K. S. E. 1993a. Application of digital signal processing and multivariate data analysis to vibration signals from ball-mill grinding. *Transactions of the Institute of Mining and Metallurgy*, 102, C39-C43.

Zeng, Y. and Forssberg, K. S. E. 1993b. Application of vibration signals to monitoring crushing parameters. *Powder Technology*, 76, 247-252.

Zeng, Y. and Forssberg, K. S. E. 1993c. Monitoring grinding parameters by signal measurements for an industrial ball mill. *International Journal of Mineral Processing*, 40, 1-16.

Zeng, Y., Zheng, M. and Forssberg, K. S. E. 1993. Monitoring jaw crushing parameters via signal measurement. *International Journal of Mineral Processing*, 39, 199-208.

Zhang, B. T. 1994. An incremental learning algorithm that optimizes network size and sample size in one trial. *Proceedings of the IEEE Conference on Neural Networks*, 1, 215-220, Orlando, FL, USA.

Zhang, J., Martin, E. B. and Morris, A. J. 1996. Process monitoring using non-linear principal component analysis. *Proceeding of the IFAC World Congress*, International Federation of Automatic Control, San Francisco, CA, USA.

Zhang, J., Martin, E. B. and Morris, A. J. 1997. Process monitoring using non-linear statistical techniques. *Chemical Engineering Journal*, 67, 181-189.

Zhang, J., Morris, A. J., Martin, E. B. and Kiparissides, C. 1999. Estimation of impurity and fouling in batch polymerisation reactors through the application of neural networks. *Computers and Chemical Engineering*, 23, 310-314.

Zhao, Y. and Atkeson, C. G. 1991. Projection pursuit learning. *Proceedings of the Joint Conference on Neural Networks*, 1, 869-874, Seattle, WA, USA.

Zurada, J. M. 1992. *Introduction to Artificial Neural Systems*. PWS Publishing Company, Boston, MSA, USA.

Zurada, J.A. and Cloete, I. 2000. *Knowledge-based Neurocomputing*, MIT Press, Cambridge, MA, USA.

INDEX

APPENDIX: DATA FILES

Autocatalysis.xls	10 000 observations on 3 variables; time series data for autocatalytic reaction (see 9.2 and 10.3).
BICores.xls	47 observations on 13 variables (see 6.8.3).
CopperFroth1.xls	497 observations on 10 variables; features extracted from digitized images of froth from a copper flotation plant (see 6.8.6).
CopperFroth2.xls	1 234 observations on 9 input and 3 target variables; representing the effect of collector addition on a copper flotation plant on froth stability.
GoldLib.xls	Experimental data for 9 different South African gold ores, each described by five compositional variables on two or three conditions (milled or unmilled) (see 5.8).
GoldLoss.xls	77 observations on 14 input variables and two target variables, viz. dissolved and undissolved losses on a gold plant.
GranularFlow.xls	Five independent time series, representing the avalanching behaviour of heaped grains of Portland cement, cake flour, maize flour, quartz and iodised table salt subject to a steady flow of material from the top.
InducedAeration.xls	211 observations on 9 variables; aeration data for 5 different impellers and different fluids (see 7.2.4., as well as Aldrich, C. and Van Deventer, J.S.J. 1994. Observations on the effect of medium viscosity and density on the rate of induced aeration in agitated vessels. *Metallurgical Transactions B*, 25B, 303-306; Aldrich, C. and Van Deventer, J.S.J. 1994. Induced aeration in slurries in agitated vessels. *Chemical Engineering Journal*, 54, 199-205).
MillAcoustics.xls	63 observations on 34 input variables (frequencies) and 10 target variables (particle sizes of Merensky sulphide ore); (see 4.9.2) and 39 observations on 34 input variables (frequencies) and 10 target variables (particle sizes of UG2 sulphide ore); (see 4.9.2).
MillData.xls	1 830 observations on 3 input and 3 target variables; operational data from a gold milling circuit.
PGMflot.xls	297 observations on 6 variables; features extracted from digitized images of a froth from a platinum flotation plant (see 3.9.4).
SakLeach.xls	54 observations on 8 variables; consumption of an additive on a leach plant (see 4.4.2).